THE NEW
DYNAMIC
PROJECT
MANAGEMENT

THE NEW DYNAMIC PROJECT MANAGEMENT

Winning Through the Competitive Advantage

Second Edition

DEBORAH S. KEZSBOM, PhD

KATHERINE A. EDWARD

A Wiley-Interscience Publication

JOHN WILEY & SONS, INC.

New York · Chichester · Weinheim · Brisbane · Singapore · Toronto

Library of Congress Cataloging-in-Publication Data:

Kezsbom, Deborah S.
    The new dynamic project management : winning through the competitive advantage /
Deborah S. Kezsbom, Katherine A. Edward.
        p. cm.
    "Wiley-Interscience publication."
    Includes bibliographical references and index.
    ISBN 0-471-25494-0 (cloth : alk. paper)
    1. Industrial project management. I. Title. II. Edward, Katherine A.

    HD69.P75 K515   2001
    658.4′0421--dc21   99-045548

10 9 8 7 6 5 4 3 2 1

To Sam and John, without whom we would not be.
Our love and thanks to you.

DEBORAH AND KATHERINE

CONTENTS

7 Optimizing the Schedule / 217

8 Leadership in a Project Environment / 249

9 Planning for and Utilizing Conflict / 285

12 Quality in the Project Environment / 404

PREFACE

This is a book about projects and about the people who make them happen. It is a practical, comprehensive guide for managing today's competitive and demanding technical projects. It addresses achieving quality and profit goals; but it also deals with people and motivating the professional of this new era. Although directed at software, engineering, and other "technical" projects, the processes recommended in this book are applicable to most projects most of the time.

As the authors of this book, we represent a small *project team* of our own. We have brought together to share with you, the reader, our innumerable experiences and collective understanding that we have gained working on and within technical development projects and business management organizations of all nature. What you will read in this book is not merely a compilation of academic teachings or training. It is an *actual* account of what we have practiced successfully and have seen practiced well. To maintain the confidential and proprietary nature of these businesses and our client's organizations, we have chosen to use what appear to be generic examples and have taken author's license with our examples and case studies. Any direct references to people or organizational practices are, therefore, cited as appropriate.

Those of you familiar with our 1989 edition, *Dynamic Project Management: A Practical Guide for Managers and Engineers*, will recognize the extensive revision of our original text. It is a total update that reflects the rapidly changing practices of running large and small projects in business and industry today. Although the basic concepts explored in the 1989 edition remain unchanged, to address the challenges facing today's project managers brought about by how the world and the economy has evolved over the last decade, we have added new content and taken a fresh approach to the topics of:

· Schedule optimization and resource allocation
· Quality management
· Procurement and contract management
· Project management information systems
· Software project management, including managing project infrastructures according to the Software Engineering Institute (SEI) Capability Maturity Model
· Teams and team management
· Leadership in the new millennium

This edition also includes more case studies, questions, and exercises that will challenge the college-bound student as well as have relevance for the business manager leading a project team charged with bringing a product or service to market.

We believe that we have addressed all aspects of project management that project managers will need to succeed in this new millennium. Further, the book is structured according to the requirements of the international Project Management Institute and its *Project Management Body of Knowledge (PMBOK)*. Project management professionals will find that this text is an excellent reference to study when preparing to take the Association's Professional Project Management certification examination.

In Chapter 1 we examine the special challenges of managing in today's rapidly changing, globally competitive, high-technology environment that have encouraged the continued wide-span acceptance of applying project management processes, structures, and procedures. We further explain, in Chapter 2, how properly designed organizations can be major contributors to project success, and describe a variety of methods for introducing project teams to the organization.

We set the context for business and project planning in an international marketplace in Chapter 3, by specifically addressing the relationship of strategic business planning to project planning. We also illustrate the importance of the concept of "time-to-market" when planning commercial projects, and then describe how project planning is interrelated to actual product and services development in Chapter 4. In Chapter 5 we provide an overview of some of the fundamental tools used to plan a project.

In Chapter 6 we describe the principal scheduling techniques that we have found to be effective for planning and later in tracking and controlling project performance. We also introduce a methodology to address the risks associated with estimating a project schedule given the uncertainty of research and development projects in the high-tech environment. In Chapter 7 we explain how to manage the variables that constrain a project's finish, such as the tight deadlines typically imposed on most organizations and the limited availability of specialized project resources.

Project leadership and the "new" role of the project manager are discussed in Chapter 8. In this new edition we examine the dimension of leadership in light of the concept of self-directed teams and the issues associated with effective empowerment. In Chapter 9 we illustrate how, if assessed and managed properly, conflict can actually be constructive and improve product development.

In Chapter 10 we discuss how applying the techniques used by commercial- and government-sponsored program efforts to exercise control over cost, schedule, and performance helps guarantee the successful achievement of sought-after objectives. Skills needed to foster good communication in today's virtual cyber-project environment are presented in Chapter 11. We further expand upon and stress the skills required to build and maintain team performance in Chapter 13.

Chapter 12, which is new in this edition, is devoted entirely to the topic of quality management and its relevance to today's projects. We cover the tools and current internationally and nationally recognized quality assurance practices that project managers can use not only to assure the quality of software and hardware development projects but that of business processes as well.

Most high-tech projects today are involved in the solicitation, negotiation, and awarding of contracts to procure needed resources to produce and deliver products and services to market. In this edition we introduce a separate chapter (Chapter 14) devoted to the topics of procurement and contract strategies which include many different specialists from across a variety of organizations.

Finally, in Chapter 15, we take a totally new approach to the topic of project management information systems, which reflects the prevalence of Web-based technology. We encourage our readers to educate themselves in the distinct capabilities of the software programs available on the market today, and match those capabilities to the distinctive characteristics of their projects.

Our intent is to create a practical text that can immediately be applied by managers and project specialists alike. We also want to promote the use of certain methodologies in project management that we believe are not only basic to project success but critical to any business's survival. For this purpose we have followed the format from many of our hundreds of seminars in project management, and have included a variety of case studies, practice exercises, and examples.

We further encourage our academic colleagues to consider the text, or portions of it, as an excellent source of information for a one- or two-semester course in project and team management. It can be used successfully at both the graduate and undergraduate levels.

ACKNOWLEDGMENTS

We would like to thank the many people who have given so generously of their time and energy to this book. Special thanks to our colleagues at Lucent Technologies—Edward Manning, Judith Gordon, and Sylvia Manna-Czaplicki, whose review and many suggestions inspired the successive rewriting of the text. We would also like to acknowledge the guidance and assistance provided by John Edward, whose engineering experience proved invaluable to this new edition, and Joseph Saez, Patrick J. Monahan, and Martin Lewis, whose help preparing and revising the manuscript was invaluable and very much appreciated. Finally, many thanks to Cindy Juette for patiently revising the artwork and to Michele Cox, a colleague and friend, whose artistic talent helped envision a cover design for this edition that reflects the dynamic nature of managing projects in today's challenging and ever changing business environment. It has been a pleasure to work with each of you.

DEBORAH S. KEZSBOM
KATHERINE A. EDWARD

August 2000

THE NEW
DYNAMIC
PROJECT
MANAGEMENT

1

PROJECT MANAGEMENT: INTRODUCTION AND OVERVIEW

And then there was light!
GENESIS 1:3

1.1 INTRODUCTION

"It was the best of times. It was the worst of times."[1] Although this historic quote by the famous English author Charles Dickens applied to the economic and political conditions of Europe during the aftermath of the late-nineteenth- century *industrial revolution*, it may mirror the economic, political, and social climate resulting from today's *information revolution.*

In Dicken's view, nineteenth-century Europe was faced with the consequences of the political and social upheavals, created by an economy that may have altered or destroyed as many industries as it helped to create. Entering the twenty-first century, we are similarly faced with an era of unprecedented prosperity created by the *information revolution.* Yet the very forces that bid well for our economic future raise concerns about the nature in which we do business today and what is required to build and maintain organizational success in the near and not-too-distant future. The same impetus that gave rise to flourishing, worldwide markets and international alliances foster the need to *innovate* and create management strategies that increase our competitiveness and chances of survival.

---

[1]Charles Dickens, *A Tale of Two Cities.*

1

Companies may be likened to living organisms, changing continuously to deal with both internal and external pressures. These pressures require each of us to *adjust and adapt* perpetually in business, as in our personal lives. We are *always* changing. Sometimes the adjustment or change is minor and may be somewhat unnoticeable, as when we switch to a new vendor or try a new supplier. Other times, however, the change may involve a variety of other people, other skills, and other technologies, as, for example, when we purchase a piece of property or buy a new home. You can do it alone, but it is much easier and more effective if you consult with a licensed real estate agent, an attorney, and a mortgage broker. Our experiences within the organization may also require a more integrated *systems approach* to managing change.

Project management is a process that offers the contemporary organization a unique vehicle for change. It is a *methodology* through which we may create more flexible, adaptive, yet *accountable* corporations, unbound by the more traditional policies and procedures originally designed to maintain the status quo. The development and installation of project management systems within more traditionally stable, hierarchical organizations will not occur without substantial resistance. For many organizations, building and maintaining the essential capabilities for survival into the twenty-first century will require major change. Meeting the challenges of change, however, may be beneficial to both the organization and the individual. Project managers, leaders, and sponsors who position themselves and their employees "right" will reap priceless opportunities and countless rewards. These new, global, *boundary-spanning* managers will be those with the initiative to assume personal responsibility for their own organizations', and their team members' futures. Today's competitive environment requires project managers and specialists to be aware of the many challenges that lie ahead and equip themselves with the strategies that assure the continued growth and very existence of the organization.

Adopting project management processes and procedures requires a penetrating examination of existing organizational and management structures, current planning, scheduling, and tracking techniques. Most of all, efforts need to be focused on involving the personnel who are the cornerstone of any project effort. Project management increasingly *is* team management!

1.2 WHY PROJECT MANAGEMENT?

In 1989, Jack Welch, CEO of General Electric, noted that one major reason for the stagnate productivity in the United States at that time was the oppressive weight of the corporate bureaucracy.[2] According to pollster Daniel Yankkelvitch,[3] secondary factors were employee alienation and declining motivation. Curiously, in Yankkelvitch's poll,[4] American workers indicated that they believed the American work ethic to be alive and well. Yet these same people reported that they tended to perform their jobs to the minimum level of performance. This tendency was attributed to an underlying long-term resentment about the way in which their work was structured, managed, and subsequently rewarded.

As a result of these, among other indicators of change, organizations worldwide are seeking ways to reduce bureaucracy and increase productivity and job satisfaction. Many are making the transition from hierarchical structures to leaner project-oriented, team-based organizations. These newer state-of-the-art organizations encourage employees to take on responsibilities, which until recently fell squarely in the managerial realm. Increasingly, project management processes are being used to create highly integrated organizations, controlled by project teams responsible for planning, controlling, coordinating, and improving their own work.

Today, corporations and government agencies are investing in efforts to extend employee involvement into the domain once dominated solely by management. Several factors are influencing this shift toward more project team-based methods of managing and accomplishing work. These factors include:

1. There is a focus on quality and customer satisfaction. Quality management processes emphasize the importance of *each* employee's contribution to the project and to the final product.
2. The philosophy that people support that which they create, involves the growing awareness that the autocratic, paternalistic management style no longer creates neither productive nor committed employees.
3. Advanced technology requires *specialized personnel* who must work in leaner, resource-limited organizations.
4. Specialization in today's leaner structures demands increasingly the need for opportunities for job sharing and job rotation.
5. There is an increased awareness that employee involvement is the key to productivity. This is based on the concept that those who must implement the work should make a major contribution in designing the work.

---

[2]J. Hoerr, "The Payoff from Teamwork," *Business Week*, July 10, 1989, pp. 56–62.
[3]Ibid.
[4]Ibid.

6. Global market competition decreases the margin for error with regard to schedule slippage and/or technological risk.

7. There is an increased realization that the project manager cannot plan, schedule, track, or control a project alone. This realization is based on the fact that in view of today's more complicated products and systems, project managers cannot possibly have all the knowledge, expertise, time, or resources to bear alone the burden of project competition.

Global competition requires managers, leaders, and professionals to think of ways to change their organization *continuously* to gain the competitive advantage. Within the past decade, managers have led the following changes:

· Todays workforce demonstrates significant changes in its values and needs. This requires adjusting one's leadership techniques to meet the needs of such a changing and diverse workforce.

· Organizations have restructured and downsized to become leaner, with fewer layers of corporate hierarchy. This requires the typical manager to become more of a team leader, a coach, and a mentor.

· Flexible work systems have enabled many companies and organizations to meet the need of an increasingly professional workforce. This requires manager's to adapt to "knowledge workers"[5] and their special needs.

· Reengineering within organizations has reduced steps in work processes and helped organizations to focus on their *core competencies*. It falls to managers to initiate the adaptation to organizational transformation, through the application of unique business processes.

· Quality management has given the workforce more power, including involvement in the decision making, planning, and implementation processes, and in achieving customer satisfaction. Increasingly, managers have had to learn to "innovate" as opposed to "operate."

· Information technologies have increased worker productivity. With accelerated development of technologies, managers have been required to adapt to the call for greater technological skills among the workforce.

· New *virtual teams* are comprised of members dispersed across varying

---

[5]Peter Drucker, *Management: Tasks, Responsibilities, Practices* (New York: Harper & Row, 1974).

international locations. These "digitally connected" teams must find ways of working together, apart. Information must be both readily available and accessible across global borders.

The changes taking place in business and industry can be summarized into four major elements: (1) globalization of competition, (2) restructuring and delayering of organizations, (3) growth of computerization and computer networks, and (4) the emergence of the information highway. As a result of these enormous changes, companies that wish to remain competitive in the twenty-first century will need to adjust in order to survive. New companies continue to arrive, and older organizations, once perceived as more stable, must heed the signs within their environment, or perish.

The challenges of this era force managers and personnel to consider how their present management practices and processes address the needs of the workforce and the organization and add value to the customer. Shortened time windows, increased costs, and tight competitive markets all require more *efficient, systemic,* and *systematic management approaches.* No longer can projects, large or small, be run by "seat of the pants" management practices. Nor can traditional practices, originally conceived to generate and control repetitive tasks within outmoded hierarchical structures, meet the demands of this politically sensitive, rapidly changing environment. Management processes must be tailored to accommodate the unique, dynamic and diverse project needs, while maintaining control over cost and quality. Project management processes have been shown to meet this decade's unique challenges.

1.3 PROJECT MANAGEMENT FUNCTIONS

It is always possible to learn from the past. As far back as 1917, Henri Fayol[6] and the leaders of the scientific and classical management movements identified functions or tasks involved in the management process. Some processes from this classical view continue to exist and are now known as *traditional management practices.* But strong differences may be discerned between the project and functional managers' roles. In fact, differences can be cited that exist between the project manager's role today and that of only a decade ago.

As indicated in Figure 1.1, project management functions begin with the process of integrative project planning. In today's competitive markets and

---

[6]Henri Fayol, *General and Industrial Management* (London: Sir Isaac Pitman and Sons, 1948), p. 8.

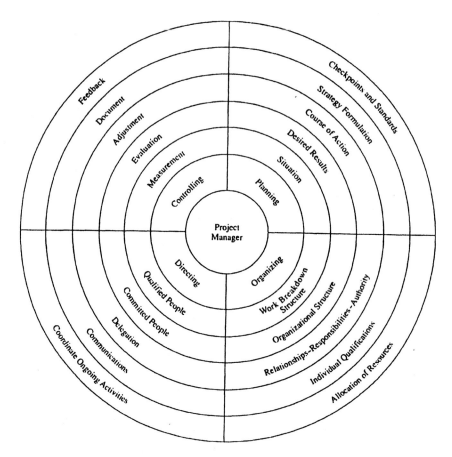

FIGURE 1.1

Project management functions.

short deadlines, it is no longer effective to design first and manufacture later. Now we look to the processes of *design for manufacture* and *design to build* that strengthen the quality of a product and assure more timely implementation. Under the leadership of the project manager, all project specialists and their functions (e.g., manufacturing, design, development, installation, operations, documentation, customer support, etc.,) determine project scope and define and communicate project mission, objectives, targets and milestones. In unison, as an integrated *team*, they develop and agree on a project plan for the entire team to track. Repeatedly, we see that one of the primary reasons that plans fail is that they have not been developed by those responsible for their implementation. *Integrative project planning* (IPP), discussed more fully in Chapter 5, involves better use of resources, encourages greater team interaction, and follows the natural flow of project work within *any* organization.

When planning a project, it is imperative for the team to discover that the best course of action in getting from point *A* to point *B* is to determine jointly those checkpoints and standards that define *when* and *if* the point of destination has been reached. These milestones, or well-defined *deliverables*, communicate to all involved what is expected and how well it is to be achieved. Such communications enable the project manager and project team to track and control a project with optimal precision.

Project control should not and *cannot* be characterized by unilateral control. *Effective control* involves joint determination of the tasks to be performed and the time required to accomplish them. Effective and, more important, successful project tracking and controlling involves joint participation in scoping or defining project activities, estimating the time and costs required, and determining quality metrics.

Above all, effective, successful project management requires the recruitment, development, leadership, and commitment of people, who truly make the project a success. As project efforts cut across what were once traditional lines of authority, the ability to influence individuals, decisions, and events becomes less of an issue of formal authority within the organization and more a matter of interpersonal effectiveness and informal bases of power. No longer are collective, participative approaches to problem solving and decision making considered a "nicety," but are a necessity as project personnel increasingly find themselves operating in a complex web of organizational relationships. Project goals can only be accomplished when they are perceived to be in accordance with the goals and objectives of the managers and specialists who comprise the project team.

1.3.1 Project Managers: Then and Now

The existing forces that drive an organization toward a more project team orientation, are (1) increased interdisciplinary tasks, (2) organizational integration, (3) complex organizational relationships, and (4) the need for synergy. Regardless of the organizational structure, or whether or not teams exist, work still needs to be planned, activities assigned, and resources allocated and tracked. The fundamentals will always remain intact. Every project calls for goal setting, handling performance problems, conducting effective meetings, tracking and rewarding accomplishments, and coaching or mentoring obligations. Although we hear much about "self-directed" teams, contrary to popular belief, the roles of the project and the middle manager are far from obsolete—they are simply changing.

Old and New Paradigms. Historically, the project manager's role developed out of the need for someone to supervise, coordinate, and engineer work related specifically to projects. In other words, the project manager's job was to plan, control, organize, and direct the work of several individuals or departments so that the project could succeed. In some cases, teams were necessary; in other cases, maybe not. Frequently, team members would be delegated project activities with minimum interfacing with other members of a project team. Operating like a systems engineer, the project manager would maneuver pieces of the project puzzle into a more or less coherent whole.

Several old paradigms have supported this past concept of the project manager:

- The project manager is the technical expert.
- The project manager makes all the decisions.
- The project manager defines the job and how it is to be done.
- The organization is structured hierarchically.
- Teams are formed only when needed.
- The focus of the organization is on specialization.
- The workforce is homogeneous.

Certainly, much has occurred that illustrates the limitations of these former paradigms. With the advance of high technology, project managers cannot possibly be experts in *all* project-related technologies. Today, project managers must rely heavily on the expertise and decisions of others. The strong focus on specialization in the past has given way to integration and an even greater need to solve problems that span several disciplines and organizations.

Cutting-edge paradigms are encouraging the creation of an increasing number of team-based organizations, with employees who are educated in the processes of group dynamics. These new paradigms are:

- Employees are considered experts who possess unique knowledge and skills.
- Controls are set collectively.
- Employees participate in goal setting and in defining how work should be organized and accomplished.
- The organization is customer-driven.
- The workforce is heavily diverse.
- Employees participate in continuous improvement of product/project methodology.

These paradigms indicate that organizations are experiencing a more flexible and knowledgeable workforce, which expects to be a vital part of the problem-solving and decision-making processes. Characteristics of project managers of the recent past and those of the future will differ. Project managers of the 1980s and 1990s:

- Are better educated
- Are open, friendly, people-oriented
- Are better listeners
- Are quality conscious
- Are receptive to new ideas and the ideas of others
- Are more participative
- Are more involved in socioemotional communication
- Allow more independence
- Encourage greater cross-functional interaction

Project managers of the new millennium:

- Will be facilitators
- Will be skilled at group process and group dynamics
- Will encourage others to participate in plans and decisions
- Will understand how to coach, inspire, and motivate
- Will no longer be perceived as the "expert"
- Will work to gather resources for the team
- Will be comfortable relying on the expertise of others
- Will be *boundary spanners*[7]

1.4 CREATING MORE FORMAL DEFINITIONS: MANAGERIAL IMPLICATIONS

Project management is the combination of people, processes, techniques, and technologies necessary to bring the project (or program) to successful completion. Measures of success certainly depend on the particular project; however, most projects are measured on what may now be considered the *Quadruple constraint* of time (schedules), cost (on, over, or under budget),

---

[7]R. G. Donnelly and Deborah S. Kezsbom, "Overcoming the Responsibility–Authority Gap: An Investigation of Effective Project Team Leadership for a New Decade," *1993 AACE Transactions*, pp. Q2.1–2.12.

accomplishing performance criteria (specifications), and last, but certainly not least, meeting or exceeding customer expectations (quality).

A more formal definition of *project management* is:

> Project Management is the planning, organizing, controlling and directing of company resources (i.e., money, materials, time and people) for a relatively short-term objective. It is established to accomplish a set of specific goals and objectives by utilizing a fluid, systems approach to management, by having functional personnel or specialists (the once traditional line-staff hierarchy) assigned to a specific project (the new "horizontal hierarchy").[8]

1.4.1 Closing the Gap Between Project and Quality Management

Total quality project management utilizes the processes of project management within a *quality environment*: that is, planning, organizing, directing, and controlling within an environment in which workers make decisions. Participative techniques are used to determine the scope of work, the flow of work, the improvement of work, and to maintain customer communications, relations, and eventual satisfaction.

More traditional management approaches can still be applied to projects. The dynamic, fluid nature of today's cutting-edge high-tech projects, however, warrant a management approach that not only addresses more immediate, short-term objectives, but also takes a more *systemic perspective*. Cutting-edge technologies demand management practices that incorporate *life-cycle management* and *address the issues of team development, power, and authority structures*.

Short-Term Objectives. Projects are run for a *relatively* short-term duration. Most commercial engineering projects today, for example, typically run approximately 12 to 18 months. Some less complicated projects may be shorter. Market windows are growing increasingly short. Project management team-based strategies help managers to better manage company resources and customer satisfaction while still addressing the constraints of time, money, technical performance, and quality. Because of the dynamic nature of short-term project objectives, more traditional management processes and tools designed for vertically interlocking structures do not readily measure or provide sufficient feedback to render timely corrective action.

---

[8]D. S. Kezsbom, D. L. Schilling, and K. A. Edward, *Dynamic Project Management: A Practical Guide for Managers and Engineers* (New York: Wiley, 1989).

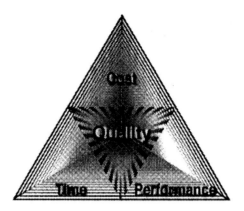

FIGURE *1.2*
Quadruple constraint.

In our earlier book we discussed the challenges of the *triple constraint*: time, money, and the attainment of performance specifications. Today, however, project managers are aware of the critical fourth constraint of quality: meeting or exceeding customer expectations and needs. In today's highly competitive global environment, one must meet the demands of budget, schedule, and performance criteria, in addition to maintaining a clear understanding of the customer's desires. Customer responsiveness and satisfaction are the final elements of the quadruple constraint. Figure 1.2 illustrates the quadruple constraint.

The project manager can critically influence and assure the quality of a project. As a liaison or *boundary spanner*[9] across a project organization, the project manager is in a position, for example, to influence design alternatives, introduce value-added methodology, and assure that customer specifications are met to ensure the quality of the project.

Systems Perspective: Synergy and Interface Management. As mentioned previously, organizations today are regarded as organic, *systemic* assemblages of people, materials, information, money, and facilities. Like any organism, organizations and projects (as organizations within their own right) must interact with both the internal and external environments surrounding them. From the external, the project organization must "breath in" customer, competitor, legal, environmental, regulatory, and other

[9]Donnelly and Kezsbom, " Overcoming the Responsibility–Authority Gap," p. Q2.8.

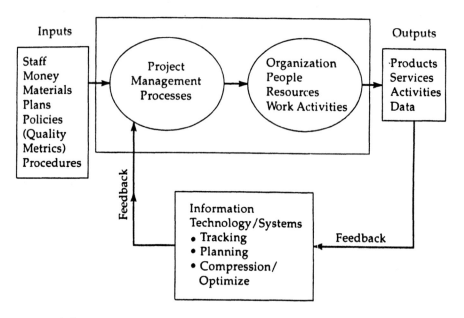

FIGURE 1.3
The systems approach to project management.

information vital to the project's success. In return, through the synchronic interactions of the internal mechanisms of the system or project organization (its people, materials, processes, and quality measures), it produces the specific product (or project) that meets or exceeds customer expectations. Such a systemic perspective to project management requires the identification of each subsystem component (hardware, software, people, facilities, processes, costs) while maintaining synergy or successful integration of interactions between each subsystem of the project to assure a *total quality project*. One must strive to make the entire system run in a manner that surpasses the performance of any individual subcomponent. This is total quality project management. Figure 1.3 illustrates this systems approach to project management.

Successful project managers experience the true sense of the term *interface management*. Originally conceived as a technique to ensure that a technical system's components matched, successful project managers now think of interface management as the integration of the organizational and managerial elements of the project. Project interface requires project managers to identify (1) the major components or subsystems of the project to be managed (2) the subsystem integration points requiring specific managerial attention, and (3) the relationships that need to be developed to manage these interfaces successfully.

Increasingly, the "way of projects" will become the way of organizations. Many of the demands of effective total quality project management translate into the characteristics that effective successful organizations will display during the new millennium. These include:

- Clarity and *congruence* of vision, mission, and objectives.
- The recognition and acceptance of constant change as the norm.
- The need to clarify, define, and manage each subsystem of the organization or project effectively.
- The need to integrate the whole. This requires identifying the interface points in order to manage them effectively.
- The importance of control techniques to monitor and to track systems and processes. This includes feedback, communication, and the gathering and transmission of accurate information.
- The demand for effective, quality-oriented leadership.
- Interface management in both a systems and organizational (political) context.

Managing the Project by Life Cycle. Effective project management re-quires *total life-cycle management.* In each phase of a project's life cycle, different levels and types of activities and procedures are required that generate their own complex set of challenges and solutions. Typically, projects go through a distinct life cycle in which a natural order of thought, action, and processes is followed. Like all organic systems, projects evolve through phases of development, marked by a beginning, a buildup, a "happening," and a finale.

During each phase of a project's life cycle, different levels and varieties of technical and managerial activities are *expected* to occur. Projects, for instance, are born when a sponsoring division, an internal or external customer, management, marketing, or the research department identifies a particular need. A variety of feasibility studies will determine whether or not an organization accepts the challenge to accomplish the goal through project mechanisms. As planning and the execution of plans occur, money, personnel, time, facilities, and other resources are committed to this challenge.

As the project's mission and resource estimates become clearer, the project moves from the concept to the definition phase. It is time for the project team to be in place. Now, specific project plans involving technical scope, budgets, schedules, and quality parameters are determined jointly by team participants. Risk areas are identified and contingency plans are mapped. The commencement of product design and development signal the need to develop *work breakdown structures*, which indicate the work to be done. Testing and quality processes are in motion that further monitor performance and assure that quality standards are met. As the project approaches execution and installation, the high-stress environment is loosened, while new efforts concentrate on the reallocation of resources.

Astute project managers, drawing on past experiences from other projects, may develop a checklist of expected events or milestones that must be accomplished to ensure that the project is on track and moving along according to requirements. Performance measurements for each phase of the project, presented in greater detail in Chapter 4, are useful in developing *these project life-cycle checklists*. Such checklists become useful tools, which serve as an inventory of accomplishments and assure that all project activities and objectives have been carried out.

1.5 POWER AND AUTHORITY: "WHO'S ON FIRST?"

Within the past two decades, we have witnessed an increased growth of the project-matrix organization. This popular organizational design *requires* project managers and team leaders to cut across divisional lines of authority to accomplish project objectives. Cross-functional teams have grown in number as customer demands have created increased need for more creative, competitive approaches to product realization. By simple observation it appears that diverse, team-based organizations and processes are capable of meeting those demands.

During the late 1980s and the early 1990s, project managers frequently complained of the challenge of accomplishing what were, at times, insurmountable tasks, while possessing insufficient authority to direct all project personnel assigned to accomplish project goals. Because of their somewhat diagonal position within the organization, a great deal of time and energy were spent negotiating with and influencing those within an organization who could provide the personnel needed by a project.[10]

With the advent today of flatter, more flexible, team-based structures, issues of project authority and power will become more pronounced and confusing than they were just a decade ago. Today we hear of "self-

---

[10]Ibid., pp. Q2.1–Q2.12.

directed" teams and wonder if managers of *any type* are required. Contemporary teams possess not only project managers, but functional, team, and branch managers. Unfortunately, in instances of several of the projects on which we have been called to assist, the analogy to the old Abbott and Costello joke of "Who's on first?" is sometimes appropriate . . . and always somewhat frightening!

As discussed in greater detail in Chapter 8, the concept of self-directed teams may be somewhat of a misnomer and undoubtedly confusing to many corporations and team leaders. First, all teams need guidance, direction, input, and structure, especially with regard to project objectives, requirements, and specifications. Moreover, with the tremendous need for communication and information dissemination required by projects, such a focal point is critical to project success. This focal point serves as an information disseminator,[11] assuring that all the data and information necessary to personnel are readily available and easily accessible. Regardless of organizational design, the role of *information disseminator* remains that of the project manager.

As illustrated further in Chapter 13, teams mature along the same lines as projects. Team maturation takes place provided the leadership, coaching, guidance, facilitation, and information available to team members help them to pass from one stage of development to the next. Project managers, through their influence, negotiation, facilitation, and communication skills, build strong working relationships with team members and the functional managers to whom they directly report. In today's highly competitive global environment, it is a waste of energy and resources to jockey for power and authority, when those energies should be focused on achieving our mutual interests, which are succeeding as a team and, of course, as a company. Through integrative efforts, project managers gain the support of their peers, their team leaders, and the managers who support them.

1.6 PREVENTING PROJECT FAILURE: MEETING THE CHALLENGES OF THE GLOBAL MARKET

The reasons that projects fail are as many and as varied as the wide variety of projects that exist. The processes of project planning, organizing, directing, and controlling may be effortless when we are focused on rather

---

[11]D. S. Kezsbom, "The Rise of the Self-Directed Team and the Changing Role of the Project Manager," *Proceedings of the 1994 Project Management Institute's Annual Symposium.*

explicit, self-contained, easily defined, and repetitive tasks. Few tasks today, however, possess such characteristics. Technology has become more complex, requiring a high degree of specialization in a scarce-resource environment. Outsourcing and international and domestic alliances are all popular methods used by companies to offset scarce resources, seek out expertise, and serve as a vehicle for market entry. But such methods do not come without a price. When work becomes more complex and spans a variety of companies and nations, complicated organizational and interpersonal relationships are required to assure a project's success.

With global competition, scarce resources, and knowledgeable, yet demanding customers, it is not surprising that all projects share some important characteristics — the *reasons for failure*. Unfortunately, while companies appear to commit *intellectually* to the concepts and processes of project management, many fail to follow up on the initiatives and support required to make projects and the project management process a success. Incomplete or inadequate definition of the project's objectives, mission, and requirements, compounded by the lack of involvement and ineffective communication across contributing entities, are sure signs of project difficulties and confusion. The confusion that results undoubtedly leads to schedule slippage, cost overruns, inferior quality, and poor customer relations. The authors' combined experiences of over 40 years working with and on a variety of projects has produced this extensive, but not inexhaustible list for project failures.

- Lack of management support
- Lack of a project focal point
- Lack of involvement of all project entities, including the customer and end user(s)
- Insufficient dedicated resources
- Too much "ownership," or a lack of ownership
- Personality conflicts
- Lack of clear shared goals and objectives
- Lack of sponsorship at the appropriate management level
- "Scope creep"
- Lack of accountability
- Intra- and interteam conflict
- Insufficient requirements
- Shifting priorities
- Inadequate involvement of team members

- Lack of understanding (or support) of the team process
- Unclear or nonexistent sense of shared mission
- Poor choice of project manager
- Lack of planning of project termination
- Poor communication
- Distance-spanning virtual projects team
- Lack of understanding of roles and relationships
- Ineffective project plans or coordination of plans
- Project manager with insufficient authority
- Project manager with poor leadership/team building skills
- Lack of a clear understanding of the project structure

The goal of project management is to provide and apply integrative management techniques to unique, often complex organizational ventures. These ventures are typically characterized by interdependent efforts, frequently both inside and outside the sponsor organization, the need to integrate a variety of specialists, the need to control time and cost; and, more frequently than not, insufficient or inadequate resources. It is critical that a common focal point brings the critical elements of the project together and *facilitates* all that needs to be done, *by those who are experts in doing it*, to offset the likelihood of failure. This is the role of the project manager.

1.7 MEETING THE CHALLENGES OF THE GLOBAL DECADES: CONCLUSIONS

Today's organizations are faced with unprecedented challenges, which hold important implications for *all* employees, managers and specialists alike. Rigorous competition, changing market environments, and sophisticated technologies all have created products with shorter life cycles than few could have envisioned. These new demands place great pressures on project teams and managers who facilitate and guide them. Today's demands to be first to market, best in class, and to remain competitive in price and quality while competing in a global market have contributed additionally to the shorter product life cycles and the stressful scheduling demands associated with such change. Firms will continue to respond to today's unique and challenging business environments by streamlining organizations and focusing on quality and speed to market. The managers and employees who survive this upheaval and advance within the organization will need to enhance and recast those leadership and management skills that proved successful for them in the more traditional organizations of the past.

The business environment calls for strong, innovative management and leadership skills. Corporations are currently undergoing fundamental changes in organizational foundation while often experiencing simultaneous rapid change in technologies. Organizationally, firms are taking steps to reduce operating costs and remain lean, placing greater emphasis on self-directed, cross-functional teams. These horizontally integrated teams consist of professionals from each discipline needed to complete the project. Team members, in turn, share certain leadership responsibilities throughout the life cycle of the project. These trends emphasize the need to increase management effectiveness and support a project management, team-based orientation.

In this book we deal with the real-world challenges and concerns of management, project managers, and employees working in this dynamic and changing team-based environment. The contents are based on actual experiences, both our own and those that we have had the pleasure of sharing with our client organizations in both commercial industry and government sectors. It is our intent to share with you, the project managers and project specialists of the twenty-first century, information to help develop the skills necessary to become more efficient and effective in today's total quality project management environment.

CASE STUDY: THE WINNING PROJECT*

It was the night of the annual project awards banquet. Steve Golden, former project manager of the XB1 processor, had been informed only two weeks earlier that the project he had managed for the past seven years was to win the national competition among other high-tech and information technology projects as Project of the Year. "Of the year," thought Steve, as he drove to the banquet hall: "more like the project of the decade!" Even now, Steve shook his head with amazement. How far they had come . . . how very far!

Organizational Background

Communication Advance Systems (CAS) was an organization in transition. Traditionally, CAS had served its internal customers or sister organizations, observing informal project management practices. This often led to adjust-

---

*This study was written for class discussion purposes only. It is not intended to reflect management practices or products of any particular company or organization.

ments of time, budget, and performance targets as the project matured. Now, with increased domestic and global competition came more formal project management systems, including the responsibility for all customer interfacing and project management with the formalization of the corporate project process group (CPPG). The CPPG had been formed in early 1987 to assure the introduction of a project into the commercialization phase of product development. Steve Golden was granted the dubious title of director of the CPPG. The group was set up to meet the traditional goals of a well-managed project: developing products on time, on budget, and within specifications. The CPPG was expected to interface with internal and external customers and to be responsible for developing the detailed concept for new products.

Steve was well known throughout the information systems, applications, and development sides of CAS. He had been with the company for over 25 years, serving in various projects and troubleshooting technical as well as management difficulties that tend to arise in an organization as large as CAS. Steve was affable and well liked, which won him the respect of management and his peers throughout his career at CAS. But with only three direct reports, Steve questioned the extent of his authority as director of the CPPG and his ability to manage the plethora of projects now on his plate. And then came the XB1 project!

Project Background

The XB1 was to become the largest telephone switching system in the United States. It was to surpass in capacity that of the XA1 system, now presently serving all 48 of the continental United States. Changes in design from the original XA1 were to concentrate on the hardware so as to minimize software glitches and make the transition from the XA1 system to that of the XB1 seamless. Millions of consumers depended, without realizing it, on the efficacy of the XA1. Therefore, the transition to the newer system had to be accomplished without a single flaw, even though the volume of callers had grown unbelievably in the last three years alone.

The XB1 was a highly demanding effort requiring development of significant new hardware, software, and process technology. This, in turn, required high levels of integration among the various phases of the XB1 project. Especially pertinent to the success of the project was the high degree of collaboration necessary between design, development, and manufacturing personnel, since the organization had a history of conflict between these areas. Second, following the need for extensive integration of project efforts, CAS had chosen to rely on the facilitating contributions of the CPPG organization and process. This was still a relatively new and evolving approach to the management of projects within CAS, especially within the manufacturing and applications areas. Third, the XB1 project was being pursued at a time of rapidly changing technologies of growing importance, such as cellular. These new technologies created a competitive environment for equipment, facilities, and management commitment and organizational resources. This created some difficulties in setting organizational priorities and allocating scarce professional and managerial resources within the organization. Finally, the project was undertaken at a considerably challenging time for CAS. Significant cost reduction pressures and mandates to freeze hiring (and, in some cases, downsizing) had affected the XB1 project as well as the overall functions or phase efforts.

To make matters more difficult, the political climate behind the XB1 was shaky, to say the least. Management had been divided in support of the changes that were to be made in the original design of the XA1. Technically, some vice presidents were dubious that the hardware changes proposed could manage the capacity of the phone calls expected to be handled by the XB1. Many directors within CAS's laboratories frequently laughed at the XB1's plausibility, and support for the new multimillion-dollar replacement of the XA1 was sporadic and inconsistent across organizational lines. What was more, the development people failed to see the need for an "outside" project management group tracking its projects. It didn't seem to matter that the CPPG was formed within a sister organization, with former development people; the CPPG was seen more as a "spy" than as a part of the project team.

The project was composed of more than five groups representing the various components or "functionality" of the XB1. The five groups were divided into three separate organizations, each reporting to different vice presidents. The groups had worked together on projects in the past, but never as a team. Steve felt that teamwork was essential to the success of this project because of the nature of the integrated technology. More important, Steve knew that the informal project tracking and reporting mechanisms of the past could no longer serve the organization. If this was to be a seamless transition in technology, the organization had to be seamless as well.

Introducing Project Management

It was early January when Steve Golden called his three project managers together. Steve had received a directive about the XB1 from his boss, Ted Lawrence. It was time to get the project moving. But Steve had several concerns and did not hesitate to voice them to his small but enthusiastic staff.

Steve Golden: I'm glad you all were able to make this meeting. With the storm the way it is, I wasn't sure that each of you could get here.

Kerry Wilson: What's the scoop on the new project Lawrence just delegated to us, Steve?

Steve Golden: Kerry, sometimes you give me the "willies" with your mindreading abilities! Seriously, I've called you all together for several reasons. First, since the forming of our group, I've been doing some heavy-duty thinking about how we should structure not only this but other projects. First, I'm not even sure that the folks in operations and deployment are speaking the same language that we are when it comes to project management.

Roger Randon: I know what you mean, Steve. When it comes to managing projects, the guys in deployment are like cowboys and shoot from the hip!

Pat Maury: Cowboys! Project management to them means they're lucky if the project gets to the customer at all!

Steve Golden: Well, I think the first thing we should do is to try to get our arms around this baby. We've got to see how these guys are planning and scoping out their portion of the work and what project management techniques they are using. Once we are aware of the situation, I think we'll be in a better position to organize and begin the planning for this project.

Pat Maury: I agree! Let's get out there and see if they know who we are and what we intend to do.

Steve Golden: That's another good point, Pat. I don't want us marching into these departments acting as if we are there to "save the day." It's important that these people trust us and provide us with *real* information, now and later as the project matures. So remember, you are there to *help* these people and the organization and get our projects out the door in a timely, quality-oriented fashion.

Kerry Wilson: Well, what's the priority on the XB1 project, Steve? Is time of the essence?

Steve Golden: Well you know management, Kerry. They haven't made it clear whether our target is time or money. We certainly know that performance outranks them all, with the importance on a seamless transition from the XA1 system to this one. Anyway, let's start at the beginning and that means seeing just what we are up against when it comes to "managing" the development, manufacturing, deployment, and installation of this baby!

At Steve's direction, the four-person project CPPG team began to meet with the various entities contributing to the XB1. Unfortunately, what they found was no surprise in most cases. Some XB1 specialists were unclear as to their specific individual objectives or did not have stated milestones against which they were working. It appeared that in those areas where personnel believed that they were planning, they were actually attempting to fit milestone dates between a given start date and a targeted end or due date. Most areas had no plan at all. Personnel, believing that since their part occurred in what they felt was the "back end" of the project, felt they had plenty of time to "plan" their strategies. Those who were honest with the CPPG team expressed difficulty in effectively "scoping out" or defining the work and understanding fully the project requirements. In many areas, no scheduling tools, such as critical path analysis, were used, eliminating the distinction between those activities of particular importance to the project's schedule and those that could more or less wait without affecting the project negatively. As a result, there were no early warning signs to indicate if slippage had occurred or how it would affect the project.

Many supervisors assigned to the project had no direct reports to work with and were already handling far more projects than they could to be effective. In conversations with various members of the CPPG, some staff also held different perceptions of the status of the various elements within the XB1 project. Many expressed concern over the degree of changes that were occurring in the original hardware physical design.

An important and not somewhat unexpected outcome of these meetings for Steve was learning the perception of CPPG held by the other organizations. With the blatantly inadequate staff that Steve had, he was only able to rely on a monthly CPPG status meeting to track and control the project. This had left the other project organizations with the feeling that the CPPG was unable to identify problems and was relying on the functional supervisors and managers to do so. Even if and when problems were pointed out, the CPPG had no ready mechanism for identifying the impact on the overall project.

Questions

1. What were the difficulties and challenges facing the XB1 project manager and his immediate project office?

2. How do these problems compare or relate to the reasons for project failure?

3. What are some of the indications that the traditional vertical hierarchy may not be adequate for managing this project?

2

CREATING ORGANIZATIONS FOR PROJECT WORK

Destiny is not a matter of chance,

it is a matter of choice.

ANONYMOUS

2.1 INTRODUCTION

In today's challenging business environment, success and innovation have become synonymous. Successful organizations have the ability to refocus and reorganize quickly and efficiently to meet market demands. The nimble, agile company clearly has the true competitive advantage.

As businesses seek to become more competitive and more efficient, they are increasingly aware of the many advantages offered by flatter, more flexible organizations and the multidisciplined project teams frequently characteristic of them. Senior managers and employees alike have become aware that project team-based organizations benefit them and their organizations in several ways:[1]

1. *Improved productivity, quality, and service.* To enjoy a competitive edge, organizations today must concentrate on timeliness, performance, quality, and cost: the *quadruple constraint.* Success in these areas

---

[1]Richard S. Wellins, William C. Byham, and Jeanne M. Wilson, *Empowered Team: Creating Self-Directed Work Groups That Improve Quality, Productivity, and Participation* (San Francisco: Jossey-Bass, 1991).

24

comes from daily work processes and product enhancements, known today as *continuous improvement*. The sense of job ownership that comes from team involvement leads to continuous improvement, or *kaizen*. Kaizen, in turn, results in quality, timeliness, service, and productivity.

2. *Flexibility and responsiveness*. Advances in service and quality depend on a firm's ability to increase responsiveness to its customers and markets. Project-oriented teams embody the flexibility to adapt and respond more quickly to changing environments and customers' changing needs. Compared to the more traditional vertical organizations, flatter, smaller, more flexible organizations can expedite more rapid and effective communication, tend to identify problems quickly, and generate and implement challenging, yet attainable action plans. Further, because of the structure and nature of cross-disciplined project teams, members tend to be more involved, better able to respond to customer needs, and more knowledgeable about the product and/or situation that confronts them.

3. *Cost-effectiveness*. To remain competitive, companies have instituted a variety of means to downsize or "rightsize" their organizations. Layers of management have been eliminated in an attempt to make the organization leaner and more flexible to customer demands. Project team-based organizations provide a means through which employees are delegated increasing decision-making responsibilities, which were once solely in the management domain.

4. *Responsiveness to technological change*. Cross-functional project teams, with their variety of specialties and disciplines, link technological processes quickly, to provide a more reliable, better-quality product. These teams mirror in structure the functionality involved with concurrent engineering, and enable back-end processes to be represented in the more front-end design and development procedures. This enables experts in manufacturing, for example, to provide their input early in the life-cycle phase of product development and to create faster, more efficient manufacturing processes. As production becomes closely linked to design, the team and all its experts become more sensitive to variations in the product and provide the communication necessary for the success of advanced technology.

5. *Reduction of job classifications*. When an organization becomes more project team-based, employees typically perform a variety of functions. Often, they will experience job rotation and the need to support and fill in for one another. Since project teams, like many

other cross-functional teams, are designed to facilitate job sharing and cross training, many organizations are responding by creating jobs that require workers who possess many skills and who can perform different job functions.

6. *New worker values.* Autonomy, empowerment, interdependence, and responsibility are the needs of today's workers. Factors such as task challenge, participation in decision making, and work design provide employees with a sense of accomplishment; which may be of greater importance than higher pay once an equitable salary is achieved.

The many success stories substantiate the powerful potential of more project team-based organizations.[2] Not only do teams improve quality and timeliness, but clearly change the way that people feel about their work and their organizations. The project team organization provides team members with a sense of ownership, which accompanies the team concept. This pride in ownership will help organizations to meet the challenges that lie ahead.

2.2 MEETING ORGANIZATIONAL NEEDS

There are many practical reasons that organizations may contemplate for installing a more project-oriented, team-based approach. Flexibility, responsiveness, cost-effectiveness, new worker values, and enhanced recruitment measures are just some of the reasons that organizations are moving toward project structures and team-based cultures. Regardless of the reasons, it is imperative that for teams to succeed, they must fit into the strategic direction and external market environment facing organizations. Regulatory influences, manufacturing processes, changes in technology or product mix, and competitive pressures are some of the potential influences that may affect an organization's decision to institute project team-based organizations.

Teams do not occur apart from an organization's work or business objectives. In fact, projects and the teams that make them a reality will have a major effect on some of the reasons for doing business. Bottom-line indexes, customer service ratings, and cycle times are all affected by the adoption of a team-based organization. Adopting a more project team-based organization will mean improvements in each of these bottom-line functions. Management must predict the results of the project team approach to

---

[2]Ibid.

the overall success of the business and to the competitive position of the organization.

Teams not only affect some of the more direct costs of doing business (e.g., costs that are observable and measurable) but will have invaluable indirect effects, such as improvement in employee morale, reduction in turnover, greater motivation, and fewer union grievances. Senior-level management must carefully investigate these effects and goal expectations when considering and supporting a true project team-based organization.

Finally, and perhaps most important, management must determine the overall "cost" of investing in the project team approach and the return on such an investment. Like any organizational change, senior management must support team-based concepts outwardly and philosophically in a variety of ways. For instance, senior management must concern itself with assessing the organization's long- and short-term business needs and in defining the role and importance of projects and project teams in accomplishing business objectives. Strategic and business plans must be consistent with a team philosophy and recognize the variety of levels of team involvement in accomplishing business goals. Appropriate action to ensure that the organization's strategic direction or mission aligns with necessary processes that enable project teams to operate effectively within the organization must be taken.

2.3 IMPLEMENTING A NEW ORGANIZATIONAL FOCUS

A more project team orientation must be part of the organization's business strategy. As such, teams must fit into the organization's business plans and senior management's assessment of how to deal with competitive markets, changing customer needs, the environment, regulatory influences, new product developments, and new technology. Project teams must fit neatly into the structure and culture of the organization, and not merely be implemented simply because it is "the thing to do."

When considering a more project-oriented team-based approach, it is important to consider the effect not only on employees, but on managers as well. Managers who prefer to maintain their traditional authority may resist the more participative approach that teams require. Management philosophies may need significant changes if wide-scale implementation to a more team-based operation is imminent. In some instances this may not be met with the most favorable reaction. Managers may feel threatened by the more project team-based orientation and become opponents of the system.

Contrary to what seemed to be a popular notion just a few years ago, *managers are needed by teams!* Teams not only require a manager's experience to guide them through their development process (as discussed further in Chapter 13) but also require the direction, information, and organizational connections that managers provide. Such information and direction prove critical to successful team functioning in a project environment, where timely, accessible information is critical to project success.

To explore the readiness of an organization in moving toward a more team-based project organization, a *steering committee* may be established to investigate the feasibility of project teams. It may also define necessary changes that the organization needs to make to assure a successful transition. The feasibility phase of organizational design is the time to explore the various types of projects in which the organization engages and determine the different structures that benefit it. It is also a time for data gathering and analysis. The size and makeup of the steering committee is contingent on the organization's specific mission. Its members should be high enough within the organizational hierarchy not only to influence change but to be supportive of employee involvement. Members must be visible and interact with all levels within the organization. The committee is charged primarily with determining if project teams are appropriate for the organization, for organizing the teams, and for overseeing the transition to a more team-based organizational culture.

2.4 BASIC REQUIREMENTS OF ORGANIZATIONAL FORMS

Project management involves the integration of company efforts and activities within time, performance, budget, and quality constraints to accomplish a unique and dynamic organizational goal. This activity takes place in an environment that is customer focused and customer responsive. The organization must be designed to facilitate the work of projects and teams. Minimally, the structure chosen must satisfy some of the following basic requirements of effective organizational design:

1. *Clarification of roles and relationships.* The organizational design or structure must indicate to each employee, manager, and specialist where that person belongs in the organization, where to go for information or cooperation, and to and from whom one escalates and/or seeks additional information or clarification.

2. *Efficiency and economy.* The organization should be nimble, encourage teamwork and communication (both internal and external), and should

actually *minimize* the need for elaborate reporting mechanisms. It should, therefore, minimize input effort and maximize output effort.

3. *Classification of company mission, goals, and objectives.* The organizational design should clearly indicate the strategy and direction of the company and how project performance leads to a specific result. The more people understand the vision or mission of the organization and the project, the more goal-directive their behaviors become. Unless the vision is understood clearly by all, employees will tend to cling to old methods and procedures, reducing efficiency and profitability.

4. *Clarification of individual and organizational tasks.* The organization must be structured so that all team members not only understand the nature of their own tasks and responsibilities but are also aware of the responsibilities of their counterparts or teammates on the project. They must also see how their tasks fit into the overall project (and organizational) mission.

5. *Encouragement of employee decision making and rapid customer response.* The organizational structure should strengthen customer focus and reaction time and encourage employee decision making. To accomplish this objective, management must minimize the impediments that may block the communication process. Structures that rely too heavily on "procedures" to get the work done may become cumbersome and make decisions difficult to attain. Decisions need to be made rapidly at the designated level and converted quickly to activities and accomplishments.

6. *Sense of stability and adaptability.* The firm must also be able to adapt to the changing market demands that occur around it. Organizations whose processes and structures support a sense of stability are at a major advantage. Open channels of communication and participative involvement by every employee help to develop a sense of employee ownership of the organization as well as that of its mission.

7. *Encouragement of innovation and development.* In today's competitive environment, characterized by megamergers and global competition, an organization's *survival is* based on innovation and change. Its structure must encourage the development of future leaders within all its "ranks" who contribute to the organization through innovative concepts. Innovation must address not only today's marketplace but tomorrow's as well. Through facilitating, mentoring, teaching, and serving, leader's within today's creative learning organizations guide the innovative contribution from all employees.

These seven basic organizational design requirements apply to *any* organization or firm, including that of the project organization.

2.5 EVOLUTION AND ORGANIZATIONAL DESIGN

To create a more project team-based organization, two approaches tend to predominate. One approach is more evolutionary in nature. It comes, in part, from successful experiences that emerged from what used to be referred to as the *quality of worklife movement*, which incorporated the original quality circles and problem-solving, decision-making teams.[3]

During the progressive movement of the 1970s and 1980s, organizations formed problem-solving teams sanctioned with improving organizational quality, productivity, and overall effectiveness. There was an increased effort to train supervisors and managers in participative approaches and to involve employees, to a somewhat limited degree, in organizational diagnosis through problem identification and solving processes.

As problem-solving teams began to reap the benefits of their training, quality and process improvement teams began to form. From this eventually grew the work team movement. Teams are now accomplishing a wide variety of activities that were once only in the realm of management. These activities include setting work schedules, dealing directly with the customer, setting production quotas, training, purchasing materials, and in some cases, hiring and firing.

Changing organizations follow an evolutionary process. It is difficult to restructure or change an organization in one fell swoop. Change of organizational design takes more than study and mandated implementation. It requires the gradual education and *demonstrated* commitment of all involved. Similarly, just as one would study the feasibility of a new technical concept or design, it is just as important to study organizational change carefully. Organizational design change involves both study and evolution.

A newly created organization has the luxury of designing its structure and work processes in whatever manner it chooses. Organizations that have made the decision to move toward team-based approaches must implement processes, procedures, and structures that support the team orientation. The majority of our efforts in this book concentrate on changing or improving existing organizations and implementing project management approaches. You will find, however, that the overall concepts apply to newly formed organizations as they do for existing ones.

---

[3]James H. Shank, *Team-Based Organizations: Developing a Successful Team Environment* (Homewood, IL: Business One Irwin, 1992).

2.6 THE IMPORTANCE OF ORGANIZATIONAL SUPPORT FOR MODIFICATION

In many instances, organizations attempt a more participatory project management approach and then wonder why such approaches never quite evolve or are fully accepted. They ponder why providing supervisors and project managers with more authority or employees with more input does not create the change they anticipate. Usually, attempts are made to modify the managers' behavior through companywide training programs. These programs may involve either information dissemination concerning employee participation and teams, or may attempt to teach the skills needed to create more employee involvement within the planning and scheduling processes. But trying to create change without modifying the organizational structure is like learning to ski without snow or skis. One lesson we have all learned is that if organizational change is to be permanent, organizational practices, policies and values must reinforce on-the-job behavioral changes. Many of us have had that awful experience of attempting to implement change after a positive training experience only to find the organization unsupportive. What usually takes place after such an occurrence is that we (the employee) either revert to our old behaviors or become so frustrated and mistrustful that we may even contemplate leaving the organization. Effective transfer of the skills involved in managing projects and programs requires continuous reinforcement and management support.

In *Beyond the Stable State*, Donald Schon argues that to truly change the nature of an organization, one must alter the way it functions.[4] Allowing changes in dress or work hours will not fundamentally alter how corporations function. Changing the hierarchical structure of the firm and the distribution of authority would have a tremendous impact.

2.7 STRATEGIES FOR IMPLEMENTING ORGANIZATIONAL CHANGE

One strategy for changing the core structure and nature of the organization has been through use of the *parallel organization*. Through this method, a steering and implementation committee is created. The committee exists in parallel with the formal organization. The formal organization continues to produce the stability and products known to the organization, while the

---

[4]Donald Schon, *Beyond the Stable State* (New York: Random House, 1971).

parallel organization, with its project and problem-solving teams, is responsible for (1) handling activities that deviate from the norm; (2) encouraging increased communication on an upward, downward, and/or diagonal basis; (3) identifying problems and developing and implementing solutions; and (4) discovering techniques for improving the organization and changing its manner of operation.

The task of a parallel organization is, therefore, to explore new methods of operating and to institute change.[5] It attempts new methods, such as managing by projects, through which the organization may be innovative and perhaps energize and rejuvenate itself. Project-oriented, team-based organizations allow for experimentation with increased problem-solving and decision-making techniques. Ultimately, the project team-based portion of the organization becomes the vehicle for transferring what is learned to the more formal structure.

Implementing a team-based parallel structure does not mean that the formal organization is neither adequate nor failing to accomplish its job. On the contrary, both the formal structure and the parallel organization should share in its successes, as they operate in congruence with one another. One structure supports the existence of the other. Stability and predictability of the formal organization give structure to the uncertainty and risk characterized by the parallel project team-based organization. One reason for creating a more innovative, project team-based parallel structure is to improve present operations, not necessarily eliminate them, and certainly *not* to condemn them.

2.8   RESTRUCTURING FOR PROJECT DEMANDS

Up-front restructuring of organizations to create a more project team-based environment will require a variety of changes. Such changes may include fewer levels, organizing around projects and project teams, changing reward structures, communications, decision-making and planning systems, and, in general, attitudes and behaviors toward people. In a telecommunications project on which one of the authors consulted, the manufacturing phase of the project moved toward cross-functional process improvement teams. These teams were the first attempt to bring development and manufacturing closer together, in an organization that had been traditionally organized around discrete, strongly functional lines. The *process improvement teams* were able to develop more horizontal lines of

---

[5]Shank, *Team-Based Organizations.*

communication and institute regular problem-solving sessions, that brought the one-time distinct organizations together. This "team within a team" concept is common for organizing along project lines. Frequently, the subportions of the work breakdown structure or project plans (discussed in greater detail in Chapters 4 and 5) represent portions of the work. A *work group* or *team*, consisting of personnel or specialists with similar job functions or tasks, must accomplish this work.

2.8.1 The Executive Steering Committee

One important ingredient in the reorganization, organization, or efficiency of a company is the *executive steering committee*. This committee is composed of key executives within the organization who have decision-making authority and responsibility for instituting corporate change and direction. The major function of the executive steering committee is to plan the corporate direction and the change strategy needed to fit new processes properly. It further monitors how change is being implemented and with what success. It provides continuity and long-term support to the project management team-based efforts, on either a division- or companywide basis. It plays a vital role in initiating project team efforts, in understanding how the change process and project team-based organization works and in making the effort to overcome resistance to the new approaches.

An important participant in the executive steering committee in terms of project success is the project executive sponsor. This person may come from any of the lines or divisions represented on the steering committee. The project manager should, ideally, report either directly or indirectly to the executive project sponsor. The sponsor, in turn, is responsible for representing the needs of the project at the highest level possible; assuring that the project enjoys reasonable time frames, resources and appropriate management commitments.

The role of the steering committee is pertinent not only to the project's success but also to the overall organization's success. Its overall function is to identify the ideal organization to accommodate what it envisions to be the firm's future. To accomplish this, the executive steering committee is involved in the following:

- Opening lines of communication within the organization
- Eliminating the barriers and issues that thwart the organization's development
- Making decisions collectively for the benefit of the entire organization
- Identifying issues that should be addressed across organizational lines

- Determining and prioritizing future projects
- Setting strategic direction for the company or strategic business unit
- Clarifying organizational goals and mission
- Acting as the focal point for consultants and other resources in developing change strategies
- Estimating resources needed to effectively accomplish overall project objectives
- Developing and working with implementation committees to define implementation plans
- Recommending targeted end dates

2.8.2 Matching Vision and Missions to Actions

We have all heard the expression that "actions speak louder than words." In the case of the mission statement and executive visions, this old platitude certainly rings true. Such statements have little effect on what people do if the actions of the executives who articulate them do not match the words. It is important that early in the implementation of a project management effort, the steering committee allocates a good deal of its time to establishing and implementing the philosophies and behaviors associated with increased involvement.

Higher management must set the strategic direction of the organization and oversee economic matters. They must restructure the system, to support the management of projects and the team approach, and reinforce positive behaviors on the part of managers, project managers, and team members. Training must begin from the top down and provide the support needed by project leaders and those working to implement a more project oriented team-based organization. Over time, the executive committee's main role should be to assure that the appropriate management philosophies; work processes and organizational structure are in place in order for the organization, and its projects, to succeed.

2.9 ORGANIZATIONAL FORMS AND PROJECT WORK

Arthur Chandler has expressed a notion that truly captures the requirements of organizing for project work.[6] Chandler's "form follows function" hypothesis quite simply observed that organizational structure in many of our more successful corporations is driven not only by the nature and

---

[6]A. D. Chandler Jr., *Strategy and Structure* (Cambridge, MA: MIT Press, 1962).

characteristics of the company's products and services, but by the pressures, challenges, and changes in the marketplace or environment surrounding them. This hypothesis was also held, almost a century ago, by Joan Woodward, one of the founders of the earlier concepts of management science.

Sound organizational structure is basic to effective project control. But just what makes an organization "sound" for project work? Are the newer, flatter, more flexible team-based organizations, characterized by lateral decision making, truly more efficient than centralized structures in which decisions and processes are determined by higher levels of management that possess responsibility for project cost and schedule control? Debate regarding this issue has persisted for decades. There is no perfect organization, nor is there a perfect organizational form. Regardless of whether the structure is layered or flat, if it is to be effective, the organization should offer the project several advantages. First, an effective project organization should assure that all functions and resources required by the project are assigned efficiently. Second, duplication of effort, or overlapping of roles and responsibilities, should be eliminated. By carefully considering and delineating the appropriate reporting relationships and lines of authority, organizations may avoid the costly *dysfunctional effects* of conflict and help to build more cohesive and productive teams. Moreover, efficient organizing helps an organization to be more responsive to both *project and customer needs*. It helps people to concentrate more on the external environment, where the true competition lies. The advantages of effective project organizing are as follows:

- Resource assignment parallel to project scope and technical requirements
- Clear definition of project authority and responsibility relationships
- Minimal duplication of effort
- Better understanding of market and customer needs
- Faster customer response
- Productive and cohesive teams
- Minimal conflict between team specialists
- Professional and personal needs of employees addressed
- Objectives and priorities clarified by communication and feedback mechanisms
- Early detection of problems or challenges leading to communication, reporting, and feedback mechanisms

2.9.1 Project Form, Project Function

The unique characteristics that render projects unresponsive to the more "traditional" management practices and policies have tremendous implications for creating an organization that facilitates the accomplishment of project objectives. As indicated in Chapter 1, projects typically are dynamic one-of-a-kind undertakings that have a specified beginning and end with a specific, well-defined accomplishment. Projects exist for finite periods of time and move through a sequence of activities, requiring constant surveillance and adjustment of people, materials, time, and money. This horizontal, integrative effort encourages a variety of technical and nontechnical opinions, which in turn creates a breeding ground for conflict. An organization engaged in projects must be structured to encourage and facilitate the accomplishment of project objectives. This requires planning, scheduling, and tracking activities that have a specific beginning and a well-defined goal or end. The systemic, integrated, multidisciplinary nature of projects calls for some form of horizontal coordination of talents, materials, and reward procedures to parallel the nature of project work. Organizing for project work must therefore, incorporate the following objectives: (1) to provide a structure in which people are encouraged to accomplish their specific tasks within the framework of the entire project, (2) to encourage collaborative efforts across the organization, and (3) to create an organization that minimizes the duplication of effort and maximizes cost efficiency. Figure 2.1 illustrates the systemic project environment.

2.9.2 Creating the Right Match

It is difficult to create an organizational structure that will address all project needs. There are certain factors to consider, however, when contemplating the right match. These factors include:

- The scope or magnitude of the project effort
- Geographic dispersion of the project team
- Government regulations and requirements
- Typical product line(s) and/or services
- Outside environmental changes and future operations and direction
- The customer
- Organizational culture, attitudes, and norms

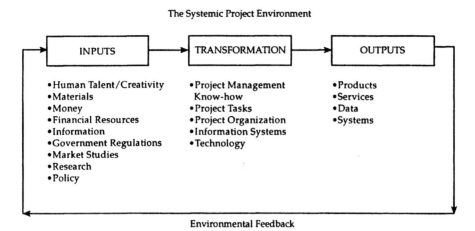

FIGURE 2.1

The systemic project environment.

· The professional and sociotechnical needs of the people who will perform the work

An organization that is effective for planning, tracking, and controlling projects must be responsive not only to the technical characteristics or specifications of the products or services it provides, but must also be aware of the social and professional relationships that its employees require. Restructuring an organization to meet the challenges and demands of today's competitive environment means more than analyzing and monitoring activities or tasks. It means assisting the informal "people" side of the organization to adapt to the changes that are occurring in the structure or form of the organization, and in the roles, relationships, and responsibilities that follow. Conflict, anxiety, and fear frequently arise from the confusion associated with organizational change. This exacerbates the existing pressures of meeting time, cost, performance, and quality constraints. A project organization that provides sufficient channels of communication and opportunities for people to meet and see the nature of their integrated efforts creates an environment that builds morale and forges cohesive, productive teams.

Modifying the organization means much more than merely changing the reporting structure or regrouping functional tasks. People enter into an organization with a variety of expectations. In today's workforce these typically entail the need for respect, professionalism, and personal growth. They also call for at least some degree of consistency. Organizational change must include the systems, processes, and procedures that address the social requirements of its primary resource — *people*. Resources such as materials, equipment, and facilities must be balanced against the human need for respect, involvement, and recognition to achieve a fully integrated organizational and project system. Such a system would consist of the processes and training that encourages increased interaction and minimizes the barriers to creating a team.

2.9.3   Symptoms of Inappropriate or Inadequate Project Organizing

No organizational design or structure can ever be perfect. As Peter Drucker aptly points out, at best an organizational structure will not cause trouble.[7] Some symptoms of inadequate organization or indications of flaws within the organizational design are:

- Little product pride or ownership exists among team players.
- Attention is devoted to one particular technical function or aspect of the project/organization, to the neglect of other components.
- Fingerpointing occurs across technical groups.
- Slippage is common, while customer responsiveness is negligible.
- Project participants appear unsure of their responsibilities, the mission, and their objectives.
- Considerable cost overruns exist, perhaps as a result of the duplication of effort or unclear delegation of responsibilities.
- Project participants complain of a lack of job satisfaction, recognition, or appropriate rewards.
- Lack of follow-through on customer complaints is common.

It is unfortunate, but when symptoms of inadequate organizing appear on project ventures, some companies respond quickly by throwing more time, money, or resources indiscriminately to an already weakened and inadequate organization. If one problem projects face is inappropriate organizational design, simply addressing the symptom while ignoring the underlying cause of the problem may leave the organization and its people

---

[7]Peter Drucker, *Management: Tasks, Responsibilities and Practices* (New York: Harper & Row, 1974).

frustrated and demoralized as projects continue to slip and conflicts continue to grow.

On the other hand, there are some companies that reorganize at the first sign of confusion. Such *organizitis,* as Drucker[8] refers to it, is a favored response of companies that have not clearly come to grips with certain fundamental changes in the complexity of the business and in the strategy and objectives that accompany the innovative organization of today.

A properly designed organization will facilitate project success. The most important factor to consider is that *the form of the organization must parallel the dynamic, systemic, and changing nature of the tasks, products, and services with which it is dealing.* Figures 2.2 and 2.3 outline the contingency

---

[8]Ibid.

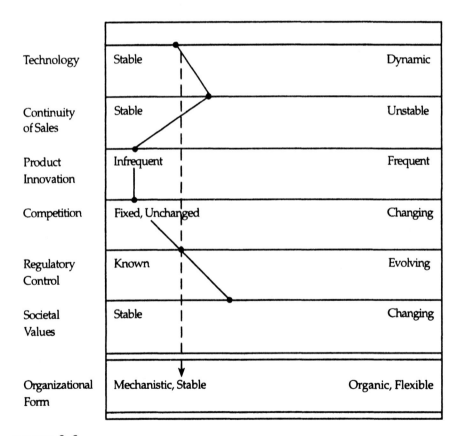

FIGURE 2.2

Contingency considerations in organizational design: public or regulated industry. (Adapted from Howard M. Carlisle, *Management: Concepts, Methods, and Applications,* Science Research Associates, Chicago, 1982.)

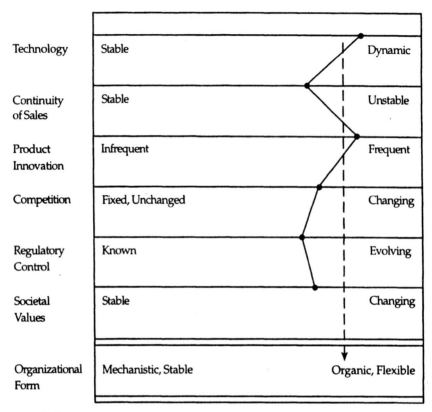

Technology	Stable	Dynamic
Continuity of Sales	Stable	Unstable
Product Innovation	Infrequent	Frequent
Competition	Fixed, Unchanged	Changing
Regulatory Control	Known	Evolving
Societal Values	Stable	Changing
Organizational Form	Mechanistic, Stable	Organic, Flexible

FIGURE 2.3

Contingency considerations in organizational design: high-tech information company. (Adapted from Howard M. Carlisle, *Management: Concepts, Methods, and Applications,* Science Research Associates, Chicago, 1982.)

considerations in contemplating organization design for different types of companies.

2.10   MODELS OF ORGANIZATIONAL FORMS FOR PROJECT WORK

There are a variety of designs that may be used effectively to structure and organize projects of various forms and functions. Since relations with customers, subcontractors, vendors, management, and other "stakeholders" (or parties interested in the success or failure of an organization) are affected by lines of authority, communication, and reporting relationships, it is wise for all project team members to be aware of the methods used to organize for project work. Then they may be better able to understand their roles and the roles and expectations of others, in order to meet project objectives with greater efficiency.

Earlier in this chapter we discussed in some detail the more team-based approaches to organizing a project. We shall refer to some of the concepts of cross-functional teams, which is a requirement of any project, throughout our subsequent discussion of some of the predominant organizational forms. Since it is virtually impossible for us to describe the plethora of organizational forms in which project management processes can or are being applied, three basic approaches are discussed here:

1. *Functional organization:* characterized by stratified levels of management, vertical lines of authority, and work that is partitioned according to specialties, disciplines, or functions. The main objective of organizing functionally is to emphasize the technology or expertise in each of the organizational lines.

2. *Project organization:* characterized by pooled or "chunked" groups of resources or specialties, banded together *temporarily* for a unique effort (a project), for a finite period of time, and under the *centralized authority* of the project manager. The object of the pure or direct project organization is, as the U.S. Department of Defense would call it, "command, control, and communication.

3. *Matrix or hybrid organization:* a mix or combination of structures, incorporating the functional, project, and *team-based* approaches. The objective of the matrix is to maximize the strengths of each of the contributing structures. The overall goal of the matrix is, therefore, project orientation, control, and technical excellence.

We discuss each of these organizational forms in greater detail in the following sections.

Organizing for project work may, therefore, assume a variety of forms, from the traditional functional grouping of work to the more complex, weblike relationships and structure of the matrix or project team-based organization. Each form has its own advantages and disadvantages. As stated previously there is no best organizational form. In designing the right approach, one must consider the typical size of the projects; geographic dispersion; number, complexity, and duration of projects; customer interface requirements, managerial and organizational philosophies and ideologies; and market demands. Only through the investigation and study of organizational and project requirements can the right match be made.

2.11 THE FUNCTIONAL ORGANIZATION

As depicted in Figures 2.4 and 2.5, the functional organization is characterized by the classic division of labor and services, and by vertical lines of authority and reporting relationships. Organizing along functional lines was once a common practice, especially among companies in which manufacturing was the predominant line of business. In these functional organizations, most company activities were more or less devoted to the assembly and production of standard products. Specialists, in turn, were grouped according to the functions they contributed. Each function or discipline created a pool of expertise, where knowledge and experience were easily shared. As part of their efforts to maintain more or less standard products, the functional departments are involved in a large volume of structured and repetitive tasks. Since activities are repetitive, uniform policies, procedures, and standards are applied more readily. When, at times, a project appears to be uncomplicated or of relatively short duration, the manager or supervisor of the lead or dominant function may be delegated the additional responsibility of managing the project. For example, in the case of a product that appears to be clearly electrical or hardware engineering in nature, the manager of electrical engineering or systems may be asked to pull the project effort together.

By now it should be obvious that to develop a total product or system, project efforts can rarely be contained within a single function. To develop anything, from a radio to a complicated switching system, hardware engineering is required to solicit the support of all functions (i.e., software engineering, mechanical engineering, documentation, drafting, and testing) to fulfill project specifications. The functional manager, wearing the hat of both the electrical engineer and project manager, must cut across organizational lines and obtain the required resources and commitment necessary to accomplish project objectives. This integration of project disciplines and activities becomes a more difficult task when delegated to any person in addition to his or her everyday functional responsibilities. Moreover, conflict is created frequently as contributing departments vie for project power and control. Technical decisions may, unfortunately, be made to enhance the power position of each function rather than for the good of the project. Decision making becomes a slow and tedious process, as commitment to what may be perceived as a low-priority project must be developed continually.

The advantages of the functional organizational structure in meeting project demands are as follows:

· Reflects logical organization of functions and disciplines

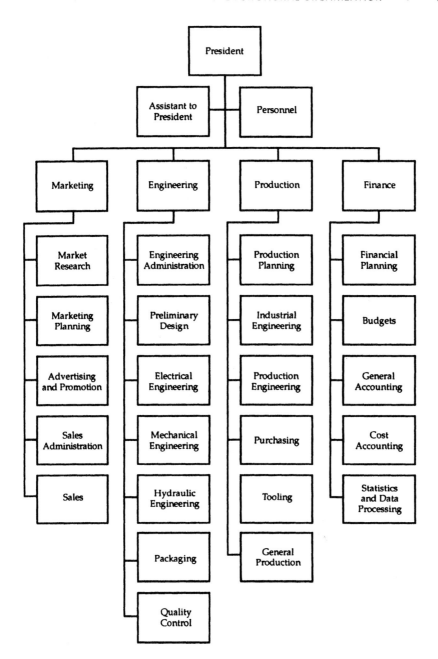

FIGURE 2.4

Functional organization.

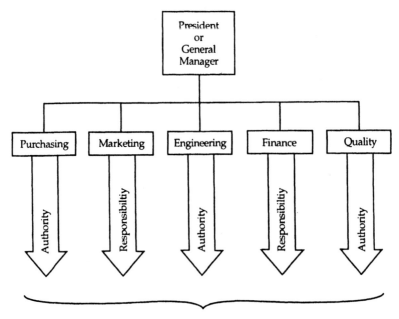

FIGURE 2.5

Vertical flow of authority and responsibility in the functional organization.

· Maintains power and prestige of the primary organizational functions
· Promotes continuity in functional procedures and methodology
· Allows control over personnel to remain in the hands of the functional manager
· Provides clearly defined communication channels
· Follows the traditional principle of occupational specialization
· Maintains a pool of specialists and talent
· Encourages technical and professional growth and advancement
· Develops technological expertise and future leadership within the specific technology

Disadvantages of the functional organizational structure are:

· Lack of a big picture or total project orientation
· Less concerned with project objectives and goals than with functional activities
· Decisions that tend to favor the predominant or lead function or the strongest group

· Lack of a project management emphasis with regard to scheduling, tracking, and cost accounting techniques
· Slow communication and decision-making processes
· Lack of flexibility and responsiveness to changing and dynamic project environment
· No singularly devoted customer focal point

2.12 THE PURE OR DIRECT PROJECT ORGANIZATION

As shown in Figure 2.6, the pure or direct project organization is also referred to as a *vertical project organization*. It frequently emerges from the traditional or functional organization in response to a high-priority project. It is, at times, likened to a task force, in which all players or team members

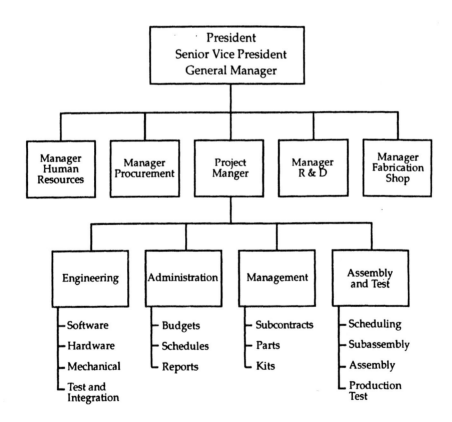

FIGURE 2.6

Pure project organization.

report directly to the project manager, for the duration of the project life cycle. The project manager's position is clearly separate from that of the functional manager, with the project manager now being granted complete line authority over project activities and project team members.

Complete line authority over project efforts affords the project manager strong project controls and centralized lines of communication. This may, in turn, lead to rapid reaction time and impressive customer responsiveness. Moreover, project personnel are retained on an exclusive rather than shared or "as needed" basis. Project teams develop a strong sense of product identification and ownership, with deep loyalty efforts to the project. Little doubt exists as to the nature of project activities, mission, or goals.

Pure project organizations are more common among technologically complicated unstructured projects, such as those in aerospace or commercial construction. These large-scale, multibillion-dollar projects can more easily absorb the cost of maintaining an organization whose structure encourages the companywide duplication of effort and the less than cost-efficient use of resources. In fact, one major disadvantage of the pure or direct project organization is the costly and inefficient use of personnel. Project team members are generally dedicated to one project at a time, even though they may rarely be needed on a full-time basis, over the life cycle of the project. Project managers may tend to retain their star personnel long after the work is completed, preventing their contribution to other projects and their professional development. Limited opportunities exist for technological exchange between personnel and projects. This, in turn, is a catalyst for frequent complaints among team members concerning the lack of career continuity and opportunities for professional growth. In some cases, project personnel may experience a great deal of uncertainty, as company priorities shift or project cancellation seems imminent.

Temporarily projectized organizations (or teams) are another variety of the pure project approach. This organization consists of a project team pulled together temporarily from their functional home base, led by a project manager or team leader. This project manager has been given a charter that establishes his or her authority to accomplish a particular objective. Lockheed-Martin Corporation has used this task force approach in the past; it was the Lockheed Corporation that dubbed this team approach "skunk works." IBM uses a similar approach to this temporary pure project organization when it "chunks" the organization and creates innovative cross-functional *design teams*. In fact, most software development companies frequently use this approach for short-term, high-priority projects.

Specialists who are involved in accomplishing project tasks and activities view the projectized task force quite favorably. It is usually not hindered by normal organizational restrictions and has many of the advantages of the full-blown pure project effort. Unfortunately, since such task force(s) occur in response to new business ventures, the advantages that they tend to carry do not occur without some drawbacks. Since the nature of the task force is *temporary*, the people who serve frequently are not relieved of their other functional responsibilities. Moreover, as in the case of the pure or direct project organization, mobilization of personnel to a central location is difficult and quite costly.

The advantages of the pure project organization are as follows:

- Centralized project planning
- Accountability clearly placed in the hands of a highly visible project manager
- Close coordination across project disciplines
- Maximum control over resources
- Commitment to project schedule, technical, and cost goals
- Strong channels of communication
- Early identification of potential problems
- Rapid reaction time
- Clearly defined customer focal point

Disadvantages of the pure (direct) project organization are:

- Lack of a "big picture" companywide orientation
- Increased organizational costs due to duplication of effort
- Limited technical communication and development across projects
- Possible conflict between projects over allocation of resources, including personnel
- Project personnel limited to a single project effort
- Tendency to retain personnel longer than needed
- Difficult to assign individuals to new projects at project completion
- Lack of career development and opportunities for project personnel

2.13 THE MATRIX ORGANIZATION

It is rare to find an organization that is strictly functional or purely project in form. More often, organizations represent a combination or hybrid of the two. The matrix organization is an attempt at creating a hybrid structure that captures and blends the advantages of the functional approach with

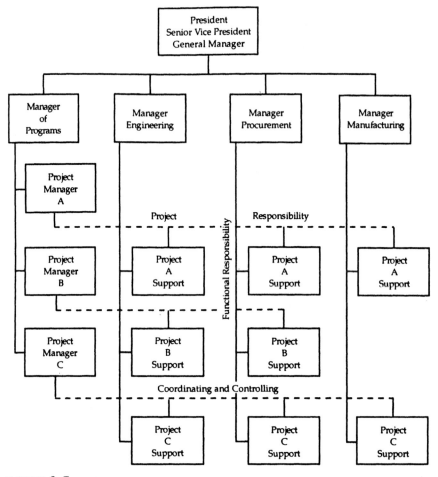

FIGURE 2.7

Matrix organization.

those of the pure project structure. Figure 2.7 illustrates a typical matrix organization. The matrix organization is an effort to maintain the consistency and excellence of technical or functional groups while addressing the need for a person designated as a specific project focal point. Matrix project managers are delegated the responsibility for engineering the project system across all contributing disciplines, departments, or sections. Functional managers, on the other hand, retain their authority over the technical specialists, ensuring that technical performance and quality standards are met, and overseeing the general technical integrity of the project. Matrix project managers are therefore delegated total responsibility and accounta-

TABLE 2.1 AUTHORITY AND RESPONSIBILITY RELATIONSHIPS IN A MATRIX (HYBRID) ORGANIZATION

Project Manager	Functional Manager	Project Specialist
Lead team in project scoping and scheduling	Participate in the development of project plans, including definition of support personnel	Prepare detailed plans consistent with overall plan
Represent project objectives in terms of cost, schedules, performance, and quality	Participate in determination of project resources	Execute tasks
	Define, with project personnel, actual project workload	Represent particular function in project reviews, cost estimates, and quality parameters
Lead project team with regard to definition of project mission, project schedules, risk assessment, and general direction	Assign personnel from unit consistent with project needs	Communicate
Request allocation of necessary resources with respect to technical needs	Maintain technical excellence of assigned human resources	
Serve as focal point/liaison for customer and vendors and alliance organizations	Recruit, train, and manage people within the organization	
Integrate and communicate project information in a diagonal, lateral, and upward manner	Assess quality of technical services and/or activities	
Facilitate the resolution of conflicts	Provide technical guidance and direction	
Report/communicate to management, the customer, and key stakeholders	Communicate	

bility for overall project success or failure, while maintaining a minimal amount of formal authority over the specialists who perform the work. Typical authority and responsibility relationships among the project manager, the functional manager, and the project specialist are outlined in Table 2.1.

Despite what can be, at times, confusing and overlapping authority and responsibility relationships, the flatter matrix organization has many advantages that make it a desirable structure for project work. A pool of project personnel, and resources is available especially for project work while maintaining their functional positions and standards derived from their home unit. Functional managers who ensure that project personnel are appropriately allocated, more readily meet shifting talent demands generated by the competitive global market and multiproject environments. Costly duplication of effort is minimized, as knowledge and experience are readily shared across multiple project efforts. More efficient use of resources is accomplished as project specialists are used on an as-needed basis. In fact, the shifting combinations of project personnel, characteristic of the matrix environment, challenges the organization to attempt to capture its form in any formal way.

Interactions in project matrix organization have often been likened to those of a marketplace. Negotiations concerning personnel assignments, priorities, equipment, and facilities are constant. Often, matrix team members complain of the continuous interruptions caused by numerous meetings they attend as they venture to accomplish their tasks on multiple projects. It is through such meetings; however, that the characteristic decentralized decision making and subsequent team empowerment occur. Group problem-solving and consensus-seeking activities replace individual and authority-based decisions, as commitment is obtained through a collective, participative approach. Members must quickly grow to value and incorporate the diverse input if project objectives are to be realized.

Project managers working within a matrix quickly come to realize that if they are truly to manage their environment they must win the support and confidence of their functional managerial counterparts. This is not necessarily an easy task. Functional managers will often resist being perceived as "service" organizations to project efforts, and in doing such, may continually and strongly assert their autonomy and power. To make matters worse, project managers may frequently be promoted to the job from the ranks of technical specialists and may be less experienced in the negotation and politics inherent in their new organizational position. They may be, therefore, less able to handle the conflict issues that lie ahead. If project integration skills are not learned quickly, the matrix frequently (and unfortunately) becomes a battleground for "we–they" issues.

When first formally introduced, the matrix organization was highly criticized for its complex web of relationships and confused lines of authority. When first created merely by superimposing a project organization over that of a strictly hierarchical functional organization, the matrix was cumbersome and lethargic. Decision making was time consuming, as the information required to achieve decisions were forwarded up, down, and across a top-heavy hierarchical structure. With the advent of flatter, more flexible restructured organizations, the matrix (when implemented through proper planning and scrutiny of present operations and senior management support) may provide a more flexible means of adapting to an ever-changing, dynamic, and competitive environment. It may further be the vehicle which creates the more team-based structure that is necessary for meeting project timeliness, quality, and overall success factors.

The advantages of the matrix organization in accomplishing project work are as follows:

- The functional organization and project efforts operate simultaneously.
- Advantages of the functional organization are maintained in an environment that is more project responsive.
- Functional managers maintain supervision over their own functional talent.
- Functional managers fill shifting talent needs of various projects.
- Functional managers promote desired technical and quality standards.
- A reservoir of talent is maintained.
- Personnel are used flexibly across projects.
- Opportunities exist for technological exchange.
- Project team members maintain a home base.
- Project planning, scheduling, and cost control techniques are emphasized.
- Project costs for personnel are minimized.
- Provides an appropriate customer focal point.

Disadvantages of the project matrix organization are:

- It violates the traditional one-boss algorithm.
- Authority and responsibility lines are, at times, unclear.
- Conflict frequently arises as a result of shared resources, multiple reporting relationships, and confused lines of authority.
- Initial adjustment of managers and personnel may be difficult and requires time.
- There is great demand for effective communication.

· Effort and time are needed initially to define policies and procedures.

· Response time may be slow, especially on fast-moving projects.

2.14   THE PROJECT OFFICE

Traditionally, the project office is viewed more as a *concept* than that of any specific place or locality within an organization. The project office may best be considered to be the "alter ego" or "other self" of the project manager. If one considers the project manager's primary responsibilities to be planning, organizing, scheduling, tracking, and controlling project resources, the project office consists of those key individuals delegated with the full-time responsibility for gathering and tracking the information necessary to support the project and ensure success.

Another view of the project office is that of a more centralized group, established to provide project management support services to all projects within a given division or strategic business unit. In either case, the project office is staffed with experts skilled in all aspects of project cost, schedule, and quality management. Project office activities therefore include:

· Customer communication
· Project management integration with functional entities through coaching and training
· Administrative support to project manager(s)
· Project forecasting and planning
· Project scheduling, tracking, and cost control
· Reporting to management
· Postproject evaluations

The structure and size of the project organization (e.g., matrix versus pure project), as well as the degree of autonomy the project office possesses, is dictated by the size of the company, the type of business in which it is involved, its number of locations, its quality processes and complexity of projects, and its management philosophies and technological orientation. In the case of a more centralized project office, one primary reason for its creation is the *institutionalization of project management*. A centralized project office, that is, one that supports a variety of projects within, for example, a particular strategic business unit, may become the functional owner or champion of project management development, including the implementation, monitoring, and continuous improvement of polices and processes associated with projects across the organization. Moreover, a more centralized project office is better able to provide a more global view of the status of projects in progress, which may prove to be critical to the strategic planning process. This global view may further enhance the utilization and subsequent optimization of resources across projects.

Regardless of size or location, the project office offers the organization a multitude of project-related benefits. The addition of a project office provides the overall organization with the following assets:[9]

- Improves delivery times (time to market)
- Results in more systematic process/product development
- Creates a project-focused environment
- Facilitates more proactive management
- Improves project and organizational communication
- Helps to identify, minimize, and mitigate risks
- Aids resource planning across multiple projects
- Provides a vehicle for anticipating problems as opposed to merely reacting to them
- Enhances "what-if" analyses and subsequent corrective action
- Defines both resources and the timing of resources across projects
- Improves business management skills throughout the organization

Typical roles (or functions) of personnel who are part of the project office are planning analyst, administrative coordinator, project planner(s), issues/change coordinator, and risk coordinator. The number of roles needed by a typical project office should not be confused with head count. One or more people, depending, of course, upon the workload, may accomplish assigned responsibilities. The actual number of staff needed to serve the project office can fluctuate with the size, complexity, and priority of the project. Key personnel in the project office may include:

1. *Project manager.* Frequently, the project manager may be considered the manager of the project office. There are instances, however, where a project management office may be superimposed upon a project matrix. In such instances, the manager of the project office may be lateral to other managers or directors and supervise several project managers, as well as project management support staff, as described above. This is discussed more fully in Section 2.15.

---

[9]Dennis Bolles, "The Project Support Office," *PM Network*, March 1998, pp. 33–38.

2. *Systems or project engineer.* The project engineer may be responsible for the technical integrity of the project and for the cost, schedule, and performance of all engineering phases of the project. This may include preparing technical scope documents, assisting in the development of the engineering plan and budget, and developing the project design.

3. *Project controller.* The controller is assigned primarily to a project office to assist the team and the project manager in the scheduling, tracking and controlling functions. This person may also be responsible for evaluating trends, predicting their effect on ultimate job completion and advising the project manager of possible remedial measures.

2.15 LOCATION OF THE PROJECT MANAGER

A question that frequently arises during the discussion of organizing for project work concerns the placement or level of project managers with regard to organizational hierarchy, and their power, prestige, and authority within the company. Naturally, the specific level at which a project manager is placed may not only determine the project manager's organizational effectiveness but also indicate the skills needed to generate commitment to project management procedures and methodologies across the organization.

Several factors determine the project manager's reporting position. Size, scope, importance, nature of the project contract (government versus commercial), and timeliness of the project are all critical elements. A basic rule of thumb, however, is that the project manager should report directly to the line executive or manager who has the authority to resolve conflicts that may occur across project divisions and who may be the advocate or sponsor of the project at a more senior level. This person may therefore assure the presence of sufficient resources and talent for the project. Thus, depending on the size and scope of the project and the organization in which it lies, the project manager could feasibly report to a general manager or to an executive vice president of a major division or strategic business unit, who occupies a role on the executive steering committee.

In most companies that are organized around a project matrix structure, project managers continue to report to a functional manager within their sphere of influence and expertise. For a large-scale project, involving several divisions, the project manager would probably come from the division with primary contractual responsibilities. The project manager's project-related activities, however, may be supervised or overseen by the executive sponsor or, in the case of a centralized project office, the manager of the project office.

Regardless of where the project manager is located within the organization, one fundamental means of achieving high-performing project teams is to establish a clear position of authority in project situations and a well-defined direct relationship with the customer. The need for clear and strong project manager authority is supported not only by common sense, but also by a variety of empirical studies. Many of these studies report not only a positive relationship between high managerial authority and perceptions of project group effectiveness, but also a strong relationship between ambiguous authority assignments and problem projects.[10]

2.16 PROJECT INTEGRATION

Quality project management *is* integrative management. It is a highly collective integrated process that requires the application of a variety of participative approaches that blend all the technical and nontechnical components into a cohesive and systemic whole. Determining the most efficient structure of an organization in today's systemic environment is only part of the task of organizing efficiently for project work. Effective, quality-driven project management begins with the integration of activities on a top-down, bottom-up, horizontal, and diagonal basis across all organizational boundaries or functions.

As depicted in Figure 2.8, the project may be conceived of as a complex, multifaceted jigsaw puzzle. Its pieces are derived not only from such functional areas as product planning and development, systems engineering, contracts and procurement, and so on, but from affiliate or outsource organizations and from the customer as well. These pieces must be fitted together neatly to create the totality or product. Anyone who has spent any degree of time working on jigsaw puzzles will immediately realize the difficulty, frustration at times, and the time-consuming nature of the task. Organizationally, this fitting together of the projects contributing pieces or functions, known as *horizontal integration* is best achieved through a participative integrative planning process, which is discussed in subsequent chapters. In fact, how projects are planned or the process of planning, may prove to be more important to successful completion of the project than the actual plan itself.

Vertical integration (Figure 2.9) refers to companywide efforts to link corporate strategic plans with functional goals and operational, or daily, project and unit activities. Successful vertical integration can be achieved

---

[10]Thomas I. Peters and Robert H. Waterman, *In Search of Excellence: Lessons from America's Best-Run Companies* (New York: Warner Books, 1982).

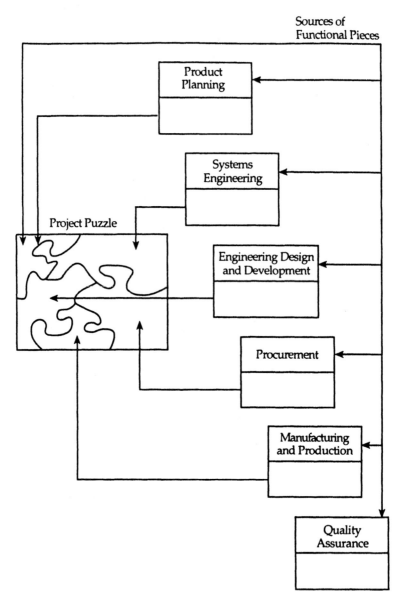

FIGURE 2.8

Horizontal organizational integration. (Adapted from William C. Wall, "Integrated Management in Matrix Organization," *IEEE Transactions on Engineering Management,* Vol. EM-31, February 1984.)

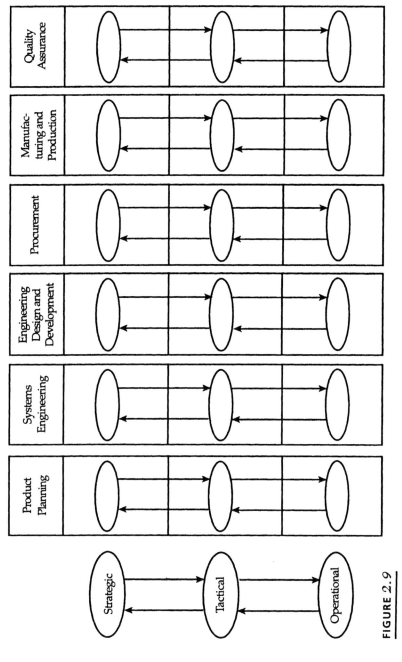

FIGURE 2.9

Vertical organizational integration is derived via integrated planning. (Adapted from William C. Wall, "Integrated Management in Matrix Organization," *IEEE Transactions on Engineering Management,* Vol. EM-31, No. 1, February 1984.)

through management practices that encourage downward communication concerning, for example, the nature of the project mission, customer expectations, fiduciary or budget constraints, and competitive practices. Moreover, downward communication must encourage and reinforce honest upward feedback and reporting mechanisms. Upward communication must pinpoint not only project successes, but potential risk, pitfalls, and problems. A "kill the bearer of bad news" philosophy serves only to bury problems that eventually lead to project slippage or technical failure.

Total project integration is therefore concerned not only with securing requisite resources in proper proportions, but also with achieving a unity of individual and group effort—a shared vision. Unfortunately, in our efforts to integrate the organization through such strategies as streamlining, restructuring, downsizing, and the design of flatter, less hierarchical structures, a project environment has been created that is highly competitive with less than sufficient resources. It is often characterized by more than an ample number of projects and an organizational structure and strategy that by its very nature is a breeding ground for conflict. By emphasizing the importance of consensus-seeking activities, responding positively to conflict and learning to apply a variety of communication and negotiation skills, project managers may derive more meaningful explanations and accurate predications of risk, for example, and establish more effective and productive project controls.

2.17   PROJECT INTERFACE AND AUTHORITY

Regardless of which organizational form or structure is decided to be most appropriate, conflict, slippage, and poor quality will eventually occur unless lines of responsibility, authority, and reporting relationships are clearly delineated. In the past, large, cumbersome matrix organizations developed poor reputations due to their confused multiple lines of authority and complicated communication patterns. Today's more contemporary, streamlined matrix organizations may prove to be more flexible than its predecessors were. If, however, reporting relationships and project responsibilities and authority continue to remain ambiguous, doing more with less will prove to be only a cliché. Effective team planning, coupled with ongoing communication, is truly the key to any organization's success. Difficulties that may naturally arise from complicated project-organizational interfaces may be overcome if clearly defined lines of authority and areas of responsibility exist.

			Team Member			
Project Breakdown Structure	Jim	Pat	Jerry	Lynn	Ted	George
1.0 Camera						P
1.1 Forward System		P				
1.1.1 Transport	S	P				
1.1.2 Structure						
1.2.3 Lens			S	P		
1.1.4 Shutter		S				
1.1.5 View Finder		P				
1.2 System Mount					P	
1.2.1 Frame					P	
1.2.2 Pedestal					P	

FIGURE 2.10

The linear responsibility chart.

The Linear Responsibility Chart (LRC) depicted in Figure 2.10 is a tool that we have found helpful in clarifying and documenting project team responsibilities and task relationships. It comes directly from the analysis of project tasks known as the *work breakdown structure* (WBS), discussed in Chapter 5. Although it may appear that the LRC states the obvious, it is our collective experience that sometimes the obvious needs to be stated. The LRC assures the project manager not only that every element in the WBS is properly assigned and accounted for, but that the work has been distributed equitably and that no one person or resource is overloaded. Like all project planning and scheduling tools, the LRC is an additional vehicle through which functional managers can buy-in, or commit themselves and their resources to a project.

2.18 CONCLUSIONS

A critical aspect of organizing for and implementing project management strategies successfully is not only to integrate project management approaches into an organization's structure, but to fully integrate these approaches and techniques into the organization's *culture*, daily experiences, or way of life. Despite the positive aspects of organizing in a more project-oriented, team-based approach, many organizations may not be ready for such an approach or for the processes that accompany it. The team concept, for instance, must fit comfortably into the organizational experience and be consistent with its values, objectives, and philosophies. Like any other organizational change or movement, project management processes must be supported in a variety of ways, by senior-level management. The decision to implement a more project-oriented organization and management approaches must take into consideration the strategic direction of the organization over the next three to five years. Project management approaches must be part of the organization's business plans. The approaches should fit into management's understanding as one critical way in which to deal with competitive global markets, changing customer needs, environmental and regulatory influences, new product development, and new technologies.

When organizing around projects and utilizing project management approaches extensively within an organization, it is important to consider how such changes may affect not only staff, but managers as well. Managers who prefer to maintain traditional lines of authority may resist the more team-based, project-oriented approaches. Changes in management philosophies may be needed if wide-scale transition to a more project team-based operation is imminent. Functional managers may become opponents of the system if they feel threatened by the more team-oriented philosophies and innovations. The best tool to overcome resistance to change is to involve those who will, in fact, implement the change to become an integral part of the planning of the change process.[11] Facilitative groups, in which attendees feel free to verbalize their concerns, provide an opportunity to have these concerns aired, addressed, and assigned to someone with the ability to see them through to resolution. Involvement in the actual change process by all who are affected, including unions, employees, and managers, will intensify the sense of ownership and lead to more likely successful implementation.

CASE STUDY: THE WINNING PROJECT (PART 2)*

We continue with the case study of Chapter 1.

---

[11]Deborah S. Kezsbom, "Implementing the Vision: Management's Role in Successful Project Team Management," *Proceedings of the 1994 Project Management Institute's 25th Annual Symposium*, pp. 684–687.
*This study was written for class discussion purposes only. It is not intended to reflect management practices or products of any particular company organization.

Creating a Project Organization

Knowing quite well that a project management system could not survive without the active, continuing, and visible support of top management, Steve lobbied vehemently and successfully for three new additions to staff. Now the CPPG could truly help the XB1 project not only to define the work properly but also to track and control project efforts effectively. Steve now was able to assign a project manager to each subdivision of the project. Further, he was able to delegate to the XB1 a full-time project manager, Roger Randon. Since Roger had worked with Steve on several other projects, the two men were highly aligned on their project management philosophies. Roger could now lead the XB1 project WBS (work breakdown structure) branch managers, as they were called, and report all occurrences to Steve in a timely and open fashion.

The first thing that Steve and Roger agreed upon was to assign a branch project manager from the CPPG to each function or project component of the XB1 project. These branch managers were to gather information and serve as liaisons between the project office and individual projects efforts in, for example, development, manufacturing, installation, and deployment. If difficulties were to arise in any area, the branch manager was to roll up his or her shirtsleeves as a member of the team and pitch in to help alleviate the limited resources.

It was just six weeks after the establishment of his new project organization when Roger approached his boss, Steve Golden, in Steve's office one chilly Tuesday morning.

Steve Golden: Chin up, Roger! It looks like your face is about to hit the ground. What's up?

Roger Randon: Steve, I'm sorry to say that the more I and my people work with these guys in manufacturing and deployment, the more we realize that we may as well be speaking a foreign language.

Steve Golden: What gives?

Roger Randon: Well, Steve, with the exception of some of the development people, I'm pretty sure the majority of the XB1 team have never had any formal project management education or training. It would help us a heck of a lot if we could run them through some sort of program and get them together to plan out this project jointly. You know, as a team!

Steve Golden: That's a good idea, Roger. If we can get everyone on the same wavelength, it should get easier for us to get the information we need to track and monitor this project properly.

Roger Randon: That's what I was thinking, Steve. In fact, if we get them comfortable enough with the process, we may be able to get them to commit to regular meetings like the training one and to regular brief reports.

Steve Golden: I was just speaking to Irving Pall in the New Jersey office. He was raving about some project management training and planning work they did out there. Let's give him a call and do something similar.

With the good of the project and the efficiency of their own jobs in mind, the CPPG team lobbied for a project management training and planning program. With the blessings of their management, they now had to convince the other areas of the project of the benefits of leaving their jobs for three full days and joining the planning workshop. After much scheduling and rescheduling, a session was finally planned for the spring and an April date was set. The session was to be held at a central location off site, in a hotel. Everyone was to stay at the hotel and since it was far enough away from the offices of any of the participants, no one would bother to leave the training premises to play "catch up" with their work. It was the CPPG's desire that in addition to training and planning for the project, some team building among the various project phase personnel would take place.

With the help of an outside management training and consulting firm, the CPPG team hosted a three-day project management training and planning session, specific to the needs of the XB1 project effort. The objectives of this three-day intensive session were: (1) to provide an overview or tutorial of project planning and scheduling methodology, (2) to clarify a project team mission statement, and (3) to assist project participants in generating a project work breakdown structure and in determining project scope and responsibilities. It was hoped that as a result of the training, a clear and uniform understanding of project and team management processes would be reached.

Overcoming Organizational "Silos"

Although resistance could be felt during the first day of the training session, it wasn't long into the first evening's work that functional barriers across the project began to crumble. Projected team members began to fully grasp the application and help of the work breakdown structure and assisted one another to define the project's scope more clearly. By the morning of the third and final day, although tired from the intense work effort each had experienced, branch managers were able to get team members to commit to a follow-up session. This session was planned to develop a formal master schedule of the entire project, as well as to gain commitment to other continuous processes that would help to assure a higher-quality project. These included regular monthly cross-functional meetings at the CPPG's headquarters in Chicago and weekly status reports to be given to the branch managers and eventually collated for Roger and Steve, who would report back to management. Roger wanted the subbranches to take a more formal project approach, including specifying short-term milestones and measurements to guide the process.

So as to keep the momentum going and to assume that the project's WBS was fleshed out to the level that the project office needed to track and monitor the project effectively, Roger's branch managers visited each of their assigned locations. They spent days with each subteam and the team leaders to define the WBS and its tasks more fully in terms of responsibility, duration, and risk. To keep conformity regarding estimation, the team quantitatively defined risk on an arbitrary scale of 1 (low) to 5 (high). The method helped the teams to focus more concretely on the riskier tasks and come up with contingency strategies.

Certain concerns that presented project risk were mentioned repeatedly at the planning meetings. The key risk areas among these were (1) risks related to the testing and production of the system, (2) risks related to the deployment, installation and startup of the system, and (3) risks related to the interface role of the CPPG and the customer organization.

Regarding development, personnel from several project areas voiced concern over the fault recovery software. Moreover, the circuit boards, which comprised the system, were of a size and complexity greater than manufacturing had ever dealt with. The aggressive rollout schedule and the technical competence and relevant experience of those who were to install the system were also of great concern. Each of these items was documented and assigned to an appropriately responsible party with a follow-up due date to report on the action item. In this way, Roger and the team actually built risk into the XB1's schedule.

Roger encouraged each branch team leader to voice his or her concerns and the concerns of colleagues. Without this time, how could the CPPG ever get a handle on the risk involved in accomplishing the project and installing the system flawlessly? If Roger had learned anything in his career, it was to listen not only to customers but also to the experts around him. He trusted their judgment about the systems and their own capabilities. What he wasn't sure of was whether when they were in trouble, they would trust him enough to tell him and the CPPG that trouble was on the horizon.

"We will just have to demonstrate to these folks that we're on their side," sighed Roger to himself.

The CPPG's presence at the various project locations proved to be invaluable. While now seeking higher staffing levels and beginning to have some success, each functional area was affected heavily by the limited resources. Just the presence of the CPPG branch manager helped the other team players to focus more on the XB1 project and more readily define the work. In addition, two (and sometimes four) extra hands (or heads) that were familiar with the technology were more than welcome.

Moving Towards a Sense of Teamness

Roger knew that a program of this magnitude and political climate could survive only if each phase or function of the project saw the great dependency that existed among them. He had learned the process of "tracking through participation" and believed that it was the only way to go to get from his peers full commitment and honest input into the design, cost, resources, and/or schedule needed. With the help of an outside facilitating company, Roger, Steve, and the CPPG team held a second planning session to determine the total duration or time required to complete the project. Although the project was large (over 500 tasks), the team took its time to integrate every task with its successor and predecessor tasks. They were also able to talk to the person who depended upon them to accomplish their end of the project and on those whom they themselves depended upon. Although Steve and Roger knew that as a project management office, they would only be interested in tracking the third- or fourth-level WBS tasks, the XB1 team brought the schedule down to the fifth, sixth, and in some areas seventh level of the WBS. In this way, team leaders were able to track their own branch areas to a more detailed task level.

The three days were intense, but by the end of the second evening together, the XB1 team began to understand each other's responsibilities more fully and saw more clearly the extent of the job they had in front of them. If this project was to be a success, by any measure, it was clear that communication among the various technologies had to occur on a regular basis. The dependencies were so great that any change in design would greatly affect manufacturing. Since integration between the new hardware, software, and manufacturing technologies was critical, especially in light of the anticipated high level of design changes, it was imperative that development and manufacturing personnel establish a collaborative relationship.

Questions

1. What problems continue to plague the XB1 project?
2. What factors helped the XB1 team to integrate project objectives and specifications?
3. Describe the culture of the original organization and its changes. How were these changes in culture initiated and sustained?

3

BUSINESS AND PROJECT PLANNING IN THE GLOBAL MARKETPLACE

*Small opportunities are
often the beginning of
great enterprises.*
DEMOSTHENES

3.1 INTRODUCTION

In the twenty-first century, as international boundaries blur and businesses become more global, organizations must engage in more rigorous and formal planning activities. When capital becomes tighter and the time required to realize projects shorter than ever before, more clarity will be needed in respect to missions, values, and strategy and in defining the results expected from long- and short-duration goals. More today than in the past, strategic planning is influenced by macroeconomic events of global proportion. As such, the shorter time frames to accomplish long-range goals may not allow for recovery from a failed product introduction when entering the new, worldwide markets. Even the mistakes made at home in realizing business and project plans only a decade ago, when our U.S. markets were captive, are no longer permissible. International competitors may already have introduced a comparable product here and won over potential customers.

In today's highly competitive, global environment, we must plan even short-term ventures lasting, for example, only a few months. Without a proper understanding of the *planning process*, organizations will fail to execute the formal strategies necessary to achieve overall business and project plans. However, to succeed, each planning team, from the higher-order strategists to the project team executing the next-generation product, must be empowered to perform and improvise to affect results. No longer will a few persons at the top attempt to master the complexity of the world economy. As management guru C. K. Prahalad admonishes, senior managers may define the "sandbox" but not every game that is played within.[1] As such, managerial intents must be shared with all contributors, especially among operations and project personnel who are closest to the customers in each international market.

In this chapter we review how top-down business planning approaches used previously are being changed to meet the challenges imposed by international markets. To improve a project's chances for success in this new environment, we explain the interrelationship of strategic and business planning to project planning and recommend nine essential planning activities to organize projects effectively. We also discuss procedures to confine product and project scope to reasonably manageable variables in light of the quadruple constraint facing project teams today: the concept of time to market and the impact of speed on international product development efforts.

3.2 TYPES OF PLANS

3.2.1 Integrating Higher-Order Plans

In the new millennium, *strategic planning* or *strategic improvising*[2] will continue to be the capstone of all organizational planning by establishing strategic directives that guide the organization's operating philosophy. While strategic planning still establishes a firm's relationship with stakeholders, employees, customers, shareholders, suppliers, governments around the world, and the community at large, strategic planning processes are no longer an activity in which only upper-level managers participate.

Like past planning efforts, strategic plans will still answer such questions as: What is our current business? What has it been in the past? What should it be in the future? While these plans will continue to project future business strategies based on technological forecasts of business activity as far off as 5 to 10 years, today they must position an organization within its

---

[1]C. K. Prahalad, "The Work of New Age Managers," in Peter F. Drucker (ed.), *The Organization of the Future* (New York: Peter F. Drucker Foundation, 1997, p. 166).
[2]L. T. Perry, R. G. Stott, and W. Norman Smallwood, *Real-Time Strategy: Improving Team-Based Planning for a Fast-Changing World* (New York: Wiley, 1993).

environment. Strategic plans must become more flexible and reversible during short-duration projects simply because there is more uncertainty associated with achieving long-range goals in the international business environment.

We are in a period of turbulence and dramatic change, from the balance of world economic power to rapid advancement of new technologies ushered in by the information age. Gone is the era where the United States was responsible for nearly 50% of the world's production.[3] So are what were once captive buyers for our products. Throughout the 1960s and 1970s, U.S. corporations were giants in their respective industries. This meant that they established the rules and essentially controlled the world's supply and demand. Their forecasts were reliable and the results highly predictable because there was no serious competition.

It was during the post-World War II production boom in the United States that strategic planning gained its popularity: hence its initial focus on capacity planning and the identification of large-scale macroprogram-level resources needed to fulfill economic obligations.

When in the 1980s and 1990s, strategic planning diverged from financial asset management of the past decades, these planning efforts broadened to encompass strategic management practices to improve company performance through organizational restructuring. These early efforts involved managers from all levels of the business, because all levels were needed to design the short-range goals and objectives to measure the effectiveness of these early improvements. But as these planning teams began to position the firm's operations around core service or product technology competence, the work that project teams would eventually carry out had to be linked to these strategies. Now, with less reliance on the formally disseminated financial plans prepared by top management, the cross-functional task force creates the action plans that drive the corporate vision.

In the new millennium, the world that project teams face will be quite different. It will be characterized by more rapid technological change and accelerated, global competitive pressures. As companies form more alliances to create new generations of technologies to advance the information age, project teams from different companies and countries will work together. Production procedures and tariffs, depending on the country of origin of the manufacturing facility, will differ. The realization of these planned efforts will be more risky and less certain than ever before. The best planning strategies in the new millennium take into consideration how

---

[3]Ibid.

different the world has become for project teams, where a quick response to accelerated, competitive pressures from companies that are simultaneously an organization's competitor and its partner is required.

What must emerge in this new era of rapid technological change is a more integrated business and product line planning process that allows for quick responses from all levels of the enterprise closest to the competition, no matter where corporate headquarters or control operations is located. Successful organizations readily communicate executive management's directives as input to its business plans and share this information with project team members to stimulate the ultimate introduction of more innovative market solutions tailored to individual customers needs around the world. There is little distinction between strategic and business planning among the following best globally practiced processes:

- Strategic intents or stretch goals that drive out-of-the-box thinking and shift planning away from business as usual
- A continuous improvement philosophy characterized by flexible and evolving planning processes
- Formal communication of the strategic plan viewed as a significant element of the planning process and a measure of quality planning
- Emphasis on action plans and strategic thinking
- Blurred distinction between strategic and business planning
- Orientation toward the future
- Vision- and market-driven planning generated by line management rather than a corporate planning office
- Reliance on task forces and multidisciplinary teams of personnel who ordinarily would not have a role in strategy formulation to brainstorm and initiate alternative courses of action
- Less emphasis on financial analysis and more focus on qualitative analysis
- Linkage of strategy to incentive compensation, the business plan, project budgets, and resource allocation

Because the span of control for project members has increased, the *tactical planning* process must also become less hierarchical. In the past, tactical plans spanned intervals of six months to two years, were performed by middle management, and involved program or project elements more micro in nature. For instance, tactical plans detailed company policies so that general guidelines for decision making and individual action could be reasonably coordinated among projects and made known to all project personnel.

Tactical planning will continue to be futuristic since it involves determining how the corporation will realize long-term strategic plans. In the future, however, tactical planning will have to be more flexible and responsive to changes in the organization's internal and external environments. It must also be flexible enough to allow an organization or even an entire enterprise to respond to changes within the confines of its long-range plans. For instance, changes in a firm's procedures for implementing engineering design changes to improve product performance and be competitive seldom need long lead times to complete and should be accomplished by engaging those who understand the impact of the procedural change. In such as a case, top-down long-range planning efforts would be ineffective.

Similarly, *operational planning* must come under the jurisdiction of project team managers who can determine how specific tasks can best be accomplished on time given the resources available. Traditionally, these plans are separated fiscally from the higher-level processes of strategic and tactical planning with the assignment of operational budgets to functional organizations or individual departments. In many enterprises, today, operational planning is still performed by functional management rather than by the project team. When operational planning for the organization is performed by functional management, these managers must allocate proper resources to the project managers to satisfy answers to such questions as: What do we need to do? When do we need it? Can we do it? How long can we take? How do we monitor progress toward achieving these commitments? In some enterprises, implementation of these plans is formalized through contractual arrangements spanning the project's duration. To assure commitment to the plan, some companies require project teams to negotiate and sign formal agreements with the operating units and functional departments that provide the resources that must be allocated to complete project assignments.

Usually, operational planning is performed as a calendar-based exercise. As such, the process limits both the functional and project organization's ability to be receptive to market opportunities and threats—challenges that can be met by short-duration program efforts requiring immediate funding and personnel. Operational plans vary in duration from one week to one year. They must be revised at frequent periodic intervals. This type of planning is typical of such high-tech endeavors as telecommunication research and development as well as the integrated-component, device design, and software applications industries. Each of these businesses is characterized by a high degree of uncertainty and rapid technological change. Typically, operational program or project plans focus on product development, process engineering, and integrated manufacturing operations.

Even operational market planning has moved down the organizational hierarchy today, because business strategy is tied to those who have greater knowledge of the customer. These short-range plans provide the detailed objectives and budget targets each departmental team leader or project team manager must now manage. The generation of corporate line-of-business operational planning has also moved down in the enterprise. Still established according to calendar-determined sequences of events, success today depends on its implementation across global boundaries and full integration of the effort at all three planning levels.

To summarize, Table 3.1 contrasts the characteristics of each planning effort and identifies who in the corporate hierarchy is responsible for the integrated execution of all three plans.

3.2.2 Business Planning in the Project Environment

As project environments become more and more complex, a project management team must understand how to finance its engineering and new product development efforts. As we enter the new millennium, all projects — military, civil, and commercial manufacturing — must be strategically driven from a financial perspective because competition around the globe is keener than ever before. *Business plans* capture the reason a business enters into product development. In some cases, competitive pressures will persuade a company to invest in the development of a product from which it does not expect to earn a profit because its purchase will entice customers to buy other core product or service offers that are more lucrative. Otherwise, the project will be expected to return a financial reward over its lifetime.

Projects must be managed to follow a firm's business strategy, and project finance mustn't clash with these objectives. Typical misunderstandings can be minimized when both project managers and product developers (1) understand their firm's business planning processes, (2) learn to assess a project's financial performance using pertinent accounting principles, and (3) use these principles to evaluate and take alternative courses of action when necessary.

A fundamental rule for the project team to learn is that a good investment must have a positive net present value (NPV) and bring value to customers. In other words, during its lifetime, the costs of developing

TABLE 3.1 CHARACTERISTICS OF THREE TYPES OF ORGANIZATIONAL PLANS

Characteristic	Type of Plan		
	Strategic	Tactical	Operational
Hierarchical level	Senior management • Chief executive officer • President • Vice presidents • General managers • Division heads	Middle management • Functional managers • Product line managers • Department heads • Program managers	Lower management • Team leaders • Coaches • Project managers
Time frame	Long-term time horizon (generally, 5–10 years)	Intermediate near-term time-horizon (6 months–2 years)	Short-term time horizon (1 week–1 year)
Content	Vision of the firm • Corporate philosophy • Mission Identification • Strategic business units Budget consolidation of strategic and operational funds Allocation of all resources	Formulation of: • Corporate performance objectives • R&D strategy • Planning guidelines • Broad action programs Budgeting at the business level Deployment of all resources	Work authorization • Monitoring of functional level cost and control system Accountability for performance standards of program activities and events
Approach	Formal, inflexible, but becoming less rigid, generally irreversible	Formal, less rigid, reversible	Formal, flexible, subject to constant updating

and producing a product must not exceed the price charged to customers in exchange for this value. Before any commercial project or civil program is funded, this question must be answered: Will the value of the product or service proposed generate enough volume with targeted customers at a price that exceeds the costs to supply?

There are five basic steps in the business planning process for a project:

Step 1: Form a business concept evaluation team. This team includes members from marketing or finance and the future project team. The charter of the concept evaluation team is to determine if the proposed project investment brings value to the corporation and its customers. This effort begins by calculating the present value of the cumulative future cash flows the investment will return.

After all annual costs and taxes have been deducted, including depreciation, the project's income is known as the *cash flow*. Cash flows are discounted based on the principle that capital earned or spent in the future is worth less than capital earned or spent today. The rate at which the cash flows are discounted is known as the *required, internal return rate* (IRR), or *discount rate* (DR). The return rate or discount rate is the rate the firm could obtain from a comparable investment of its capital. This rate is determined subjectively by each firm and often varies depending on where in the world the capital investment will be made. The NPV of the investment is determined by discounting all of the project's cash flows using the company's discount rate. The planning committee, or decision board reviewing the concept evaluation team's business plan, is able to use this measure to judge whether the project will yield a return commensurate with its risks. This selection committee will also use the project's internal rate of return to compare the investment program's return to that of alternative projects.

Step 2: Prepare the business plan. The plan must describe the business opportunity or problem the proposed project or the program addresses and discuss how it fits strategically into the existing business. All real and perceived market and technical risks and uncertainties must be identified. Good plans include proof of the business concept. This may be an actual system or prototype, field test results, or documented market research where product or service concepts were presented to appropriate market segments and the prospective customers willingness to purchase was tested. Business plans should also contain an analysis section that presents, for each alternative proposal, a *cash flow model* and *sensitivity analysis*, a view of how the known risks associated with realizing the project will change its NPV.

Step 3: Present the business plan for review. Discuss the proposal informally with all stakeholders: the members of the decision body (corporate planning review board or steering committee, venture capitalist, export–import bank, or end users commissioning the plan) who have the authority to finance it. By discussing plan details informally with various representatives from the potential funding organizations and obtaining their buy-in to the concept, the team will learn how to answer specific concerns that individual stakeholders may raise later before actually funding the project. Review the outline of a typical business plan presented in Appendix 3.1. This summary provides very useful information that concept teams may need to supply to a decision body when seeking funding for a program or project.

Step 4: Prepare a preliminary program or project plan. Once potential and future projects are matched with the firm's market strategy, some funding is usually allocated to prepare a *preliminary program* or *project plan* summarizing the project. This initial plan gives the organization's senior management a high-level indication of who will carry out the actual project work, how and when it will be completed, and at what cost. The team charged with writing the preliminary program or project plan may be the same team responsible for the business plan or a new team selected by senior management. Several of the original team members who championed the business plan usually stay on the team. Representatives from every discipline who may be involved in the project should have input to the plan.

Step 5: Present the preliminary program or project plan for review. When the preliminary plan is complete, the team will probably be required to present and review it with senior management before additional funding is assigned. The confidence the team instills in the plan often influences management's decision to fund or not fund a project. Project teams that develop the plans in process are rarely successful. Later, as more details become known, the team must revise the preliminary plan. A prescribed process for creating the more comprehensive *project summary plan* is described in Chapter 5.

3.3 CONSEQUENCES OF POOR PLANNING

As we have seen, the purpose of higher-level strategic and tactical planning is to identify the efforts that managers at the project level must make to move an organization toward realizing its strategic goals and improving its market position. For this reason, all the planning activities for the project — its scheduling, tracking, and team building approaches — must be tightly coupled to the project's business objectives. Also, requirements to create a project team place great demands on organizational resources and personnel needed to complete project tasks. Therefore, to facilitate the utilization of these resources in a timely and cost-efficient way, project plans must document not only the immediate team members' work, but all required commitments from other organizations. Moreover, a good operational plan must identify not only required resources (human, capital, and equipment) needed to staff a project, but also development schedules and the range of associated durations for each major project activity. In addition, it must delineate managerial accountability and authority. Finally, a good plan will provide for a system for measuring the results of the planning process as well as the work performed through the development of a test plan.

A good plan also documents the strategies used to structure the project and achieve the successful execution of project objectives for future reference and adaptation by others. Unfortunately, this is not always the rule. Postproject audits, conducted over the past several years in organizations around the globe, reveal that many programs often do not succeed. Usually, such negative results are the consequence of a planning effort that failed to match the characteristics of the project to the organization's overall managerial approach. Postproject audits also caution teams to examine the critical nature of each the following conditions as factors that can potentially influence their overall success or failure:

- Disagreement over the scope, length, and/or technological magnitude of the project
- Aggressive time-to-market demands, inadequate budget to expedite project work, or product introduction schedules that are perceived by the team to be too tight
- Geographic dispersion of project specialists
- Government regulations and/or technical requirements that vary from country to country
- Standard manufacturing and quality processes from which deviations are not allowed
- Organizational structures and/or reporting relationships that hinder communication among specialists who need access to project information

Projects that fail to be responsive to these challenges never really get off to a good start. Personnel quickly become disillusioned when chaos prevails at the project level. Project objectives are misunderstood, resource allocations become inadequate, schedules slip, and budgets are grossly overspent — all because project requirements were not sufficiently or clearly defined. Sometimes the various functional departments have developed their own planning documentation, and a newly assigned project management team may discover that no effort was made to integrate the plans across functional or operational divisions. Furthermore, the organization's failure to assess project management requirements periodically or to use a systematic decision-making process to monitor any apparent deviations from a baseline plan almost always results in uncertainty, discontinuity, and seat-of-the-pants crisis management. In short, progress is impeded, which results in a product that is delivered late or released into an unresponsive market. These occurrences may also succeed in undermining the authority of project leaders who manage the team.

3.4 WHY PLANS FAIL

No matter how hard project management tries, planning is not perfect, and plans sometimes fail. There are several reasons that even the most systematic efforts go awry:

- Business strategies are not understood at the lower organizational levels.
- Strategic or long-range planning objectives are not redefined at timely and regular intervals when technology or external forecasts change.
- Competitive pressures to generate "a me, too" product or service outweigh management's best judgment about the firm's knowledge of the technology or ability to finance the investment.
- Resource allocation is inadequate and corporate commitment for funding new projects shortsighted or insufficient for developing technologies.
- Planning is performed by a planning group without direct input from functional or line managers and other lead project personnel who will complete the project activities.
- No standardized planning process is in place, or those assigned to complete the plan do not follow the process because the benefits of using it are not known.
- Planning stops short at the master schedule level; those who must complete the detailed planning and control work are not consulted until the project is well under way.
- Plans are based on insufficient data.
- Plans are not updated at periodic intervals to let personnel in new projects know what staff is available to be assigned other work.

- Not enough time is allowed for proper estimating or scoping of the work.

- Estimates are best guesses not based on benchmarked standards for the industry, on research, or on historical data gathered on similar projects in the firm.

- Staffing for multiple project efforts in the same organization is done in isolation.

- No one knows the project's initial staffing levels or overall requirements; typically, qualified personnel are available only part-time or simply are unavailable when needed.

- The major work activities and tasks that must be performed have not been defined, nor has anyone scheduled their associated milestone dates.

- End-item specifications are nebulous and project deliverables not outlined in contractual agreements.

- Project personnel come and go constantly, and little regard is paid to the resolution of the resulting schedule and resource conflicts.

- The project manager does not have the authority to negotiate scope, budget, or resources with project stakeholders, departmental management, or the customer.

- Top management neglects to establish an organizational structure designed to fit specific projects or direct the information-flow requirements of the project.

- There is no global exchange of product ideas — projects do not co-locate R&D resources with production facilities to permit daily, concurrent engineering.

- Financial estimates are poor; the value of the project may erode due to inflation, new taxes, labor actions, or unforeseen forces of nature and strife.

- The lack of maintenance-of-value clauses to protect the project from currency fluctuations in an international contract causes the project to overrun its allotted budget.

- Exposure to foreign currency transaction costs jeopardizes the project's ability to budget accurately. Also, if the project's financial resources cannot be repatriated or converted from a local currency because of exchange controls or other government action, failure to set aside large cash reserves may prohibit the project from funding relocated work.

3.5 PROJECT PLANNING TECHNIQUES

A project's success depends on *planning, monitoring,* and *tracking*; these are the techniques of project management. Refining these techniques produces optimal results: meeting predetermined objectives through the best use of available resources in the least time and at the least cost. Integrated project planning must take into account the following concepts:

- Formal process whereby a truly cohesive team is formed that commits collectively to carrying out mission-critical project goals
- Measurable project objectives and project management strategies to realize these collective commitments
- An evolutionary life cycle of project planning
- Preparation of a work plan that describes all the project deliverables and summarizes the estimated costs and schedules to complete them

Thus, to be minimally effective, project planning must include:

1. A project management requirements analysis to determine a valid methodology based on project requirements for organizing the functions needed to manage the project
2. Overall project specifications and statement of work (SOW), including technical performance, usability, reliability, and quality requirements, as well as target completion dates and cost objectives
3. Strategies for accomplishing measurable project objectives, including the use of a work breakdown structure (WBS) to delineate each of the tasks that need to be performed
4. Development of contingency plans and risk mitigation strategies for high-risk portions of the project
5. Forecasts of all project completion dates from all project support organizations and a statement indicating the degree of reliability of these estimates
6. An organizational structure and design for the project accompanied by a detailed description of the authority, responsibility, and corresponding duties of all personnel: managerial, professional, and support staff
7. A schedule and budget for accomplishing the work activities that cover every organization responsible for project deliverables, including contingencies, meeting, and report due dates

8. Policy statements for making decisions that balance product quality and technical standards against the cost for completing by specified dates

9. Standards or quality process assurance guidelines that define the criteria for acceptable and unacceptable individual and team performance

3.5.1 Project Management Requirements Analysis

The *project management requirements analysis* is an orderly and detailed technique for outlining project methodology and procedures.[4] Its primary purpose is to (1) assess and validate present strategies used in accomplishing project objectives; (2) determine specific project requirements contingent upon scope, timeliness, quality criteria, regulatory requirements, and organizational structure; and (3) pinpoint strategies that may be improved upon, eliminated, or added to increase the likelihood of project success given the special characteristic and demographics of the project.

Project teams have varying and changing needs that reflect the wide array of the members' professional backgrounds and interests. The very fluid nature of project work essentially dictates that it be managed using approaches that are quite different from those used in functional organizations. Since there is tremendous potential for conflict in the project environment, the management strategies must encourage team work, frequent human interaction, a participatory spirit, and a reward structure to celebrate milestone events and major project accomplishments. Strategies that are effective for managing projects must be sensitive and responsive not only to the nature and challenges of the specific product technology, but to the social and professional relationships that the structure creates.

The approaches determined by the project management requirements analysis will also vary, depending on the needs to communicate and facilitate group cooperation. For instance, long-term projects will require regularly scheduled status meetings to review and report on progress, while short-term projects may demand even more frequent but brief meetings of key team members to address problems and find solutions quickly.

Management may also conduct a project management requirements analysis to identify appropriate strategies to foster professional growth and a sense of affiliation among team members. Since the project structure will create new reporting relationships as team participants are regrouped to

---

[4]D. S. Kezsbom, "Match Strategies to Structure with a Project Management Requirements Analysis," *Industrial Engineering*, Vol. 23, No. 4, April 1991, pp. 56–58.

work on project assignments, the kind and frequency of rewards sought by the team will be out of sync with the annual performance evaluation cycles conducted in their functional organizations.

We have used the following project management requirements analysis model with a diverse group of international management teams and recommend it as a first step in planning a project. The analysis process allows a new project team to match project needs to the planning approach.

Step 1. Conduct an assessment of critical issues and/or needs of the project's sponsoring organization.

(a) Meet with the sponsor to establish a clear vision of what is required to manage the project.

(b) Document the perceived planning, scheduling, tracking, and team building needs.

Step 2. Create a project requirements interview document that will help:

(a) Pinpoint present project management strategies.

(b) Determine subsequent strategies to plan schedules, allocate resources, track progress, and so on.

(c) Define responsibilities, reporting relationships, budgets, project management scheduling software, and other higher-quality management tools that will be used to manage the project.

Step 3. Identify project management subject matter experts or team partners to be interviewed based on the project's technology or functional mix; these players will probably become the key implementers and core team of specialists involved with the project.

Step 4. Use a team of objective interviewers to collect the data.

Step 5. Verify the interviewer's observations through independent synthesis of the data collected.

Step 6. Invite personnel involved in the interviews who will lead the project in a feedback session, where they will document a strategy to manage the project.

3.5.2 Project Objectives and Contingency Planning

There are several techniques used in commercial and civil program management to assure that a project realizes its intended outcome and that the organization achieves its intended market strategy. First is the creation of a set of *quantifiable project objectives*. These are criteria that must be met for the project to be considered successful. In Chapter 5 we recommend a process to involve project team members in the writing of these objectives. Their participation in the preparation of project objectives is imperative since the team's performance is measured against these criteria. Objectives for a new product or service offer may include, for example performance targets (usability, serviceability, reliability, and quality), customer satisfaction ratings, financial targets (return on the investment), and market strategy (share growth, segment penetration, recovery, etc.).

It is also good project management practice to get the project participants' buy-in to mission-critical project objectives. The team members' collective understanding of project objectives helps optimize project decision making and lets the project manager readily call on the affected organizations anytime for advice when choosing among alternative courses of action. When analysis is carried out beforehand and documented in the project objectives, a project manager is in a better position to help senior management make more informative trade-off decisions. Especially when considering, for instance, whether to reallocate budgets to develop a leading-edge product that may not be immediately profitable or to continue the funding of another that uses less advanced technology but appears to satisfy all existing market requirements.

Another technique to avoid failure is to assess risks associated with the project and plan contingency allowances. *Risk assessments* should be conducted not only at the project's outset but also continuously during all subsequent phases. Because anything is likely to happen to take the project off track, good *contingency planning* is dependent on the project team's constant monitoring or tracking of critical external factors. This surveillance permits the team to react as quickly as possible and to control deviations from the original baseline by putting alternative plans into place.

The following list of contingencies, although not exhaustive, includes most of the major risks that the project manager must plan for:

- Errors in estimations of the money, time, and personnel needed to complete the project
- Management of the project's *transaction exposure*: potential losses from the sales of securities or property used to finance the project in case cash problems arise due to fluctuations in foreign *currency exchange rates* (FX), *interest rates* (IR), and *commodity prices* (CP)
- Design changes in the product that necessitate additional budget for fabrication and final production
- Omissions from initial estimates to control for changes in scope specifications, technical requirements, and estimated costs
- Cash reserves to purchase additional resources for expediting the work

- Variations in the averages used to estimate resources required for project completion
- Risks associated with changes in regulatory procedures decreed by government policy
- Price adjustments to the value of a contact that are owed to subcontractors on U.S. government and international production contracts to pay for variations in the features delivered

The rule of thumb for estimating reserves for contingencies is 10 to 20% of the total project budget. Yet because the level of risk and uncertainty is greatest during the preliminary startup, design, and development phases of most technical projects, these estimates may need to be larger. The amount may also depend on the level of risk associated with the technology under development. The project team may obtain their best forecasts from analyzing past performances of similar development projects and research endeavors.

3.5.3 Common Development Processes

Throughout the late 1980s and 1990s, commercial industry embraced the quality management practice of establishing *common* or *structured development processes* to expedite time to market and avoid failure. Common development practices set the framework for all project activities shared by team members. Because the entire organization is familiar with the same process, individual teams do not spend time reinventing how they will proceed. Standardized steps to complete activities are consistently applied to all projects, ensuring higher compliance to original customer specifications within the initially estimated completion intervals. End-of-phase reviews are completed, and the decision to rework the plan or to proceed on course is publicized to the entire organization. These end-of-phase reviews are an excellent way to gauge the quality of the endeavor and are a market-driven strategy for managing the entire product life cycle. In many companies they act as gates for filtering out inferior deliverables. In the new millennium, businesses will continue using structured new product development processes, as they are pressured in the global market to compress the time required to complete the development of a product or service. Many organizations will streamline how they implement the models they introduced in the 1990s as they strive for success through product and cycle time excellence (see Chapter 4).

3.5.4   Scope Management

Finally, a planning technique that the Project Management Institute expounds in its Body of Knowledge is to define the scope of the program effort. *Project scope management* includes all the processes required to ensure that a project includes all the work, and only the work, needed to complete the project successfully. Scope may refer both to the product or service and its features and functions, as well as the work that is necessary to deliver to market a product or service with the functionality specified. *Product scope* is the degree of functionality developed and incorporated into the product. Completion of the product scope is measured against project specifications, while completion of the project scope is measured against the project plan.

The product *specifications* represent a statement by the customer of what that customer wants the product to do. However, in the commercial marketplace where there is no one, single customer, marketing must inform the project team what the targeted customers require, often needing to differentiate between their "wish lists" and what they are probably willing to purchase. For example, product specifications generally enumerate expected performance and quality criteria. It would not be uncommon for customer specifications for a communications systems to state (1) that a transmitter be able to transmit at 347 MHz and (2) that spurious signals be 30 dB below and above the 347-MHz signal. In addition, the customer can specify how to perform specific tests and under what temperature, humidity, and shock conditions.

In a production contract, for instance, the specifications also include a statement regarding the number of units, say 48 of 50, that must pass the performance tests. The mean time between failures (MTBF) is another test that requires the product to be operated for a given time, for instance, 5000 hours or so, without failure. Each of these tests must be considered when determining the scope of the work to be done.

When a project is commissioned, by a customer, or done on contract, the project and marketing managers should discuss the specifications with the customer to (1) ensure that the product is developed to meet a complete set of realistic specifications and (2) guarantee that all the elements of the product scope are defined, including any constraints. Project scope is defined by decomposing or subdividing major project deliverables into manageable components. This project management practice creates a project work breakdown structure, a widely employed tool that is discussed at length in Chapter 5.

3.6 CYCLE-TIME EXCELLENCE IN TODAY'S GLOBAL MARKETS

3.6.1 Quicker Product Development

Often, projects must address two juxtaposing elements of the quadruple constraint: meeting the schedule and time-to-market demands and satisfying customer demand for value. Whereas on-time delivery is a championed objective of every well-managed project, speed to market doesn't necessarily guarantee success in the global economy. A necessary prerequisite among the many factors that influence how rapidly products are accepted in the market is the execution of a *quality-driven market strategy* that satisfies the expectations of customers in specifically targeted segments. The following factors are causes for shorter product life cycles:

· Rapid advancement of technology
· Aggressive marketing
· Continuous willingness to try new products
· Increased pace of technological change
· Rapid rate of new product introduction
· Faster information flow to consumers
· Faster information flow to competitors
· Greater willingness to invest in technology to reduce development and production costs
· Reduced product development expense
· Improved customer service, product reliability, performance, and perceived value

For two decades, the professional literature has been replete with studies citing how large and small firms benefited from earlier market entry because they have reduced their new product development times. The first of several studies, conducted by McKinsey & Co.[5] in the late 1980s, showed that high-tech products that came to market even 6 months late earned 33% less profit over five years. In contrast, profits were reduced only 4% when they came out on time but 50% over budget.

Another study by Arthur D. Little[6] resulted is this analysis:

---

[5]B. Dumaine, "How Managers Can Succeed Through Speed," *Fortune*, February 13, 1989, pp. 54–59.
[6]A. Topfer, "New Products: Cutting the Time to Market," *Long Range Planning*, Vol. 23, No. 2, April 1995, p. 63.

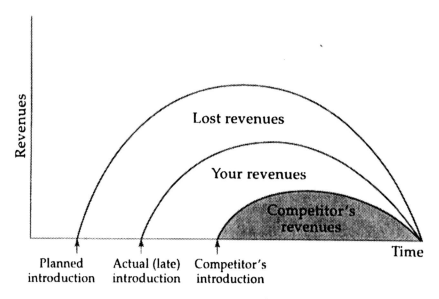

FIGURE 3.1

Cost of belated product launch.

- Exceeding the launch time by 10% (which for a development period of two years is less than three months!) reduced total revenues by 25 to 30%.
- Exceeding production costs by 10% reduced total revenue 15 to 20%.
- Exceeding R&D cost by 50% reduced total revenue only 5 to 10%.

As Figure 3.1 illustrates, projects that experience a belated product launch risk very large losses. To learn how to calculate the cost of expediting a project, see the discussion of project time/cost-trade-off procedures in Chapter 7. In a 1989 cover article, *Fortune*[7] magazine showcased companies that were superfast innovators and superfast producers. The authors encouraged U.S. industry to bring new technology to market sooner to reap a plethora of monetary rewards. Generally, as an early entrant, companies can charge a premium until others follow, provided that their products are ready for the market and meet customer needs.

---

[7]Dumaine, "How Managers Can Succeed Through Speed."

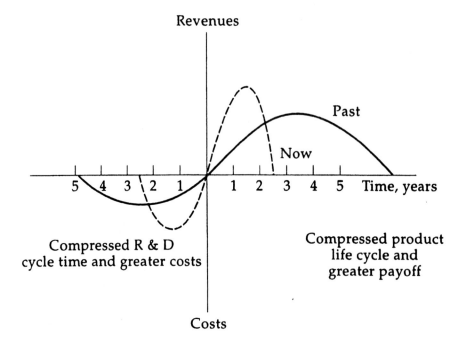

FIGURE 3.2
Costs and payoffs related to life cycles.

Throughout the 1990s, to be competitive, multinational firms stepped up the pace of technical change, narrowing the windows of opportunity and effectively cutting the lifetime of all kinds of products (see Figure 3.2). As the world's markets became more saturated, competition fiercer, and product life cycles shorter, the average total life cycle for a new product compacted to only two or three years rather than the typical five years just a decade earlier.

How beneficial is it for companies to do more with less? Do the economic rewards justify the relentless pursuit to bring technology to market in record time? This quest appears permissible because the technology that many international companies produce can be considered standard components that go into products fabricated by other international firms. Moreover, many high-technology projects result in products that easily compete in global markets without incurring additional costs for redesign or excessive localization. Yet, despite these advantages, speeding high-technology products to market does not guarantee the realization of anticipated profit margins.

Management cannot forecast the market and product life cycle of a high-technology product without considering the technology under development. For this reason, many projects never result in products that are a commercial success. Often, project teams are juxtaposed between the ramp-up of new technology products that are right for the market and the extension of older technology products that are less in demand. Consequently, these project teams never gain enough experience developing the product to reduce its fabrication costs because its technology is still immature. Manufacturing rarely achieves the desired production levels to afford the firm any economies of scale.

Consider, for example, the introduction of a new generation of photonics or optical transmission technology that is brought to market before there is a proven demand. Because the technology isn't fully developed, sales are low and creep along slowly. Potential customers must *prove-in* or evaluate the technology, usually in their own newly designed products, before sales improve. In addition, these customers often can choose among several competitive alternatives that are either already in the market or about to enter. Consequently, the technology may barely penetrate and, perhaps, never saturate the market.

The time trap is a very real dilemma for project managers. To be competitive in the global economy, a business must combine such operational strategies as reducing development cycle time for new products with market timing, product positioning, and global market program development. For this approach to be successful, it must be documented in long-range plans and made operational. For many organizations, this may require major restructuring to (1) overhaul how product development is financed so that projects are funded when a market need is uncovered, (2) adapt processes for faster new product development that emphasize concurrent engineering in both the home facility and abroad, (3) encourage teaming of development with potential customers around the globe to learn what local requirements are expected to bring value to these international markets, and (4) create between marketing and product development an interdependency to improve the marketing and R&D interface.

3.6.2 Impact of Speed

Until recently, there have been very few scientific studies measuring what affects a business's ability to shorten the time it spends bringing products and services to market. However, a few studies provide imperial data to support an organization's implementation of sound project management practices to achieve fast, user-oriented product development. The first of two studies to explore the effect of the length of product development time on actual cycle times was conducted by Abbie Griffin at the University of Chicago's Graduate School of Business.[8] Griffin examined 343 successfully completed projects from 21 divisions of 11 companies to test and explore:

1. The relationship between project strategy variables (newness and product complexity) and the company's new product development processes (use of cross-functional teams and formal processes)
2. The impact of each of these variables on three standard cycle-time variables: (a) development time (DT), (b) concept-to-customer time (CTC), and (c) total time (TT)

The new product development cycle (NPD) can be portrayed as a stepwise history of the initialization and realization of imitative and innovative products. The number of stages in the cycle may vary. Generally, most firms adapt a four- or five-stage model.

Stage 0: Concept generation. This stage begins when the ideas for a product or service first surface.
Stage 1: Project definition. This stage starts when the product strategy and target markets have been identified and the business analysis has been approved in a business and project plan. Technical specifications and product requirements also are documented during a design phase.
Stage 2: Development. Research and development are funded and physical work on the product or service is completed.
Stage 3: Integrated manufacturing and production. Technology from the laboratory is replicated and released for sale from a production facility.
Stage 4: Commercialization. This stage begins with limited production trails, model fabrication, ramp-up, and full production of the product or product components of a service.

Cycle time is measured by three intervals: *Development time* runs from the start of stage 2 through product introduction in stage 3, *concept-to-customer time* is stage 1 through product introduction, and *total time* is stage 0 through product introduction.

Although Griffin did not find a relationship between product newness and complexity or between the formal process and the strategy used to

[8]A. Griffin, "The Effect of Project and Process Characteristics on Product Development Cycle Time," *Journal of Marketing Research*, Vol. 34, No. 1, February 1997, pp. 24–35.

develop products, her work proves very meaningful to how an organization or project team approaches new product realization. Her research finds significant relationships between using a cross-functional team and a formal development process and confirms that *larger cycle-time reductions* are associated with cross-functional teams for newer products and formal processes for more complex products. These empirical results quantify the expected increase in time that more complex and newer products require for development. The research also supports claims championed by project management experts—that formal design and development processes, specifically those conducted during NPD stages 1 and 2, help organize and expedite multifaceted project work. Griffin's study also implies that forming cross-functional teams early, especially when undertaking innovative product development efforts, will improve cycle time.

A second study, conducted by University of Maryland researchers Abdul Ali, Robert Krapfel, Jr., and Douglas LaBahn, used a sample of 73 small manufacturing firms to investigate the association of faster product development with shorter break-even time.[9] *Break-even time* was defined as the elapsed time from the end of product development when the project team has launched or made the product commercially available to the point in the product's sales life cycle when cumulative product contribution has repaid development and startup costs. The research was designed to test whether businesses that pursue a market strategy to produce unique or differentiated products take longer to come to market and whether development time increases as the amount of innovativeness increases.

Analysis revealed that the mean product development cost for these firms was $350 million; the projects took a year and a half and five person-years to complete. The mean time to break-even after launch was approximately 18 months. Among those respondents who had completed projects and those nearing completion, no significant differences existed regarding innovativeness, marketing strategy, time needed to recover the initial investment, firm size, newness of recently developed products, product substitutability, and life span of new technology.

Statistical analysis further revealed that the break-even time varied with its predicted values, leading the study authors to conclude:

· Greater innovativeness and higher level of technical complexity delay a new product's arrival into the market.

---

[9] A. Ali, R. Krapfel, Jr., and D. LaBahn, "Product Innovativeness and Entry Strategy: Impact on Cycle Time and Break-Even Time," *Journal of Product Innovation Management*, Vol. 12, No. 1, January 1995, pp. 54–69.

- Faster product development leads to shorter break-even time.
- Being first to market, lower relative price, and company size help achieve shorter break-even time.
- Product advantage — adding more functionality or features to differentiate the product from others in the market — usually prolongs the break-even time.
- Better product advantage or differentiation also comes with faster product development, however.

It is important to note that technical content had the strongest delaying effect on new product development cycle time. So businesses in this study achieved faster cycle times not by sacrificing product quality but by maintaining a level of *technical simplicity* in their products. Even though R&D managers take professional pride in advancing the state of the art of their product endeavors, unless this sophistication satisfies some unique customer requirement, the delay coming to market may not pay off. But as these authors caution, while shorter cycle time is desirable, faster product development must be judged against product performance and development cost — familiar quadruple constraint components. Further research is also needed to understand the delay experienced in NPD stage 0 and the impact that speedy delivery to market has on consumers since their resistance to continued innovation can exhaust profit margins.

3.7 CONCLUSIONS

The new millennium poses tremendous challenges to businesses: global competition, restructured processes, leaner and geographically dispersed organizations, and the need to consider customers' expectations and cultural values when designing products and services for multinational consumption. To meet these challenges, we introduced several of the requisite project planning techniques to help project teams succeed in this rapidly changing and competitive environment:

- Aligning project and program plans with overall organizational goals and business objectives
- Planning for the unknown and expecting the exception
- Using a common or structured development process
- Analyzing project management requirements and building a sense of trust at the onset of the project among team participants accomplishing the work

· Speeding up product development to benefit the organization while balancing customer expectations and the project's technical requirements for usability, performance, reliability, and quality

In Chapter 4 we examine specific development methodologies and design tools that project teams can use to enhance project and organizational performance, meet market demands for on-time delivery, and provide value to the customer.

BIBLIOGRAPHY

Barwise, P., P. R. Marsh, and R. Wensley, *Must Finance and Strategy Clash?* Harvard Business Review Paperback 90053. Boston: Harvard Business School Publishing, 1991.

Berry, D., "Strategic Processes for Organic Growth: What We Need Versus What We've Got," *Planning Review*, September–October 1996.

Byars, L. L., L. W. Rue, and A. Z. Hkaker, *Strategic Management.* Homewood, IL: Richard D. Irwin, 1996.

Cassimatis, P., *A Concise Introduction to Engineering Economics.* Winchester, MA: Unwin Hymann, 1988.

Drucker, Peter F., "Toward the New Organization," in Peter F. Drucker (ed.), *The Organization of the Future.* New York: Peter F. Drucker Foundation, 1997.

Hayes, R., "Strategic Planning: Forward in Reverse?" *Harvard Business Review*, Vol. 64, No. 6, November–December 1985, pp. 111–119.

Hayes, R., "Too Much Dictating from the Top Why Strategic Planning Goes Awry," *New York Times*, Sunday, April 20, 1986.

McGrath, M. E., M. T. Anthony, and A. R. Shapiro, *Product Development: Success Through Product and Cycle-Time Excellence.* Stoneham, MA: Butterworth-Heinemann, 1992.

Mockler, R. J., "Strategic Management: The Beginning of a New Era," *International Review of Strategic Management*, Vol. 6, 1995, pp. 1–35.

Panfely, P., and L. A. Sonnier, "Learning from Best Practices in Strategic Planning," *Planning Review*, Vol. 24, No. 5, September 1996, p. 48.

Porter, M. E., "What Is Strategy?" *Harvard Business Review*, Vol. 74, No. 6, November 1996, pp. 61–76.

The Powerful Business Plan Writer — Success Inc. Irvine, CA: Virgin Melbourne House, 1989.

Project Management Institute Standards Committee, *A Guide to the Project Management Body of Knowledge.* Upper Darby, PA: Project Management Institute, 1996.

"Strategic Planning," *Business Week*, August 26, 1996, pp. 46–52.

Topfer, A., "New Products: Cutting the Time to Market," *Long Range Planning*, Vol. 28, No. 2, April 1995, pp. 61–78.

Vernon-Wortzel, H., and L. Wortzel, *Strategic Management in the Global Economy.* New York: Wiley, 1997.

Wheelwright, S. C., and K. B. Clark, *Revolutionizing Product Development: Quantum Leaps in Speed, Efficiency, and Quality.* New York: Free Press, 1992.

EXERCISES

3.1. Classify each planning activity as either a strategic, tactical, or operational process.

   (a) Planning the five-year research and development program for a major laboratory

   (b) Planning a project budget

   (c) Scheduling the task durations for a software development project

   (d) Defining the architecture of a new integrated switching system

   (e) Defining the mission of a new business venture

   (f) Documenting procedures to distribute a new financial service offer to customers

3.2. Describe briefly how project management has accountability for realizing the business strategy of the organization

3.3. Which of the following statements is not true?

   (a) The purpose of strategic and tactical planning is to identify specific efforts that project-level managers undertake to achieve strategic goals.

   (b) Business objectives and the project's planning, scheduling, tracking, and team building approach are interrelated.

   (c) Adoption of the project plan requires acceptance by those who complete actual tasks and by those who interface with them.

   (d) Once a project baselines a planning document, changes cannot be entertained.

   (e) A well-managed project is planned by a team of cross-functional specialists, who actually contribute time and expertise toward completing the scheduled tasks.

3.4. Review Section 3.3. Explain why each of the following is a potential cause of project failures.

   (a) Only the senior-most levels of the organization understand the business strategies of the project.

   (b) A planning group performs all the planning functions, directing how functional organizations complete project activities.

(c) A software applications development project relies solely on a conceptual agreement with a joint venture partner rather than negotiating a plan defining project responsibilities and deliverables.

(d) A project rolls out a new product offer to several countries simultaneously because there appears to be little need to customize the offer for these markets.

3.5. Good project planning includes nine required steps. List at least six of the nine steps used to plan a project.

3.6. Contingency planning is an amount of money, time, personnel, and other resources added to each activity estimate to cover all of the following except:

(a) Errors in estimating

(b) Changes in design

(c) Expediting the work

(d) Uncertainties in estimating resources

(e) Fluctuations in currency exchanges

(f) Omissions

(g) Insurance policy premium against natural disaster, labor action, strife, and unrest

3.7. Your project team is estimating the cost of developing a new applications program to manage a communications device. The loaded cost, including personnel and all resources, is estimated at $1 million.

(a) List at least five probable risks that a project this size might experience.

(b) How much money should be set aside for contingencies?

3.8. Explain possible drawbacks to contingency planning and what other management techniques might be used as strategies to control inherent risks associated with realizing project objectives.

3.9. Project scope management includes the processes required to ensure that a project includes all the work, and only the work, required to complete the project successfully. Which of the following are not inputs to a scope management plan for the project?

(a) Product specifications

(b) Product description

(c) Critical assumptions about resource availability or constraints on the project start date

(d) Strategic plan

3.10. The amount of planning performed on a project must be commensurate with:

(a) The scope of the work

(b) The information available to size and estimate the work

(c) The project manager's authorization

(d) (a) and (b)

(e) All of the above

APPENDIX 3.1: SAMPLE BUSINESS PLAN OUTLINE

Table of Contents

Confidentiality Statement

    I. Title Page

   II. Executive Summary

       A. Business Description

       B. Current Position and Future Outlook

       C. Management and Ownership

       D. Uniqueness of the Professional Service

       E. Funds Sought and Use

       F. Financial Summary

  III. Objectives

  IV. Strategies

   V. Business Description, Status, and Outlook

  VI. Management and Ownership

 VII. Product or Service Description

       A. The Unfulfilled Need

       B. Uniqueness of the Professional Service

       C. Pricing and Value

       D. Trademarks, Copyright, Licensing, and Patents

VIII. Market Research and Assumptions

       A. Market Overview and Size

       B. Market Segments

       C. Customer Profile

       D. Competition

       E. Market Share

      F. Geographic Market Factors
      G. Barriers to Market Entry
  IX. Marketing Strategies
   X. Operational Plan
      A. Sales Forecasts
      B. Services
      C. Advertising and Promotions
      D. Selling Methods
      E. Service and Delivery
      F. Supply and Stocking Plans
      G. Facility Plans
      H. Design, Technical, and Engineering Plans
      I. Proposal Plans
      J. Project Backlog and Management
      K. Administration Plans
      L. Employment and Personnel Plans
      M. Critical Issues

Confidentiality Statement

Include a statement that reads as follows: "The information, data, and drawings contained in this business plan are confidential and are not to be disclosed to third parties without prior written consent of _____ company."

I. Title Page

State the name, contact person, address, and telephone number of the business.

II. Executive Summary

A. Business Description. The business description will consist of two or three sentences describing what your business does and must include a description of the industry. Give a brief history of the business and the corporate structure (i.e., partnership, sole proprietor, or corporate affiliation).

B. Current Position and Future Outlook. Describe the current position and future outlook. In a short five-sentence paragraph clarify:

1. Business strengths
2. Potential for growth of your industry or area of project work (e.g., technology, operations)

3. Industry strengths
4. Plant facilities, equipment, other assets (e.g., patents, reputation)
5. Business goal and objectives
6. Plans to accomplish or achieve these goals

C. Management and Ownership. Describe the management team and their experience in the industry. If seeking venture capital for your project or commercial funding, list the owners of the business.

D. Uniqueness of the Professional Service. Answer the question: What makes this business opportunity unique? You must describe the uniqueness of the proposed project/program/venture. Compare your proposed features, pricing, company image, and distribution system (location) to those of your competition.

E. Funds Sought and Use. State the funding being requested for your project/program and how the funds will be obtained.

1. Will the amount of funding be raised through debt (e.g., collateral, stock, loan) or equity (venture shares)?
2. Explain how the project will use the funds (e.g., research, new product development, production expansion, marketing).

F. Financial Summary. The financial summary highlights the financial forecast for your project. It is the summation of the projected income statements and may forecast three to five years into the future. Include a chart or spreadsheet of your calculations.

	Actual Estimates[a]				
	20xx	20xx	20xx	20xx	20xx
Net sales − cost of goods sold	—	—	—	—	—
Gross profit − selling, general, and administration expenses	—	—	—	—	—
Other income	—	—	—	—	—
Net income	—	—	—	—	—
Income (after taxes)	—	—	—	—	—
Annual net cash flow	—	—	—	—	—
Discounted cash flow	—	—	—	—	—
Cumulative discounted cash flow	—	—	—	—	—
Net present value	—	—	—	—	—

[a]Thousands of dollars or other currency denomination.

III. Objectives

Objectives are project targets that your proposed project will satisfy in at least three areas:

1. *Profitability.* State the effect the project/program/venture will have on sales growth, assets growth, and return on sales. State the projected sales from the investment in the future.
2. *Market positioning goal.* What position would your investment in the project allow you to occupy in the marketplace?
3. *Other goals.* Are there employee and customer satisfaction goals that completion of the project will allow you to realize? Are there public safety and environmental objectives that the project/program will satisfy?

Briefly list two or three objectives in each of the areas above.

IV. Strategies

Discuss in one or more pages the strategic objectives of your business, organization, or company. Project strategies are the avenues your project will use to achieve the objectives stated previously. Later, you'll describe your plans for implementing them. Strategies, generally, will relate to more than one of the objectives and may include financial, marketing, manufacturing, technology, and employment or resource allocation (materials and personnel) methods.

V. Business Description, Status, and Outlook

Describe the business and industry, including current and future positions. Describe how the project or program proposal relates to the business. Include:

1. Type of business or corporate structure; your company's history completing similar projects or programs.
2. Type of industry the business/project/program represents and its potential for growth.
3. Strengths and growth potential offered by the project or program. List any other intangible benefits for completing the activity.
4. Facilities/resources needed to complete the work.

VI. Management and Ownership

Name the key managers involved in the project and describe their experience managing similar ventures.

1. List the company owners, if relevant.
2. List the key managers; describe their qualifications and experience.
3. Append a company/project/program organizational chart. Include a partial chart here.

VII. Product or Service Description

A. The Unfulfilled Need

1. Describe the market's unfulfilled need—the niche your project's product or service will fill.
2. State the origin of the idea (e.g., technological advancement, research patent, service opportunity, etc.).

B. Uniqueness of the Professional Service. Describe how the project/program is unique in the industry, how it meets the unfulfilled need, and how it differs from competition.

1. How do your operations meet the unfilled need?
2. What unique features and benefits do you offer the marketplace (e.g., image, location, price, merchandise lines, advanced technology, etc.)?

C. Pricing and Value. Summarize the rationale behind your pricing policies, comparing the value of the project's/program's product or service to other competitive offers in the market.

1. Describe how the price was determined.
2. Compare price to completion.
3. What are the projected gross profit margins?
4. Is the product or service price or volume sensitive?
5. What is the expected reaction from competitors?

D. Trademarks, Copyright, Licensing, and Patents. Describe conformance to government regulations. State whether you are protected by copyrights and/or patents in countries where they are valid.

VII. Market Research and Assumptions

A. *Market Overview and Size.* Define your industry and market. Measure their size in money ($, £, ff., etc.), annual capital investments, geography, demand.

1. Provide an overview and history of the market.

2. Describe overall product/service demand.

3. Cite total revenue by geography and growth projections for regions where you expect to sell your product/service.

B. *Market Segments.* Markets consist of segments that focus on different types of customers, products, and services. Describe the segment or niche that your project/program supports. How does it compare with other segments of equal size? What makes the chosen segment unique?

1. Describe your segment.

2. What are the growth projections that make this niche attractive?

C. *Customer Profile.* Who are your customers? What type of people are they (managers, government purchasing agents, military generals, political bureaucrats, foreign government planning officers)?

Gather demographics and sociographic information about them (age, income, and occupation) to draw a profile.

Learn what motivations they might have to purchase or what affiliations or relationships must be developed to close a sale (psychographics). For instance: Are they an innovator or early purchaser of technology who will pay a high price to acquire the latest features, or are they content to stay the course until prices drop and their current systems are totally obsolete.

Are proper introductions and endorsements needed in high-context cultures (e.g., Asian and Middle Eastern societies) before you can be considered as a potential contender?

1. Describe the type of customers you will target and explain how you will service their needs.

2. Profile them.

3. Cite market test results and success with similar introductions in this or other market segments.

D. Competition. Describe your major competitors; give their market share and financial position; explain what marketing strategies they employ.

1. Cite the market volume (i.e., monetary volume) for each competitor.
2. List the strength and weaknesses of each.
3. Describe their current and anticipated strategies and marketing niches.
4. Give their relative financial position (i.e., P/E ratio).

E. Market Share. Where does your company rank?

1. Explain where your business stands versus competitors in each segment and geographical market (e.g., North American home computer market segment versus European business segment).
2. Cite market share data.

F. Geographic Market Factors. Describe your distribution strategies.

1. What areas do your retail outlets cover? Do you have direct channels in the segment locations where you propose selling?
2. What other regional factors might you need to consider (i.e., favorably or unfavorably opposed political factions in the regions where you will market)?

G. Barriers to Market Entry. Will it be easy to enter the market when the project/program is complete?

1. Give any financial barriers (high fixed costs or additional capital investments required to realize potential savings over the life of the product).
2. List what technical training sales, services, and manufacturing personnel may need to sell, service, and maintain, or to tool the factory, to complete the project.

IX. Marketing Strategies

Draw on your knowledge of the market research conducted by your organization so that your team can market successfully the product/service when the project is completed.

1. Describe which target customers your research identified.
2. Describe their unfulfilled need.
3. Identify how to contact the customer (promotion, channel — direct or indirect, etc.).

Briefly explain sales strategies. Pay special attention to how to sell in various low- and high-context cultures around the globe.

1. Explain how your sales team will interact with the customer, close the sale, or take the order.
2. How would the order be filled and final payment be accepted?
3. What credit terms will be extended?

After closing, state your policies for:

1. Delivery
2. Service
3. Repair
4. Maintenance
5. Ongoing support

X. Operational Plan

A. Sales Forecasts. Work with your product management team to list the sales and pricing objectives of your project's commercial offer. Forecast them for each targeted market segment for the next 1 or 2 years.

1. List overall sales projections one to two years out.
2. Analyze each segment's profitability; forecast by geography area (e.g., North America, Asia) or customer group.

B. Services. Detail how you will manage and price the service offer. Explain how it relates to your uniqueness and how your policy differentiates you from competitors.

1. Position the service offer for your product.
2. Forecast sales (with profit) for each service line.
3. Set the price in terms of customer value.

C. Advertising and Promotions. Explain your promotional plans. Remember that advertising must be localized for each geographic region where you sell.

1. Choose a media mix and budget for it.
2. Forecast results.
3. Identify other mechanisms for publicizing the product/service offering to the market (e.g., public relations campaign in government publications, commercial trade fairs, etc.).

D. Selling Methods. Account for the selling expenses associated with your product.

1. What is the compensation plan, and will you use special incentives to push your new product?
2. Account for the seasonally of selling (e.g., U.S. end-of-year holiday sales peak in consumer electronics).

E. Service and Delivery. What is your execution plan for customer service, credit, warranties, and maintenance contracts once sales are closed?

1. When will you transfer product/service knowledge from R&D to your services support organizations? What will this cost?
2. What provisions are in place to collect payments?

F. Supply and Stocking Plans. Do you have a stocking plan?

1. Identify your buyers and your suppliers.
2. Establish stocking and inventory control procedures.

G. Facility Plans. Where will you administer sales and service? Identify international locations where you will provision the product.

1. Describe your facilities plan.
2. Make allowances for security, insurance, zoning permits, etc.

H. Design, Technical, and Engineering Plans. Discuss the technical plans (project, technical design plans, etc.) that support the project/ program. Create a list of:

1. Objectives for product development
2. Technical service
3. Quality assurance and reliability

I. Proposal Plans. Discuss how proposal or bid documents will be handled.

1. Explain tactics for proposal writing, bidding, etc.
2. Name the team responsible for this activity and major deliverables and milestones for the target clients.

J. Project Backlog and Management. Describe your project/program management plans at length.

1. Are there other projects that the firm has bid on, and what is the likelihood that contracts will result?
2. Describe the proposed or in-place project management structure.
3. What are the tactics for each project to meet profitability goals?

K. Administration Plans. Who is in control of daily operations? Who is the functional management, and what policies will be used to control operations?

1. Which functional organizations are responsible for completion of the work effort?
2. Cite any pending changes to the management or business structure that may affect completion of the project.

L. Employment and Personnel Plans. Describe the employment level necessary to complete the work. How will they be compensated? Is special training necessary to bring the workforce up to speed?

1. List any payroll expenses indirectly associated with the project work (e.g., insurance, overtime wages, fringe benefits).
2. List any government compliance (in the United States: EEO, OSHA, etc.).
3. Include training plans, criteria for evaluation of performance, etc.

M. Critical Issues. List the critical issues or risks associated with this venture, along with action plans to mitigate their effect.

1. Are there unproved marketing, technical, or operational activities that could delay completion or overrun the budget?
2. Describe actions that would be taken to avert the impact of any such occurrences (e.g., contingency allocations to be set in reserve).

PLANNING AND DEVELOPMENT METHODOLOGIES

A journey of a thousand miles must begin with a single step.

LAO TSU

4.1 INTRODUCTION

Using a standard and common development process can add to an organization's competitive strengths. *Operational plans* that focus on examining the systems and procedures that are used to bring products and services to market can also create a fast-cycle capacity for getting to market ahead of the competition. For a software applications business, this means detailing the methodologies followed to produce and test code. In an industry engaged in manufacturing, plans would address materials production procedures.

In today's turbulent and highly competitive environment, planning a new business venture and succeeding requires that every organization have an intimate knowledge of its response to four important questions:

1. Are we properly prepared? Is our startup plan sufficiently detailed to manage all operational activities?

2. Can we meet customers' demands and realize a solution that meets their needs? Will we be able to produce the product or deliver the service within the time required and at the quality level specified?

3. Can we manage the development and production processes? Will we maintain control over engineering and design, testing and system verification, process inventory, finished goods, and retail stocks? Can we avoid excess downtime and materials waste?

4. Are we utilizing our resources in the most expedient and timely manner and encouraging contributions from all our talent? Can we seek the best; attract the best; retain the best human capital?[1]

In this chapter we illustrate the planning and common development methodologies that not only help answer these questions but also improve overall cycle time. We explain how using certain project-level design tools will affect customer satisfaction and encourage firms to improve their time-to-market commitments by reviewing how they manage overall organizational and project team performance. Although organizations will be challenged in the twenty-first century to get the *right* product to market quickly, high performance in the marketplace will not be measured by speed alone. To succeed, products and services must conform to customer expectations, satisfy their demands for value, and perform reliably — familiar components of the quadruple constraint that cannot be traded off without considerable loss of creditability with customers.

4.2 SHORTENING PROJECT LIFE CYCLES

In the new millennium, companies will face relentless demands to shorten their R&D cycle times to stimulate a continuous flow of high-quality products and services to targeted markets around the globe. How are businesses planning and organizing today to succeed in the future in this team-oriented and demanding activity?

How a company approaches new product development determines how fast it brings products and services to market. Over the past three decades, there have been basically four widely used methodologies: (1) phased development, (2) stage/gate methodologies, (3) structured development, and (4) product and cycle-time excellence (PACE).

None of these methodologies are exclusive of the next. In fact, projects have adapted aspects of each to improve on-time delivery. *Phased development* was introduced by the U.S. National Aeronautics and Space Agency (NASA) during the 1960s and 1970s. This methodology broke up long-

---

[1]A. F. Smith and T. Kelly, "Human Capital in the Digital Economy," in Peter F. Drucker (ed.), *The Organization of the Future* (New York: The Peter F. Drucker Foundation, 1997), pp. 199–211.

duration project efforts, often spanning several years, into specific short-term phases to minimize technical risks. The approach also created a sense of urgency among project team participants to finish all phase-specific activities, since progressing to the next phase could not be accomplished without receiving approval and/or continuous funding for completing the preceding phase. Project teams would remain in their functional organizations and were managed using the matrix management structure. Although widely implemented throughout the high-tech industry and mandatory for several years on federally funded procurement contracts, commercial projects that followed phased development were disadvantaged:

- The matrix structure made it difficult for project managers to make project work a priority assignment over other technical work.
- The impact that market information might have on the scheduled development activities was not considered except during the early design phase.
- The rigid delineation of functions kept teams from engaging in concurrent activities.
- Teams frequently lost time waiting for approval and funding from upper management to proceed to the next phase.
- Often communications among the various functional and project team members do not exist. When completed, designs were, in effect, thrown over the wall to manufacturing with little understanding of the upstream requirements.

Moreover, international competition began rapidly to magnify these disadvantages, such that, by the early 1990s, U.S. industry had to maintain tighter control over project capitalization. In response to the need to speed up the processes of new product and service introductions, two methodologies were introduced in parallel: the stage/gate model shown in Figure 4.1 and the common or structured development approach discussed in Chapter 3.

Stage/gate models divide the product development process into stages. Projects do not proceed beyond review points without approval of senior management. This method requires that projects be managed by a full-time *core team* whose membership represents all functional departments. Functional specialists, who may join and leave the project team as required, support the core team.

Whereas the phased development approach could be characterized as a cascade or waterfall of project activities, in a stage/gate model, activities are

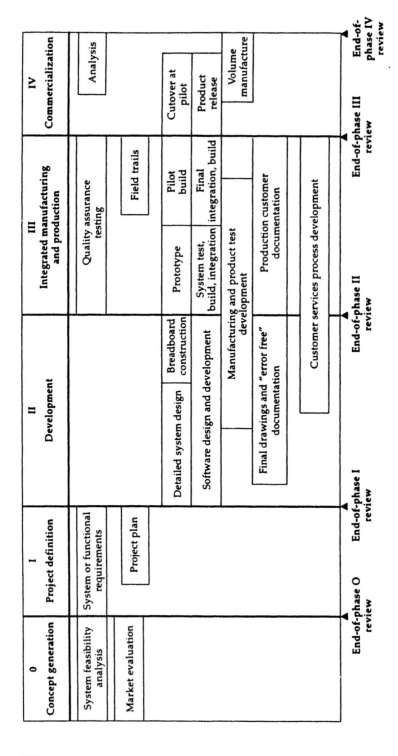

0 Concept generation	I Project definition	II Development	III Integrated manufacturing and production	IV Commercialization
System feasibility analysis	System or functional requirements		Quality assurance testing	Analysis
Market evaluation	Project plan		Field trails	
		Detailed system design / Breadboard construction	Prototype / Pilot build	Cutover at pilot
		Software design and development	System test, build, integration / Final integration, build	Product release
		Manufacturing and product test development		Volume manufacture
		Final drawings and "error free" documentation	Production customer documentation	
		Customer services process development		
End-of-phase O review	End-of-phase I review	End-of-phase II review	End-of-phase III review	End-of-phase IV review

FIGURE 4.1

Stage/gate product development model.

performed concurrently. Parallel project activities save a project the time required in the earlier model to rework the outputs of the previous phases. Teams design and develop, fabricate and test, ramp-up and tool concurrently. Frequent reviews are built into the development and production schedules to check progress and encourage communication across all the functional groups represented on the team. At the end of each stage of activities, there are additional reviews with senior management to decide whether to proceed, abandon, or redirect the project.

In the stage/gate approach, projects are more attuned to changes in the marketplace that affect both technical and market risks. Project plans are rewritten after each gate decision to reflect market reaction to the proposed development. Based on this updated information, product introductions may be delayed or accelerated, the plan of record changed, and/or resources allocated to more promising projects.

In today's fiercely competitive environment, larger engineering firms must be more responsive to pressures from small entrepreneurial companies with reputations for introducing innovations to the market quickly. The models and common development practices that the larger companies choose to adopt must help reduce the time taken to complete projects and bring new services and products into the marketplace. One such model that promises to accelerate new product and service development cycles while reducing time to market is the *product cycle-time excellence* (PACE) model. The approach emphasizes cycle-time reduction and has many components of the stage/gate model: (1) phase reviews, (2) core teams, (3) structure development, (4) market and product strategy, (5) technology management, (6) design reviews and tool development, and (7) cross-functional project management.

Each project is typically divided into logical phases. Usually, a review is required prior to proceeding to the next phase. However, this review is conducted by a product approval committee of senior managers who have responsibility for overseeing strategies for new product development. The "go" or "no go" decision is based on previously established criteria documented in the operational plan. Targets are also established to reduce development intervals and improve time to market based on project complexity, newness of the technology, and so on. PACE gives the organization the ability to carry out the strategies recorded in its business plans while controlling the direction and timing of its new product introductions. Figure 4.2 illustrates typical checkpoints, milestones, and targeted intervals for a software development project using PACE process methodology.

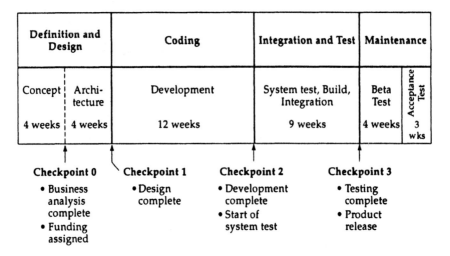

FIGURE 4.2

Benchmarked intervals and checkpoint mileposts for a software development project.

4.3 LIFE-CYCLE PHASE ACTIVITIES

Every project team member must understand, at a detailed level, the structured activities and steps in its new product or services introduction process. In fact, we recommend that a business train, at a high level, the entire workforce. This will facilitate a common understanding of how the organization manages the related marketing, production, and field support functions that dictate how employees from these functional areas, for instance, process and fill orders, support the launched products and services, repair field returns, and so on. Usually, core teams are charged with keeping project team members informed. Many organizations employ process management groups of highly trained specialists who write and maintain the documentation that describes, phase by phase, the required activities and engineering tasks that must be performed. They are available to the new product development core teams and WBS branch managers to help these groups educate the project teams and audit their adherence to the company's process guidelines. Table 4.1 to 4.5 summarize outputs typically required for each phase and standard quality guidelines associated with the development of a new product or service. We describe the overall phases in the following sections.

4.3.1 Customer-Market, Concept, and Definition Phases

Typically, the preliminary *customer-market* and *concept planning phases* of a new product or service development begin when senior management creates a cross-functional, market planning, or *concept evaluation team*. This team is established to test the feasibility of developing and producing a new technology, product, or service identified by market research (see Table 4.1). The planning team carries out these activities under the leadership of a market manager, project manager, or both. The team also calls on highly skilled groups of creative systems and design engineers for input. In all cases, the germs of product ideas defined in the early customer-market phase set off this evolutionary process.

Often, the team is directed by a managerial, steering committee. Together with the planning team, they share the following three primary responsibilities: (1) timely solution of all business problems (2) allocation of adequate resources to staff the project startup organizations, and (3) preparation of all preliminary project documentation (e.g., preliminary functional requirements, required quality standards, estimations of the targeted development/production intervals, and cost objectives to the design team).

During the *concept phase*, the market planning team evaluates the global, economic markets and the feasibility of developing the proposed technology or system in terms of corporate strategy, competitive advantage, risk, and an identified market need, while the engineering design team prepares a preliminary design. The team members accomplish this by interacting simultaneously with potential customers or whatever public or government agency has furnished the specifications. Teams dealing with vendors or other contracting entities will find that these dialogues become iterative and that they are extremely important to proper planning of the project. Since the customer usually specifies a system in an incomplete manner, ident-

TABLE 4.1 CUSTOMER-MARKET PHASE

Outputs
- Determined market need
- Preliminary product
- Requirements established based on commercial analysis

Quality process assurance guidelines
- Overall business mission and strategy
- Defined corporate technical strategy and R&D function within corporation

TABLE 4.2 CONCEPT PHASE

Outputs
- Engineering design concept
- System feasibility analysis (technical prospectus)
- Initial data as to:
  - Cost
  - Duration
  - Specifications
  - Development risk

Quality process assurance guidelines
- Approved business plan
- Proven technical, environmental, and economic feasibility
- Identified technologies in all key areas

---

ifying only the key features needed, the planning team is charged with specifying the technical, quality, and usability (human) requirements that were not thought of by the customer.

During the concept phase (see Table 4.2), the system engineers and designers attempt to define the features that must be developed and gain customer approval for them, perhaps by creating a very early prototype and seeking customer input. Later, in the *design phase*, the prototype is perfected based on this input and a better understanding of the technology being developed. This iterative process leads to a better quality product that is delivered within allotted intervals and budget. Unfortunately, when this iterative procedure is omitted, downstream phases will experience numerous specification changes that cause delays, cost increases, and reduce morale among project team members. To minimize aggravation and maximize satisfaction, we strongly recommend this first step in the development process.

In the *definition phase* (see Table 4.3), specifics are added describing as accurately as possible the technical and economic requirements to develop the technology or proposed system. This is again an iterative process accomplished in an interactive manner between the planning and design teams, where the planning team asks the engineering design team to *partition* the project into functional tasks. This technique, called the *work breakdown structure*, is discussed in detail in Chapter 5. Later, functional and project management use the partitioned tasks to determine targeted intervals for the project and to allocate personnel and other needed resources based on task duration.

TABLE 4.3 DEFINITION PHASE

Outputs
- Project summary plan
- Work breakdown structure
- Risk and other areas of uncertainty identified
- Master schedule and milestone events established
- System or functional requirements analysis completed
- Final product performance requirements and test
- Qualification (usability, serviceability, reliability) requirements prepared
- Identification human and nonhuman resources
- Initial project documents
- Policies
  - Procedures
  - Budgets
  - Product or service description
- Interface points determined for each major task

Quality process assurance guidelines
- Approved project plan
- Approved system and functional requirements
- Approved product performance and test
- Qualification requirements

Documentation. The number and types of documents produced during the concept and design phases may be different depending on the source of funding sought for the project. The primary planning documentation produced during these phases for commercial projects always include (1) the product or system technical prospectus (2) the product definition, (3) a business plan, and (4) a project plan. However, a firm bidding on a government agency proposal to obtain project funding and approval might deliver only the formal research and development prospectus in lieu of the business plan, whereas in the case of a regulated public utility, a prospectus to develop a system that supports its existing business would briefly describe, for example, the new system and the operations it affects. The document would include estimates for required resources, capital, other expenses (personnel and computing requirements), and a high-level development plan or work program outlining proposed features and their development sequence, schedule, and intended release dates.

The business plan for a commercial project, similarly, would describe the status of the new technology under investigation within and outside the company, including competitive threats. The plan would also cite expected benefits to the company or government due to successful technical and commercial development. In other words, the planning team in the business plan attempts to determine the payoff for investing in this technology; the revenue and profits anticipated; and the market share, cost reduction, quality improvement, energy savings, or compliance with government regulations expected. To learn how to write a business plan that addresses these concerns, consult company guidelines and review the sample outline in Appendix 3.1.

The activities of the concept and design phases may be the response to a solicitation from the local or national government or from either an external or internal customer. In such cases a prospectus is written that contains a description of the technical personnel and facilities available, various approaches considered, and the finally agreed-on approach for satisfying the specifications. A quantitative comparison of the approaches, including risk, is required. Duration, costs, and schedules, as well as management control techniques, should be presented in the project plan.

4.3.2 Design and Development Phases

The plan, which includes the organization and schedules required to satisfy the documented specifications, detailed testing procedures, and quality metrics is produced in the customer-market, concept, and definition phases. Firm control of the system's technical definition is essential throughout these phases. Control helps ensure the timely success of the project and will preclude eventual redesign in later phases. The standard, functional requirements should represent primarily the market or customer needs. They should be constrained only by what is technically feasible and economical to develop. However, unless these requirements are very carefully communicated to all members of the project development team, the system designed often ends up technically inadequate, user-unfriendly, late, and grossly over budget. This is particularly true of very large and complex software development projects where the programming team may rush through requirements planning and immediately start writing code.

The intent of the design and development phases is to produce working and tested models of the system using the specification and preliminary design defined in the previous two project planning phases, and to determine the inadequacies of any analyses conducted previously.

In the *design phase* (see Table 4.4) the project team simulates all aspects of the system or product and then codes and/or constructs it. The resulting system should meet operational specifications with the exception of form,

TABLE 4.4 DESIGN PHASE

Outputs
- Detailed system, component, and module design
- Software simulation/prototype
- Hardware simulation/prototype
- Breadboard construction
- Completed unit and functional test plans
- Complete data processing test plans
- Preparation of all initial drawings, schematics, and circuit descriptions

Quality process assurance guidelines
- Approved revisions to project plan
- Successful completion all unit and functional testing

physical requirement, and final testing. In the *development-test phase* (see Table 4.5), the project team codes and/or constructs the final system and installs it in the chassis. This system should look and react like the final production unit.

The project manager has a number of responsibilities during the design and development phases. These include (1) reviewing and confirming with the core, project, and senior management teams the decision to continue

TABLE 4.5 DEVELOPMENT-TEST PHASE

Outputs
- Prototype, including physical design and testing
- Proved-in design of all features, structures, and interfaces
- Final drawings and documentation:
  - Schematics
  - Circuit packs
  - Equipment
  - User training plan

Quality process assurance guidelines
- Established performance specifications
- Scheduled milestones and other deliverables
- Serviceable design
- Usability interface design
- Manufacturable product
- "Error-free" documentation

the design effort, (2) developing a prototype system for customer approval, (3) developing the final system, (4) producing test plans and ensuring that the test results meet specifications, and (5) developing procedures for installing the system at a qualification site.

It is important that project management attempt to stop development efforts that fail to meet technical specifications or that appear to lead to a product cost greatly exceeding what is needed to recover a profit. This is a strategic decision usually made by senior management based on the recommendations of the project team. Also, as we explain in Chapter 7, overtime is best spent during this period to accelerate the development interval. By crashing the project or expediting work needed by downstream processes, project management can ensure its timely completion so that maximum profit can be obtained from the on-time market introduction of the resulting product or service.

4.3.3 Integrated Manufacturing and Production Phases

The purpose of the *integrated manufacturing and production phase* (Table 4.6) is to transfer technology from the laboratory to production. Historically, planning this transaction was perhaps the most difficult in the soon-to-be-launched product's own life cycle. In the past, in large companies, this was

TABLE 4.6 INTEGRATED MANUFACTURING AND PRODUCTION PHASE

Outputs
- Timely procurement of all required resources: inventory, supplies, funds, labor
- Verification of all production specifications
- Volume production start
- Feedback to developers and manufacturing engineers
- Component and unit testing
- System integration and testing
- Quality assurance and reliability control or compliance testing
- Technical manual and all other documentation preparation and production
- Product installation/deployment support plan development

Quality process assurance guidelines
- Manufacturable product
- Established cost targets and scheduled milestones
- Established performance specifications
- Serviceable design
- "Error-Free" documentation

where the proposed system crossed organizational jurisdictions or territories and fell subject to a variety of communication barriers and related difficulties. Because the stage/gate model, used today, increases communication among members of the permanent core and project teams, many of these problems have been eliminated. Also, today's rapidly changing technology and highly competitive global markets are driving manufacturers worldwide to integrate the manufacturing phase with the chronologically earlier engineering design and development phases. The overall goal is to bring products to market sooner and collapse the lengthy manufacturing ramp-ups that were acceptable only a decade ago.

For over a decade now, *quality by design* has been the theme of such high-tech U.S. firms as 3M Corporation, AT&T, Hewlett-Packard, Lucent Technologies, and Texas Instruments, as well as the quest of most international trading houses in Japan. Systems analysts and hardware designers are encouraged to apply quality control methods to analyze system reliability early in the design process, long before the project moves through manufacture. The belief is that the engineering/manufacturing/quality interface can be improved and the total *concept-to-customer* (CTC) delivery time reduced by attending to quality considerations in the design and development phases. Errors identified here, in the front end of the development cycle, save significantly more than when found in the final product because errors incurred in succeeding phases are much more costly to fix. If the development team is to hand off a product that is *fit for use* and free of error, quality must be engineered into the product from the beginning.

As discussed in Chapter 3, one aspect of defining quality is seeing it as conformance to specifications. During development, the system must be demonstrated *manufacturable* or constructable or implementable; that is, the parts ordered should be usable in the manufacture or production of the product. For instance, often during prototype assembly, parts are selected carefully to ensure that the prototype system will function. However, such a system is not *manufacturable* in a situation where only 500 of the 1000 parts ordered work.

The quality metrics of the stage/gate methodology discussed earlier in this chapter require project management to assign to the design and development teams appropriate personnel to represent production/manufacturing/construction. It is essential that these people become integral members of the project team, so that they may contribute valuable input into the manufacturability of a product before costly design errors are perpetuated beyond the laboratory. Similarly, after transitioning the prototype product to production, members from design and development should join the manufacturing team. The exchange permits the team to discover how to improve the overall product design in subsequent iterations.

Firms have enjoyed the following benefits from collaborated design for manufacturability:

· Shortened design for manufacturability cycles and quicker product or system launch
· Reduced engineering changes after designs were introduced and frozen
· Bottleneck or logjam elimination caused by previous production cross-checks

Other quality improvement measures that quality-oriented companies incorporate into their new product development processes may include:

· Conformance to customer or user requirements
· Training staff in the use of statistical quality control measurements
· Just-in-time procurement
· Process control engineering
· Improved supplier performance
· Timely new-product introductions
· Cost saving from reduced rework

These and other specific measures for managing the incremental or continuous improvement (quality) of development and production activities are discussed in Chapter 12.

4.3.4 Installation Phase

In the *installation phase*, the primary operational emphasis shifts from planning the product or service to that of controlling the product or service introduction. Assuming that the product's design has been verified through preliminary unit and system testing, the project management team's duties now emphasize interfacing with the customer or client because it is at this point that the product begins its commercial or sales life cycle. Typically, the following activities must be planned and carried out by various services support organizations that interface with the firm's customers: (1) negotiation of qualification sites for acceptance or market pilot/early introduction testing (beta testing in the market often is done in the development phase as well); (2) preparation of all customer support services: documentation, training, installation manuals, repair services and so on; and (3) contract for delivery.

TABLE 4.7 INSTALLATION PHASE

Output
- Documentation delivered to installers and customers
- Trained installation support personnel
- Trained users and customers
- Field performance and reliability testing
- Product release/cutover at pilot (early introduction) qualification sites
- Acceptance testing of product by users and customers
- Customer feedback to product planners and developers

Quality process assurance guidelines
- Established performance specifications
- Scheduled field test and early training unit shipping milestones
- Serviceable product standard
- Accepted product qualification evaluation standards

The planning and conducting of final acceptance or marketplace tests of complex, leading-edge technologies, such as asynchronous transfer mode data switching systems, are extremely important. These qualification tests, conducted in a friendly environment, help the project team to understand the initial customers' reactions to a new product or service. The tests also equip the team with ideas for improving the product's design and adding new functionality in subsequent releases. Members of the documentation development team may also gather important feedback about how to revise the accompanying maintenance and training manuals or other packaging materials. Marketing may also be able to research the design of planned promotional materials.

Had such high-tech firms as IBM and AT&T tested their new computer workstations running the OS/2 and UNIX operating systems through formal marketing introduction qualification testing, they might have saved millions of U.S. dollars in planned R&D activity in the early 1990s. Several offers in these product families were withdrawn from production after repeated efforts by both companies to "fix" too late in the production phase what the customers told them was wrong. The output and quality assurance guidelines of the *installation phase* are described further in Table 4.7.

4.3.5 Divestment Phase

A project will live on forever if its timely completion is not planned, organized, or controlled. For this reason, some organizations often relieve the project management team of its assignment and transition to the position a new team of qualified specialists who are especially skilled in project closeout and product maintenance procedures. The primary responsibilities of the project team during project termination include:

· Developing a specific plan and schedule for redeployment of project personnel.
· Ensuring that all required actions are taken to facilitate the customer's acceptance of all project deliverables.
· Assuring that the acceptance plan and schedule comply with contractual requirements.
· Assisting marketing and/or future planning or concept evaluation teams in preparing an operating plan for the new product line. The plan should not only introduce new products resulting from continuation of the R&D effort but also enhance the market position of those products already in existence by preventing the effects of obsolescence or market saturation.
· Conducting, along with the functional management, project evaluations or retrospectives to capture "lessons learned" in an attempt to improve on future project management procedure and technique.

Table 4.8 outlines a description of the outputs and quality guidelines for the divestment phase.

TABLE 4.8 DIVESTMENT PHASE

Outputs
· Monitored and improved product quality
· Monitored and improved product market position
· Transfer of resources to product maintenance or to other new projects
· Evaluation of project lessons learned
· Performance problem assessment of product and any deviations from original specifications, planned costs, and schedules
· Recommendations for new development efforts
· Policy and managerial evaluation of techniques and procedures employed on project

Quality process assurance guidelines
· Preplan and publicize the closeout schedule
· Acquired data base of product defects, reliability, usability, maintainability, and performance measures

4.4 COLLAPSING SOFTWARE PROJECT LIFE CYCLES

In the first edition of this book, the life cycle of a software project was presented as a continuous flow or waterfall of project phases and activities that included definition, design, coding, integration, testing, and maintenance. By the mid-1980s, however, software was included in nearly all high-tech projects and represented the *critical path*. In other words, because the total project duration was set by the software life cycle, the code developments undisciplined nature threatened the preliminary completion estimates of even hardware-oriented programs. Some authors described these early days of hand-built coding as a jungle, where too often software development projects failed to deliver. Fortunately, by the mid-1990s, the growing discipline of *software engineering* had succeeded in educating a generation of new application programmers and began to introduce a number of standardized methodologies heralding improvements in both development time and overall software quality.

A new process model for software development introduced in 1988[2] is often referred to as a *spiral* because it links incremental activities across life-cycle phases. Special emphasis is devoted to specifying downstream activities to identify associated risks and to determine the software system's objectives, constraints, and alternatives. Recently, the software industry refined the model. The current approach identifies the system's stakeholders' "win conditions": the end user's or customer's expectations and willingness to pay, while progressing through the cycle and executing plans for the downstream activities.[3]

The software life cycle still consists of the phases shown in Figure 4.3. However, today, the project team is engaged in more front-end process activities during the early *definition phase*. These consist of (1) documenting requirements, (2) reconciling constraints, (3) designing the customer interface, and (4) planning the software project. The design that results is much more detailed and includes writing a *technical prospectus* that describes the design's functionality. Since a requirements document can be used to check potential customer and supplier understanding of high-level architecture and specifications, less rework and greater end-user satisfaction can be expected.

---

[2]B. Boehm, "A Spiral Model of Software Development and Enhancement," *Computer,* Vol. 21, No. 5, May 1988, pp. 61–72.

[3]B. Boehm, and P. Bose, "A Collaborative Spiral Software Process Model Based on Theory W," *Proceedings of the 3rd International Conference on the Software Process: Applying the Software Process,* Reston, VA, October 10–11, 1994, pp. 59–68.

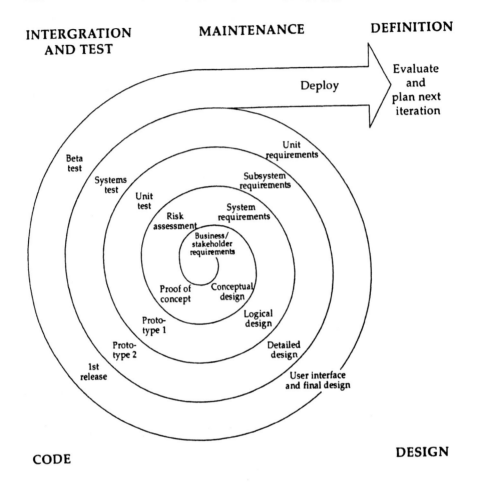

INTERGRATION AND TEST **MAINTENANCE** **DEFINITION**

Deploy

Evaluate and plan next iteration

Beta test

Systems test

Unit test

Risk assessment

Unit requirements

Subsystem requirements

System requirements

Business/ stakeholder requirements

Proof of concept

Conceptual design

Logical design

Proto- type 1

Proto- type 2

1st release

Detailed design

User interface and final design

CODE **DESIGN**

FIGURE 4.3

Software life cycle portrayed as a refined process spiral.

The main outputs of the definition phase include:

- Prioritized requirements, assessing trade-offs, and documenting business objectives
- Identified tools to trace requirements and record modification requests back to their original source and final acceptance test
- Agreed-upon acceptance criteria
- Defined development test plans, including acceptance tests that measure usability, performance, and system security
- Defined authorization policy for changes and procedures that operationalize the policy

- Documented risk management and reuse strategy for program libraries of *object-oriented code*
- Defined configuration items

The project team in the next phase, the *design phase*, tries to specify and agree on an appropriate physical solution to the potential customer's or end-user's business problem. Several appropriate solutions for the application are typically positioned against existing or potential architectures in a series of trade-off discussions before construction begins. Since past failures often resulted from "scope or feature creep," the specifications are "firmed" as early as possible. Preliminary designs are tested in prototype form with representative end users to determine the "best" alternative before development proceeds. Then the project management practice of defining a *work breakdown structure* (WBS) is used to obtain the software organization, budget, and schedules.

The next activities in the design phase are to document the program's *requirements* — write the *detailed specifications* that software developers and testers will both use to build and verify that the software system performs as intended. The major outputs for the design phase for a software application or integrated hardware/software product include:

- Architecture identified for component development of the system
- Existing components identified for reuse
- Firm design of screens, user interfaces, menus, wizards, on-line documentation, and help messages
- Completed design of the software program, including the transactions, control software, and package configuration
- Standards specified for the physical databases and file structures
- Procedural controls designed for security
- Performance and workload standards designed for each application subsystem
- Validated design against requirements through tabletop, peer reviews, and prototype construction, including end-user feedback on system usability
- Traceable test plans and unit tests to design requirements
- Completed plans for system, integration, regression, and acceptance testing
- Updated list of configuration items
- Coding standards specified

· Software and hardware platforms, development tools, and off-the-shelf components specified

To reduce development time many software applications and integrated hardware/systems today include *commercial off-the-shelf* (COTS) components. Or one of the *component development architectures* and *object-oriented programming* disciplines may be required to encourage quicker programming and testing of the code. A *component* is a reusable software element that can be used by developers to assemble an application. The *object-oriented development* (OOD) *approach* to software development is based on the reuse of existing libraries of self-contained operating units.

This creates economies for the project in the *coding phase*, where the programmer can call up large modular units that are already coded, tested, and debugged. The final module unit, of course, is tested again to verify that it meets the specifications for which it was designed. The coding phase is completed only when each module operates without error and *conforms to specifications*. The main outputs of the coding phase include:

· Components selected for reuse from a program library
· Components selected for reuse from the marketplace
· Software developed for the project, such as physical databases, files, etc.
· Procedures developed to control for security
· Unit tests executed on software elements
· Predefined hand-off to system verification for integration and systems testing
· Updated list of configuration items

In the *integration and test phase*, the project team integrates the software modules together and then integrates the modules with any system hardware. Shrink-wrapped applications for the commercial market might require more frequent or *daily build and smoke tests*, where files are compiled, linked, and combined into an executable program every day. Also, to speed up the activities in this phase, many companies use automated tests to perform routine verification and debugging of the system. The significant outputs of the integration and test phase include:

· Insurance that all software *components* are under configuration control
· Summary reports of all tests conducted: integration, performance, and regression

- Analysis and reporting of all defects, including user interfaces, on-line help, and paper documentation
- An agreed-upon well-defined criterion for handing off the program to acceptance testing

Installation at a customer site is considered acceptance testing. Many firms perform this final test just prior to, and in parallel with, the *production* or *manufacturing phase*. Commercial applications are products, so they go through a production phase to replicate and distribute quantities of the application to the marketplace.

All information systems, whether built for in-house corporate offices or commercial distribution, have a *maintenance* or *operations phase* that occurs after delivery of the final release package. A request for a maintenance release capable of performing in a different operating environment or an upgrade release with more functionality may result once customers become familiar with the system.

Maintenance development should follow the same rigorous rules as those of any new development. Changing specifications will result in changing modules and testing software characteristics not planned previously. This will result in uncovering new and previously undiscovered bugs that must then be corrected. Basically, there are four types of maintenance activities, and from a business perspective it is important to know what additional programming costs the project may incur during maintenance development. These include:

1. *Corrective maintenance:* to fix bugs and to address fault reports
2. *Perceptive maintenance:* to improve the system or make it faster, better, etc.
3. *Adaptive maintenance:* to add new capabilities
4. *Preventive maintenance:* to add needed safeguards

Software maintenance, like good programming practice, requires the following:

- Revisiting the defined business objectives and time-to-market goals. This includes reaffirming the direction for *reuse programs* for the new development. What is the company targeting: code, requirements, designs, documentation, or test data?
- Adherence to processes that require thorough understanding of *specifications* prior to design. Conversely, the new release should include only those performance requirements that are necessary (i.e., that are contained in the specifications).

- Standard design procedures and development practices that identify and specify all system components and the connection or interfaces between them.

- Required *sanity checks* and regular reviews of the code by the project team.

- Documented code that explains what the programmer is doing, records all test results, and provides an *audit trail* that relates the software to the specifications laid out in the system requirements.

- Designs that are testable, maintainable, and usable. Applications must be tested against specifications. So designs should enable tests to be performed readily.

- Restructuring the development projects into small, incremental efforts. Frequent minor releases let programmers quickly address changes in end-user requirements, processor technology, or international differences in language and culture without having to wait until the next major market introduction.

4.5 CUSTOMER-DRIVEN DEMAND FOR QUALITY

Markets dynamics are driving companies to pursue shorter product-development cycles, and so are competitive pressures to halve development costs while doubling productivity and the quality of the products or services brought to market (the all-too-familiar components of the quadruple constraint!). The earlier in the cycle — and the more rigorous and systematic the processes introduced by the project team to address these concerns — the more probable the anticipated success. The quest today is to produce products and services that are right for the market the "first time." Customers define what is right: For most enterprises around the globe, *quality* is meeting and exceeding customer expectations for value. This means that projects must define quality in the preliminary or *front-end* (concept and design) stages of new product development.

4.5.1 The QFD Approach

To understand and meet customer expectations, project teams must work together and with customers to deliver what is needed and valued and to affect changes in downstream processes (succeeding development, manufacturing, and support activities) that will help teams realize their market and project objectives.

Project teams can achieve an early definition of customer needs and expectations using the product- or service-realization approach known as *quality function deployment* (QFD), a system for translating customer requirements into appropriate technical specifications for each stage of the product or service development cycle. It is a visual, powerful planning tool, first used in the 1970s in Japan, at Kobe Shipyard, by Mitsubishi Heavy Industries. In the 1980s and 1990s, it was subsequently adapted by the U.S. automotive industry. Attributed as having helped Japanese manufacturers enjoy great competitive strength over the past decades, QFD has improved product development by (1) limiting the number of design changes, (2) shortening total cycle time, (3) identifying the "must haves" for improved engineering, improved productivity, and reduced costs within all phases of the development process, and (4) focusing the project team's efforts on these critical issues.

The term *quality function deployment* is derived from six *kanji* characters: *hin shitsu*, meaning qualities, features, or attributes; *ki no*, meaning function; and *tten aki*, meaning deployment, development, or diffusion. The translation is not exact — *hin shitsu* is synonymous with qualities, not quality.

Steps to QFD. The QFD approach begins with interviewing the customer. Continuous dialogue with the customer permits a firm to understand the *voice of the customer* to gather market-driven research that is available to make key trade-off decisions between what the market wants and what the company can afford to offer. This information can be used to map out arguments for decision making throughout the entire product development process, creating a closed loop for improving costs, quality, productivity, profitability, and market share.

QFD, typically, is a four-part process with two phases addressing product planning and design and another two emphasizing process planning and shop-floor activities. At the heart of the first phase is the product planning matrix or *house of quality*, which maps the relationship between customer needs and the technical design concepts under consideration. The house of quality provides the framework to guide a cross-functional project team through each phase of the QFD process.

As shown in Figure 4.4, the house of quality matrix contains information about customer values. These include (1) what is wanted, (2) mechanisms for determining how to address and realize customer wants in the product's or service's design, and (3) performance measures used to judge how much is needed to satisfy customer wants. *What* items may start off as a list of loose project objectives that the team can refine into statements of basic customer requirements. This refinement continues until every objective

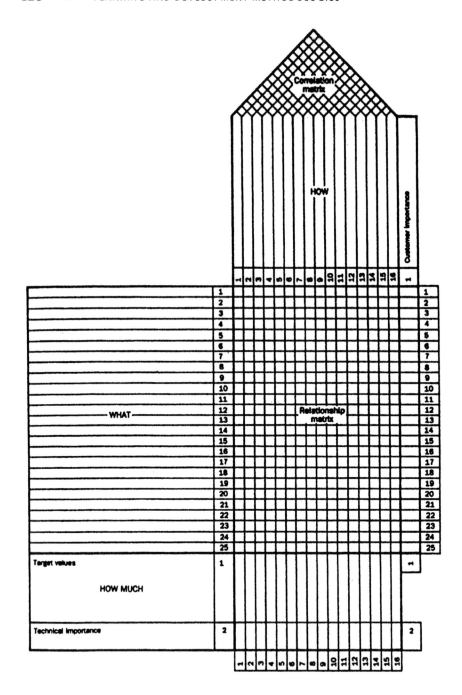

FIGURE 4.4

Schematic view of the house of quality. (From P. G. Brown, "QFD: Echoing the Voice of the Customer," *AT&T Technical Journal,* Vol. 70, No. 2, March/April 1991, p. 25.)

and/or listed item is actionable. Such detail is recorded along the bottom of the matrix. It is needed to gauge whether or not customer requirements have been met. The *how much* items also set targets for further detailed development.

In general, the QFD approach is carried out to assess competitive strengths and preliminary design requirements prior to introducing a design to the factory floor. Even though the approach may be conducted differently depending on the specific industry, the same correlation matrix is used across all industries to show the strength of positive and negative relationships between customer and technological needs. The following specific guidelines are recommended to produce a new service introduction in the telecommunications industry:

Step 1. Identify customer needs.

Step 2. Rank the priority of customer needs.

Step 3. Identify technical concepts to address customer needs.

Step 4. Map or show the relationship of the technical concepts to the needs.

Step 5. Identify technical emphasis.

However, the same generic steps would also be deployed in a more traditional product-development application: for instance, in the tool and dye industry to machine parts.

As shown there are five basic steps to the QFD approach. Steps 1 and 2 focus on the customers' perspective; steps 3 and 5, on the technical or company perspective. However, step 4 is an intersection of the two perspectives. This is where the organization must translate or interpret the customer's perspective — or *voice.*

The mapping of technical concepts to customer needs usually is depicted in a matrix with technical concepts on the horizontal axis and customers needs on the vertical. A cross-functional QFD planning or project team fills in the resulting cells according to the relative impact of each correlation (strong, medium, weak, or none). Each step is replicated until all the positive and negative correlations are recorded in the "roof" of the house of quality. QFD is a powerful, flexible approach that project teams can adapt to be successful in their design and execution of customer-driven products and services.

4.5.2   Value Engineering and Value Analysis

Value engineering (VE) is a problem-solving method practiced during the front end of product development to improve designs, reduce costs, and organize project work based on its value to the customer. Value engineering is both a strategic and a tactical planning activity that can be used to refocus or align organizational goals to improve both profits and products. Although senior management initially may introduce a VE program to cost reduce, for example, a particular manufacturing procedure, the real power of the problem-solving exercise lies in the improved communications that results at the functional or project level among engineering, operations, marketing, and finance.

Value analysis (VA) is usually performed after a particular product or service has been launched, to reduce the cost of product design by reducing component and material costs. Hence VA is done in the back end or upstream of the design and development phases.

Companies, today, are moving toward higher customer responsiveness and using VE to improve overall corporate performance and the effectiveness of both their services and manufacturing operations. VE is a particularly appropriate vehicle to help firms deliver world-class service, where every step of the service delivery process is analyzed to be superior to those of competitors. Essentially, there are two components to the VE approach: (1) determining what is of value to customers by involving employees who are close to them and (2) driving out costs by examining alternative means for achieving a particular function.

Ultimately, the team performing the VE exercise is called upon to make a number of subjective judgments. Aspects of the product or service of value to customers should not be reduced; components or features, not of value, are appropriate candidates for streamlining or elimination. To determine total value, the team must first identify elements of *use*— functions the product or service must perform. Second, the nonfunctional *esteem* or *status-value* features that prompt a customer to buy are analyzed. VE teams often find that the proportion of each of these two variables varies from product to product and service to service, so there are no hard-and-fast rules to use when determining which value functions to identify.

There are several approaches that a project team can use to collect data and become familiar with the component costs and value of the features in the products and services it offers. One technique of particular interest is known as *value chain mapping*. Like the VE approach, a value chain divides the work done by an organization based on its value to customers. Basically, there are three kinds of work:

1. *Value-adding work:* work that is done to satisfy customers' requirements

2. *Essential work:* work that is required to sustain the organization but is not valued by customers

3. *Non-value-adding/nonessential work:* work that the customer does not value or work that does not sustain the business

Strategically, resources should be allocated to value-adding projects that support only core business strengths and help a firm maintain or achieve competitive advantage. Teams should consider essential work as prime for cost reductions or possible outsourcing. Non-value-adding or nonessential work should be eliminated. Following are the typical steps to use to develop a *value chain flowchart*:

Step 1. Choose a key product or service that you provide to your customers. Develop a detailed flowchart of the process. Include and map all activities in the chart. Then note whether the process is a:
  (a) *Materials flow process:* activities that directly support the material's movement from one status to another (e.g., transporting)
  (b) *Transaction flow process:* activities used to administer the material or service (e.g., purchase orders, issuing of material drawings, etc.)

Step 2. Label each activity:
  (a) Value-adding
  (b) Non-value-adding
  (c) Essential

Step 3. Calculate the amount of elapsed time to progress through each activity in the chain.

Step 4. Determine the percentage of time spent on essential, value-adding, and non-value-adding work.

Step 5. Focus only on those activities (typically, 10 to 15) where time is spent satisfying your customer (e.g., travel to a customer site to configure a design, customized assembly, etc.).

Step 6. Place priorities on these opportunities. Adopt the top three to five non-value-adding activities and focus your efforts on improving them. This is where dramatic quality improvements with the greatest payback to your customers and your organization can be made.

Step 7. Lobby for "buy-in" from all potential stakeholders to make the necessary process changes. Publicize the potential benefits to your organization (e.g., resource savings, quicker deliveries, fewer design changes, etc.) for making the improvements highlighted in your value chain map.

A project team should chart the process from a customer's perspective. Once the data are gathered, recommendations are reported to all involved organizations or project units. This should occur before further quality improvement efforts are undertaken in the targeted areas.

Teams engaged in process improvements in a production facility, for instance, may want to start the VA exercise by attacking "big-ticket areas" that are costly to operate and waste valuable company and supplier time and resources. In the service industry, the teams might start by redesigning processes that dissatisfy their customers. Areas to consider might include streamlining the total materials flow from outside supplier to delivery to end user and shortening the order fulfillment processes for a product or service. Savings and changes in customer satisfaction levels must be measured as the plan is unveiled, to take advantage of opportunities as they arise.

4.6 ACHIEVING ON-TIME DELIVERY THROUGH PROCESS IMPROVEMENTS

So far, we've recommended several, project planning activities that can be used to maximize the profitability of an organization's new product introductions and increase the longevity of the product or service offers in the market. Among these are:

- Matching project structure to corporate strategy and obtaining individual team members' buy-in to a set of project objectives that establishes overall commitment to achieve them.
- Shortening concept-to-customer (CTC) delivery time and being first to market through cross-integration of marketing, engineering, and manufacturing.
- Improving product performance by working with customers from targeted market segments to determine fitness for use and important value-adding and non-value-adding features.
- Planning an incremental development and release of software applications to help master the timing of changes requested by customers or technical shifts to the underlying hardware platform.

While implementing these planning techniques cannot guarantee increased performance, coupling them to fundamental process changes: Specifically, how companies conduct their NPD increases their chances for success. Table 4.9 illustrates how external and internal factors can influence NPD success.

TABLE 4.9   INFLUENCES ON NPD SUCCESS

External Factors	Internal Factors
• Global competition • Changing marketplace • Increasing government regulation and/or deregulation of major industries • Timeliness of competitor introductions and exaggeraged proclamations	• Business and R&D strategy • Product superiority • Preliminary technical market assessment • Speedy and timely product introductions • Horizontal linkage such as marketing and R&D • Planned incremental development • Cross-functional teamwork • Concurrent marketing

Throughout the 1990s businesses around the globe invested heavily in process reengineering, with a large concentration of their efforts focused on the redesign of their new product development processes. What empirically gathered data exist to support the premise that project organizations need to adapt their processes continuously to meet the challenges of the twenty-first century to succeed. A 1995 study conducted by the economics faculty at Erasmus University in the Netherlands tested a hierarchy model of acceleration activities for an organization to follow when speeding up its NPD[4]:

Step 1: Simplify.
- Technical designs
- Interpersonal relationship with external suppliers
- Controls: reports, documents, etc.
- Organizational structure

Step 2: Eliminate delay.
- Unused time between NPD activities
- Access time for retooling, handling, warehousing, inventory, scrap, and rework
- Transit time between production facilities

---

[4]M. R. Millson, S. P. Raj, and D. Wilemon, "A Survey of Major Approaches for Accelerating New Product Development, *Journal of Product Innovation Management*, Vol. 9, No. 1, March 1992, pp. 53–69.

- Non-value-adding activities
- Nonproductive meetings between functions (e.g., R&D, production, and marketing) by fostering teamwork, employee empowerment, and buy-in for NPD process

Step 3: Implement parallel processing.
- Perform two or more tasks at the same time.
- Complete all basic scientific research before scheduling parallel operations as to not waste time pursuing different approaches to the problem.
- Make sure that the integrated project team attends all functional reviews to assure collective decision making.
- Eliminate overlap between process development of NPD activities and the actual product development cycle by engineering the process before development begins.

Step 4: Eliminate steps.
- Unnecessary NPD operations
- Non-value-adding activities that do not contribute to customer satisfaction
- Parts and the associated task to assembly them into the product
- Product components that customers find no use for in their designs

Step 5: Speedup.
- Use new technologies for faster communications (e.g., Internet/intranet, e-mail, fax, videoconference, groupware).
- Adapt common or structured development and customer-focused design methodologies.
- Implement management structures to support project organization and NPD activity.

This hierarchy model was introduced in 1992. The study determined that accelerating NPD in accordance with the model helped eliminate important mistakes that otherwise might limit financial results.[5]

Further research in another study supports the conclusion that product process is critical from idea to launch of a product. Researchers at McMaster University in Canada gathered empirical data on new product introductions from a sample including major chemical firms in four countries (Canada,

[5]E. J. Nijssen, A. R. L. Arbouw, and H. R. Commanduer, "Accelerating New Product Development: A Preliminary Empirical Test of a Hierarchy of Implementation," *Journal of Product Innovation Management*, Vol. 12, No. 2, March 1995, pp. 99–109.)

Germany, the United Kingdom, and the United States).[6] The performance measures used to gauge success included (1) overall profitability of the firm, (2) technical success, (3) annual sales revenues, (4) relative market share, and (5) impact on the company of the new product's sales and profit.

The study's findings relate directly to how to improve product planning and process management activities that provide a competitive edge. The researchers' findings and conclusions are easy to understand and to apply:

- Market success correlates positively with offering a *product advantage* (differentiating the product from competitors) through *product quality* and creating *value* for the customer's money.
- Implementing a *launch strategy* tied to quality of service and satisfactory technical support, as well as timing the reliability of shipments to guarantee adequate product supply, also correlate positively with market success.
- The elements associated with early and accurate *product definition* are linked to financial success. However, many companies tend to rush through or ignore front-end analysis of customer needs at the expense of other technical activities.
- The quality execution of a *NPD process* is critical with the following 10 activities correlating positively to market success:
  (1) Full business and financial analysis of the project at mileposts immediately following product development but prior to commercialization
  (2) Elevation of production ramp-up
  (3) Initial screening of product ideas
  (4) Field trials of the product
  (5) Full-scale launch that introduces the product to market with specifically designed marketing activities
  (6) Test market (beta or early introduction) sale of the product
  (7) Marketing research
  (8) Preliminary market assessment
  (9) Business and financial analysis leading to a go/no go gate decision prior to the development phase
  (10) Pilot or trial production to test manufacture

---

[6]E. Kleinschmidt and R. Cooper, "The Relative Importance of New Product Success Determinants: Perception Versus Reality," *R&D Management*, Vol. 25, No. 3, July 1995, pp. 281–298.

- *Project organization* contributes to success. The projects that were lead by a project champion and where the team is in place for the entire project, from concept formation through launch, are financially successful. Use of the cross-functional team is a leading contributor to new product success.
- In general, only three elements of *project familiarity* contribute to the firm's financial performance:
    (1) Introducing new products into an existing product category or family
    (2) Competing in product categories with familiar competitors rather than against new competitors
    (3) Selling to an existing customer

Another study of Mexican-based high-technology firms tested the perception of R&D, manufacturing, and marketing professionals regarding variables that pave the way for more effective NPD.[7] Respondents from all three groups surveyed shared the same perceptions. Each group believed that *internal forces* concerning NPD performance, such as project evaluation prior to selection, technology, and manufacturing capability, along with *cross-functional cooperation*, have greater effect than external factors. According to the same three groups surveyed, top management also has a role in the NPD process. Corporate executives must act as champions, supporting the project manager and project team, while balancing outside market forces.

As this research indicates, achieving on-time delivery and maximizing the profitability of hardware and software projects will require the restructuring of project organizations and adaptation of constantly refined processes. To improve profitability and compete and succeed, process improvements and more rapid, incremental product introductions around the globe will be inevitable. Organizations that implement processes that have more cross-functionally integrated activities, especially in the early product conception and project formation phases, not only succeed with current product and service introductions, but also build a fundamental framework for future introductions. These process improvements are necessary to sustain profitable growth, market share, and future fiscal performance.

---

[7] X. M. Song, M. M. Montoya-Weiss, and J. B. Schmidt, "Antecedents and Consequences of Cross-Functional Cooperations: A Comparison of R&D, Manufacturing, and Marketing Perspectives," *Journal of Product Innovation Management*, Vol. 14, No. 35, January 1997, 35–47.

4.7 CONCLUSIONS

In this chapter, we have discussed several alternatives to conventional new product development and why more structured methodologies, in addition to better communication among all involved project and functional organizations, are making it possible to reduce cycle time and improve performance. Undertaking the principles discussed in the chapter helps prepare organizations to meet competitors challenges and satisfy customers expectations. In Chapter 5 we show you how to navigate the rough waters of a project and chart a course using proven planning tools.

BIBLIOGRAPHY

Bernstein, L., "Software in the Large," *AT&T Technical Journal*, Vol. 75, No. 1, January–February 1996, pp. 5–14.

Brown, P., "QFD: Echoing the Voice of the Customer," *AT&T Technical Journal*, Vol. 70, No. 2, March–April 1991, pp. 18–32.

Dant, B., and S. Kensinger, "Re-engineering Engineering: More Madness," *Computer-Aided Engineering*, Vol. 16, No. 3, March 1997, pp. 60–63.

Eureka, W., and N. Ryan, *The Customer-Driven Company: Managerial Perspectives on QFD.* Dearborn, MI: American Supplier Institute, 1988.

Feiner, A., and K. A. Edward, "Development Processes and Applications at AT&T: An Overview," *AT&T Technical Journal*, Vol. 70, No. 2, March–April 1991, pp. 2–6.

Feiner, A., and K. A. Edward, "An Overview of Quality and Project Management Processes at AT&T," *AT&T Technical Journal*, Vol. 71, No. 3, May–June 1992, pp. 2–7.

Govers, C. P. M., "What and How About Quality Function Deployment (QFD)," *International Journal of Production Economics*, Vol. 46, No. 47, December 1996, pp. 575–585.

Greasley, A., *Project Management for Product and Service Improvement.* Oxford: Butterworth-Heinemann, 1977.

Jenkins, S., S. Forbes, and T. S. Durrani, "Managing the Product Development Process, Part 1: An Assessment," *International Journal of Technology Management*, Vol. 13, No. 4, 1997, pp. 359–378.

Kythe, D., "The Promise of Distributed Business Components," *AT&T Technical Journal*, Vol. 75, No. 2, pp. 20–28.

Leech, D. J., and B. T. Turner, *Project Management for Profit.* New York: Ellis Horwood, 1990.

Pawar, K., P. Forrester, and J. Glazzard, "Value Analysis: Integrating Product–Process Design," *Integrated Manufacturing Systems*, Vol. 4, No. 3, 1993, pp. 14–21.

"Managing Megaprojects," *IEEE Software*, Vol. 13, No. 4, July 1996.

McGrath, M. E., M. T. Anthony, and A. R. Shapiro, *Product Development: Success Through Product and Cycle-Time Excellence.* Stoneham, MA: Butterworth-Heinemann, 1992.

Norris, M., *Survival in the Software Jungle.* Norwood, MA: Artech House, 1995.

The Project Management Institute Standards Committee, *A Guide to the Project Management Body of Knowledge*. Upper Darby, PA: Project Management Institute, 1996.

Quality Functional Deployment for Product Definition: Guidelines for Understanding What Customers Really Want. Holmdel, NJ: AT&T, May 1990.

Rosenau, M. D., Jr., *Faster New Product Development: Getting the Right Product to Market Quickly*. New York: American Management Association, 1990.

Rosenthal, S. R., *Effective Product Design and Development: How to Cut Lead Time and Increase Customer Satisfaction*. Homewood, IL: Business One, Irwin, American Production and Inventory Control Society, 1992.

Servi, Italo S., *New Product Development and Marketing: A Practical Guide*. New York: Praeger Publishers, 1990.

Topfer, A., "New Products: Cutting the Time to Market," *Long Range Planning*, Vol. 28, No. 2, April 1995, pp. 61–78.

Wheelwright, S. C., and K. B. Clark, *Revolutionizing Product Development: Quantum Leaps in Speed, Efficiency, and Quality*. New York: Free Press, 1992.

EXERCISES

4.1. Which of the following does not have a positive influence on project performance?

   (a) Application of a common or structured development methodology

   (b) Rapid deployment of development resources during the design phase of a project

   (c) Use of a panel to learn the voice of the customer and influence design

   (d) Relieving project team members of their responsibilities when it is time to end project work

   (e) Shortening the overall development cycle by eliminating less critical laboratory tests that occur at the end of the project

4.2. True or false? The project team should wait until the project has been well defined before planning the development phase of its life cycle.

4.3. Explain why the waterfall approach is no longer an accepted methodology among hardware and software system developers.

4.4. The rapid pace of technological advancements has changed the time that organizations can spend developing products and services to introduce into today's markets. Which of the following might inherently delay the cycle time of an NPD project?

(a) Relative newness of the technology under development

(b) Use of cross-functional team of project specialists

(c) Formal processes, such as integrated project planning

(d) Complex product designs

(e) Project evaluation and selection processes

(f) Internal reward systems

(g) Product designs that provide specific advantages over competitive offers

(h) Product launch strategies.

4.5. Faster new product development can be achieved by:

(a) Eliminating dead time or delays between phases of the NPD cycle

(b) Performing tasks in parallel

(c) Eliminating non-value-adding product and service features through value engineering

(d) Designing systems using quality function deployment techniques

(e) Communicating the project's mission, goals, and objectives to team members early on

(f) (a), (b), and (c)

(g) All of the above

4.6. Explain how using common or structured development methodologies can influence the length of the concept-to-customer (CTC) interval of a new product or service development.

4.7. Quality, cost, timeliness, and product value are often viewed as conflicting values that require project teams to trade-off when reducing product development cycles. Explain why it's easier to earn customer loyalty and gain market share by bringing a desirable product to market first than to win back customers with a late entry that does not meet their needs.

4.8. E&S Computing's newly assigned core team is challenged with deciding the technical design for a new portable computer that has enough functionality to satisfy the needs of the mobile business executive. Representatives from marketing have recorded and ranked the priority of customer needs in a QFD planning matrix (Figure 4.5). Systems engineering personnel have also indicated several available technologies that would potentially support these needs. Your role, as project leader, is to direct the team in choosing and planning the development of the technologies that best support customer needs and E&S's business directives.

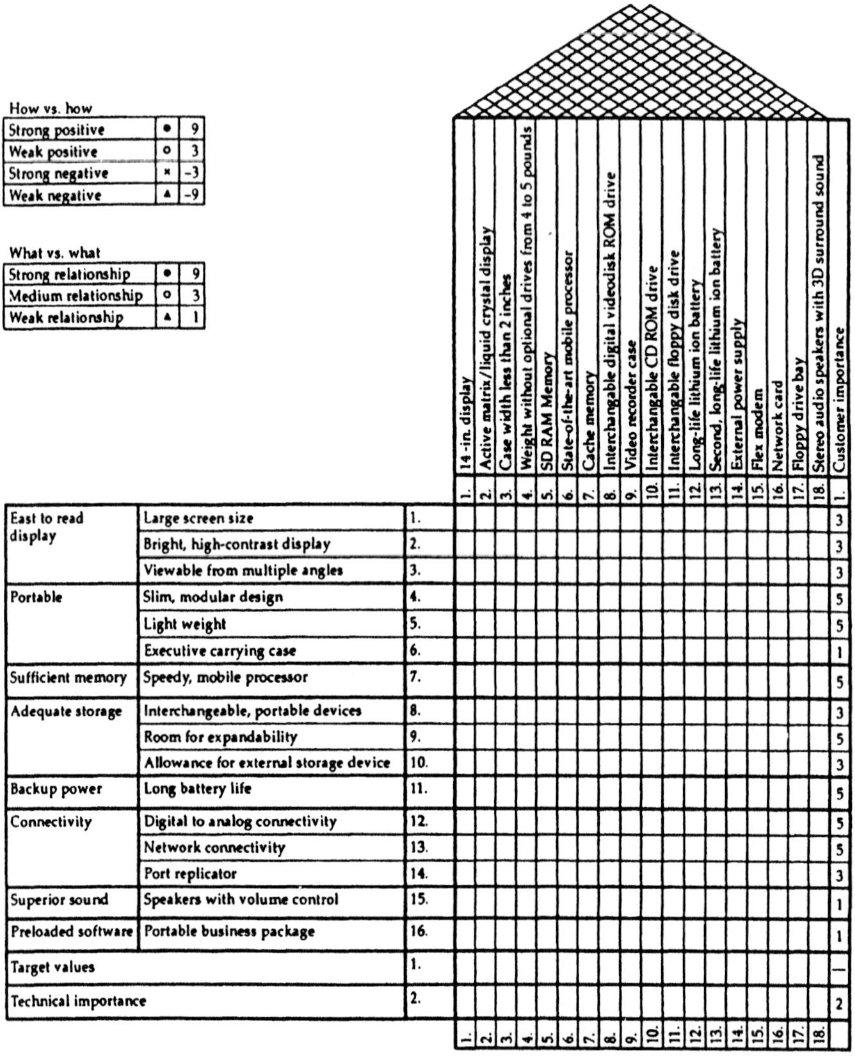

FIGURE 4.5

QFD quality planning matrix.

(a) Select the appropriate symbols to indicate the relative importance of the technical concepts (shown in the vertical columns) to customer needs (expressed in the horizontal rows) and populate the matrix.

(b) Indicate the interdependencies of the technical characteristics and record the correlations in the house of quality's roof.

(c) In the matrix, for each of the assigned relationships in the columns, write out those that customers target as most valuable.

(d) Calculate their technical importance. Multiply the numeric equivalent of the symbol selected for each of the relationships in the rows with their overall customer importance rating. Note the sum for each column.

(e) Discuss with the team which technical components would produce a value-adding feature list for E&S's business customers. Remember to consider cost, technical difficulty, and availability of resources.

4.9. Review the research findings in Section 4.6. As a senior member of Midland Bank's steering committee for new program development, what recommendations would you make to the project team developing a home banking application? The target customer market includes expert and novice users of personal computing technology who already own one or more computers that they use for business and/or educational purposes in their homes. The payback or break-even period for the investment is anticipated to be 6 to 12 months before new technology would require the bank to introduce an upgrade release of the software application.

4.10. You are a newly promoted project manager at Airlink Fare, an international carrier with aviation routes throughout the United States, Central and South America, Canada, Europe, and Asia. Your first assignment is to engage a value engineering team in the redesign of the passenger check-in and baggage-handling process for U.S. domestic departures and arrivals.

(a) Create a value chain flowchart for the current process.

(b) Indicate at least three non-value-adding activities that are customer dissatisfiers.

(c) Reengineer these activities and document the process flow of the new passenger service.

5

CHARTING THE COURSE USING THE TOOLS OF PLANNING

Luck is what happens when preparation meets opportunity!
ANONYMOUS

5.1 INTRODUCTION

It is hardly surprising to most of us that to succeed, high-tech, high-performance projects must be planned carefully. No mountain climber would ever set out to scale a summit without first selecting a route, choosing a competent team, assembling proper food, and ensuring that all the equipment is in order.[1] In fact, projects of all magnitudes require considerable time and effort to plan. The great explorer and navigator Christopher Columbus took several decades putting together the pieces of his plan that came to be known as the *Enterprise of the Indies*. Although some historians today look back at the events leading up to his 1492 crossing to the Americas from Paos, Spain and consider them fortuitous, the truth is that Columbus prepared in detail for his adventure. He skillfully altered the course of world history because he: (1) was a visionary with the ability to make theory a useful tool; (2) knew how to read and draw portalan navigational charts, which by the late fifteenth century aligned distances with meridian lines, compass barring, and reference points,

---

[1] J. M. Kouzes, and B. Z. Posner, "Plan Small Wins," in *The Leadership Challenge* (San Francisco: Jossey-Bass, 1987), p. 232.

thereby making sailing by compass navigation possible; (3) had explored the Atlantic as a youth and become intimately familiar with the trade winds; and (4) had befriended wealthy nobleman and the clergy, who arranged his appearances before the royal Spanish court.

As in modern projects, Columbus had trouble securing immediate financing. His proposal to sail west to reach east was first refused by the Spanish sovereigns in 1486. During the ensuing years until its acceptance, he sought backing from the monarchs of Portugal, England, France, and the Genose government. To assemble a crew, he needed to demonstrate that other competent pilots and captains were willing to accompany him. During the voyage these leaders actually sailed their ships in close proximity so that they would be able to caucus, plan the daily course, share navigational sightings, and keep the crew from mutinying! Columbus's vision, perseverance, and tactical planning skills would be greatly admired by project team leaders today.

Drawing on this analogy, in this chapter we provide an overview of the tools used to plan a project. These are the methodologies used not only to define day-to-day project activities but also to align team members with organizational goals. We begin by reviewing the essential elements of the project summary plan. This plan charts the course for the project and becomes a trusty navigational aid during its execution. Then we explain how to implement an integrative planning process to form and launch a *team* of contributors who are poised for success. We conclude with an in-depth explanation of how to prepare a project work breakdown structure and work package.

5.2 PROJECT SUMMARY PLAN ESSENTIALS

The *project summary plan* is a summation of all of the project's essential elements. It is prepared in such a manner that those managing the project can learn at a glance *what* is to be done; *why* and *how* it is to be done; and *when* project deliverables are to be ready, by *whom*, and at *what cost*. It is the culmination of an integrative planning effort by members of the project's *multidisciplinary* team. In fact, the preparation of the summary plan builds the foundation for how the project personnel learn to operate as a true *team*. The actual discussion and the collaborative writing of the plan elements, by the personnel who will eventually complete the work, can improve project quality and timeliness. Not only is their input obtained, and their roles and responsibilities clarified, but their commitment to completing project objectives is also discussed.

The planning process should include those team members who will carry out the project tasks and activities, since planning efforts benefit from the *collective input* of all the specialists who bring their unique expertise and experience to the group. The plan creates the basis for project control by establishing *participative procedures* for discussing difficulties and resolving problems as an integrated team. The elements of the plan can be monitored and deviations from the established baselines used as early warning signs that a project is headed for trouble. The components of a summary plan include:

- Project mission, goals, and objectives
- Project management methodology
- Product development approach (stage/gate model, design approach, etc.)
- Project critical success factors
- Business plan
- Project overview and/or statement of work
- Specifications/technical requirements
- Project milestone schedule
- Project work breakdown structure
- Master budget
- Project network
- Project task responsibility matrix
- Project risk response/contingency plan
- Master test/system verification plan
- Change management plan
- Services support plan
- Training support plan
- Documentation plan
- Procedures to control project documentation/quality records
- Project team membership; core team composition
- Qualification plan/launch plan

The *project mission statement* recounts how the project effort supports the organization's mission and aligns the project with the business entity's or unit's formal business and technical goals. A statement that represents a common vision or focus, the mission is used to guide the project team and influence individual team members' behavior in a direction that contributes to project goals. For the project plan to be successful, it is critical that the entire team accept these statements. Concerns about goal accomplishment must be aired and addressed not only during the planning process but also during subsequent project meetings. Here they may be documented as open items in action plans to be resolved at a later time by the project team or to be escalated to senior management for immediate clarification and further input. *Project objectives* are measurable statements describing what is expected at the outcome of the project effort. They may specify targeted intervals for each project phase as well as the cost to bring the service or product to market. The objectives should also communicate to all levels of project personnel the quality and quantity parameters necessary to meet customer specifications.

The *project management methodology* is a description of the procedures the project will use to plan the work activities, track progress toward their completion, and communicate, report, and document results. Data gathered during a *project management requirements analysis* may be included in the summary plan to explain how a particular management structure or procedure was chosen for the project.

The *development approach* references the organization's standard methodology for the type of activities under way (e.g., systems engineering, software design, code inspection, hardware development, production). Exceptions or changes to these standards, agreed to by the project team, should be described using as much detail as is required to comply with corporate guidelines. Frequently, one needs to reference internal memoranda that document how a specific division, department, or unit performs these functions and to include this information in the summary plan.

The project team must enumerate in the *summary plan* key conditions or *success factors* that must be present in the marketplace and project environment for the project to succeed. The team must also be aware of and note any factors that may have a positive or negative impact on the project's financials. These concerns should be documented in a formal financial or *business plan* that predicts the anticipated expenses and earnings that will incur over the life of the project. *Critical success factors* are statements of assumptions that are crucial to the project's successful completion. Some examples include events that could potentially delay the product's completion and competitive actions or nonactions that might influence acceptance by customers. To predict the likelihood that the assumptions may happen and any negative financial implications to the project, the project team often assigns a probability to their occurrence. This information may be used later by the team when determining risk mitigation strategies and assigning contingencies to cover the costs to recover the project should the assumptions with any negative repercussions prove true.

The *business plan* contains the financials for the project. These include resource expenditures, cash flows, future income, and the project's *net present value* (calculated cumulative, future income after all annual costs and taxes have been deducted, including depreciation). The business plan should also contain sections describing the market and competitive position compelling the development of the proposed technology, service, or product. Appendix 3.1 contains further information on how to prepare a business plan for a project.

The *project overview* briefly describes the work to be undertaken. It may explain the technology under development or describe the proposed feature enhancements being produced to stimulate market demand for a company's core product line. Like the *statement of work* (SOW), the project overview is a narrative describing the work activities to be completed, the funding, and product specifications. The SOW is issued, typically, by a contracting organization to its prospective suppliers. Generally, all proposals to the U.S. federal government and its agencies must be written and prepared in accordance with the specifications stated in this document. In fact, it is good practice for firms involved in any form of contract negotiation with the government to reproduce the SOW in the body of their proposal response to indicate compliance with the customer's work requests.

The project team may often need to specify and/or clarify for its customers how the project's *technical requirements* will be satisfied and even restate them in the summary plan.

The *milestone schedule* included in the plan is a high-level or *bird's-eye view* of the project schedule. It should include such items as the start and end dates of major tasks or activities, along with tangible milestone events, report due dates, and so on. Generating the milestone schedule is a critical step in building the integrated project team when accomplished collectively. Not only do the team leaders acquire a sense of interdependency and respect for each team member's complementary roles, but the players gain a greater understanding of project objectives and respect for their co-members' diverse talents.

The *work breakdown structure* (WBS) breaks down or divides the work activities into definite units or packages. These *work packages* can be assigned specific project or job numbers, expected costs, personnel required, duration expected, and required equipment. Because undefined work often results in duplication of effort and frustration, it is best to include all the team members who may perform the work in developing the WBS. Project objectives typically become clearer and understood by the team through this collective effort to define what is required to accomplish project goals and objectives.

The *master budget* is an estimate of funds required to complete each work activity over the development phase of the project and will typically include the early introduction testing of the product or service offer in the market. However, budgeting for other postproject, life-cycle activities, such as advertising and marketing collateral, are likely to be expensed as corporate overhead and often are not included in the master budget for a development or production project. The master budget usually is presented as a schedule indicating the anticipated funds to be expended as a function of time. If the project team anticipates the need to expedite project work to meet tight project deadlines, the project budget should include estimates or contingencies to cover the additional expense. The master budget should be prepared once the WBS is complete.

The *project network* is a schedule that illustrates the sequencing and interdependencies of the work activities. It also indicates whether these tasks will be accomplished serially or in parallel. The network defines the *critical path* for the project by identifying which activities, if delayed, will delay the project. The critical path is discussed in detail in Chapter 6. The generation of the network often is assigned to a team of project management specialists who are skilled with any of the many commercially available project management information systems. These specialists use the collective input of the entire project team to produce the network. The project teams' collective input to the schedule and its constant monitoring and revision, after a baseline of the initial plan is established, is essential for the project to succeed.

A *task responsibility matrix* or *linear responsibility chart* (LRC), discussed in Chapter 2, illustrates who has been assigned what tasks and illustrates how the project team will interact when accomplishing the project objectives. If roles remain fuzzy, the team will be frustrated. No one will be accountable for successes or failures. Serious barriers may arise, and the project will experience nonconstructive conflict. The resulting matrix can be used to prevent such misunderstandings and clarify roles and responsibilities.

Projects are inherently risky. The summary plan must account for and communicate to all project members the actions or contingency plans intended to mitigate these risks. Good *risk assessments* formally evaluate both the business and technical risks associated with a project. The project team assigns a probability that the risk will occur to determine the cost to the overall project for mitigating or accepting them. *Risk response* is described in more detail in Chapters 6, 7, and 10.

Another critical step in the integrated planning process is to build a *master test* or *system verification plan*. This plan describes how the system under development will be tested and the acceptance criteria for release to a customer. The test plan reports the cost to stage the tests, the resources needed (personnel and capital expenses), and intervals scheduled to detect and correct defects.

The *change management plan* documents how the project will manage *modification or change requests* (CRs) for design changes once the team baselines the project's technical requirements. In software development projects, for example, CRs that result from executing unit and system tests are assigned severity and priority levels for resolution by the developers or testers who discovered them and are published for review. Then, selected project members who represent both these communities and act as the formal *change control review board* for the project, accept or reject the proposed corrections, and finally, approve any modifications to the original project specifications. Systematic tracking of all proposed changes is a required project management activity. It is also important to describe in the *change management* portion of the summary plan any automated tools that the project will use to maintain software source code configurations and document the modifications that the review board approves.

The *services support plan* describes how to develop the service agreements (warranty and after-warranty maintenance procedures) that accompany the product to market. Often, a *business plan* is written to determine the cost and benefits to the firm for extending customer service offers beyond the warranty period.

The summary plan should also include a *training support plan*. This plan describes how customers will receive training on how to use the product. The training plan will also include a budget and schedule that is rolled up into the master plans for the project.

Similarly, the *documentation support plan* details what documentation will be prepared to accompany the service or product offer. Preparation and product costs, along with an appropriate schedule linking these activities to the master plan, are required.

The summary plan must also outline how the project team will document the formal development and product introduction processes they are following. Most corporate quality departments publish detailed guidelines regarding what project *quality records* must be kept as the team passes the various quality gates and provide templates to record the formal outcome of project assessments.

The *project team membership* and the lead *core team* nominees should be listed in the summary plan. The team will also want access to an electronic version of the membership roister where affiliations, locations, e-mail addresses, and phone and fax numbers are posted. Once the project is under way, it's important to update the list of project participants periodically as personnel go onto other assignments and new members join the team.

Finally, the summary plan should contain a *qualification* or *launch plan*. This plan describes the strategy for the product or service introduction and the agreed-upon criteria for a successful market introduction. How will the company know when project efforts are complete? How will the team know that their services are no longer retained by the project? Like project objectives, these criteria can be generated through an integrative planning exercise involving all representatives from each cross-functional discipline. Obtaining the collective team membership's agreement during the planning phase by having everyone visualize or articulate what's required to succeed helps establish a common understanding of where the project is headed.

5.3 TEAM DEVELOPMENT AND THE INTEGRATIVE PROJECT PLANNING PROCESS

Businesses must organize by adapting strategies that improve performance in both domestic and global markets to face the demands placed on them from multinational competitors. One strategy that we have applied and found extremely successful for creating major improvements in product performance is to involve all the specialists who work on a project in the front-end development and subsequent maintenance of an integrated project plan. Projects that engage in this planning process are able to define their functional interfaces and generate a more committed and dedicated team. Even the most seasoned team leader is apt to have some difficulty getting diverse participants to understand where the project is headed and what it will take to succeed. Accomplishing this with diverse staff, from a variety of technical and business disciplines, who come from different countries and cultural backgrounds, requires a back-to-basics, hands-on exercise in integrative planning.

The *integrative project planning process* (IPP)[2] is a group decision-making procedure that brings all project participants together under the leadership of a *project champion*. An excellent planning tool that stimulates collab-

---

[2]D. S. Kezsbom, "Making a Team Work: Techniques for Building Successful Cross-Functional Teams," "Experts Only," *The Linton Training Sourcebook and Buyers Guide 1995*, pp. 172–173.

orative thinking, project teams can use the IPP to decide how to turn customer specifications into a successful commercial venture by (1) determining the scope of the work effort, (2) planning the allocation of required resources, and (3) identifying how to deal with associated risks and concerns.

Full participation in this decision-making process develops commitment and ownership of project goals and procedures, the integrated summary plan, and schedule. Each specialist comes to possess a shared sense of responsibility that is essential to the team's foundation. This process demands creativity as well as a thorough understanding of group dynamics. Because of these demands, an outside facilitator may best lead the process. The chosen person, whomever he or she may be, must be perceived as an active participant who is involved with the team as it develops over the life of the project and becomes a more cohesive unit.

The following are the steps we recommend:

Step 1. Remember that the likelihood of project success is dependent, in part, on how the team development process unfolds. Communication is critical to building an effective and productive team. Start the IPP planning session by explaining its objectives and expected outcome. Then focus the team's attention on collectively developing a clear understanding of the team's *mission* and what it will take for the project to be a success. Negotiating a mission can be a frustrating and time-consuming activity, but without common agreement over where the team is headed—its *mission*—little progress will be made drawing up a plan that everyone understands. The facilitator may need to start by seeking clarity from each member of the group of his or her own interpretation of the mission. This may be the first opportunity for many team members to express their personal reasons for wanting the project to succeed. There may be others who, although already working on the project for several months, learn for the first time what is expected.

Step 2. Once the mission has been agreed upon by all concerned, the next step is to begin defining a *work breakdown* of the project that delineates all work assignments starting from the topmost levels of activities. For the purposes of this exercise, it is sufficient for the team to use Post-It notes to capture a snapshot of the work to be done so that activities can be rearranged and reordered as they are assigned to functional groups. Later, each functional group, assigned the work, will prepare a more detailed WBS, as we explain in Section 5.4.

Sometimes, to get the team started, the facilitator and project champion may have to share how they feel the overall project might accomplish the level 2 WBS functions. About midway through the session, the team should have identified within the WBS most of the major subbranches of activities required to complete the project. With this accomplished, functional subteams can be formed and asked to break down their tasks further into assignable *work packages*. Before ending the session, it is important that each of these subteams presents its conception of the work assignments to each of the other teams. The specialists must be encouraged to question each other's plan so that all ambiguities are clarified and explained. If the work is not finished, continue the exercise on another day. The duration of the sessions will naturally depend on the scope and complexity of the project effort.

Step 3. Once the project scope or definition is clear and a WBS according to each of the tasks that illustrates the project plan is developed, the next step for the team is to arrange the activities along a timeline. This process provides the team with the opportunity to define the boundaries of the project in relation to time, and to specify how work will be conducted to achieve these goals. Often, market management will impose time constraints on the schedule, due to competitive pressures and threats of new product or services entries in the market. The *integrative planning process* provides teams with the opportunity early in the project life cycle to negotiate date changes, trade off proposed functionality, or *crash* the schedule before a plan is announced. The intent is to analyze each task, consider its dependency and precedence relationships to other activities, determine duration and cost, and eventually assign someone's name to its completion. Out of this process, team roles and responsibilities are clarified.

Step 4. The final exercise in the IPP session is to ask the team to explore the project's critical success factors, as well as the areas that represent considerable risk or that may possibly place the project in jeopardy. These include both *internal* and *external factors*, such as incorporating advanced or state-of-the-art technology into the product, government action such as possible intervention into a presently unregulated market, or the firm's relative inexperience selling this type of product or service to the target market. This exercise helps teams learn how to collaboratively establish contingency and mitigation strategies associated with project risks and to identify their potential impact on the project schedule.

In summary, there are five essential components for using IPP successfully to create an integrated project plan. These are:

1. Create a positive climate that helps participants understand how important team planning is to building commitment to project success.

2. Establish a common vision by forming a mission statement that is developed jointly by team input.

3. Create a better understanding of what is expected from each team member so that valuable resources and time are not wasted duplicating roles or responsibilities.

4. Develop a sense of interdependency among team players to accomplish the schedule by providing an opportunity for specialists to visualize the Post-It scheduling process and experience how each other's diverse talents contribute to the team mission.

5. Develop the team's ability to use conflict in a constructive manner by reaching consensus on the team procedure to be used to process critical decisions.

5.4 THE WORK BREAKDOWN STRUCTURE

A properly conducted IPP results in one of the most important components of an integrated project summary plan, the project *work breakdown structure* (WBS). The WBS defines the work or tasks to be performed and is the primary planning and analysis tool used in almost all projects, including information and systems engineering, construction management, and pharmaceutical, because it answers two key questions: (1) What is to be accomplished? and (2) What is the necessary hierarchical relationship(s) of the work effort? The WBS also aids the project management process by (1) providing a complete list of the software, hardware, services, and information technology work tasks that must be completed during the development and production of a product; (2) defining the responsibility, personnel, cost, duration, risk, and precedence of each work task; and (3) providing an easy-to-follow numbering system to allow hierarchical tracking of progress.

This important planning and analysis tool was introduced in the 1960s by the following U.S. federal agencies: DOD, DOE, FAA, and NASA. Today, it is widely employed around the world by corporate research and development laboratories, software development companies, manufacturers, and civil engineering and facility management organizations. In other words, it is an appropriate and necessary tool for all forms of projects. The aim of using a WBS is to better manage investments and projects of a variety of type and duration.

The WBS partitions the project into manageable elements of work, for which costs, budgets, and schedules can more readily be established. When properly prepared and completed, the WBS satisfies the needs of both project management and the customer. Integration of the appropriate project's organizational structure with the WBS helps the project team to assign specific project personnel responsibility for the technical tasks that need to be accomplished for successful project completion.

5.4.1 Preparing the WBS

Formation of the WBS family tree begins by subdividing, or *partitioning*, the project objective into successively smaller work blocks until the lowest level to be supported and controlled is reached. This treelike structure breaks down the project work effort into manageable and independent units that are assigned to the various specialists responsible for their completion. The WBS links, in a very logical manner, company resources with the work to be performed. Figure 5.1 shows a level 2 WBS for a radio transceiver (transmitter–receiver). We see that the basic radio system listed in level 1 is partitioned subsequently with its primary functional blocks in level 2. If done properly, there should be no work that must be performed that cannot be considered to be a part of one of these level 2 blocks. Task numbering follows each task and allows not only the tracking of task completion but associated costs.

The WBS is not a breakdown by discipline or functional organization, such as hardware, software, systems, or mechanical engineering. Rather, it is a *deliverable-oriented charting* of project elements grouped as functional blocks that helps to organize the work project team members are assigned. Each of the level 2 elements in Figure 5.1 is now partitioned into a grouping of level 3 elements. Since the WBS represents the full scope of any specific effort work on the receiver might possibly be organized as shown in Figure 5.2a, and work on the controller, in Figure 5.2b.

In Figure 5.2a, we see that the receiver consists of a low-noise preamplifier, tuner, frequency synthesizer, demodulator, and audio amplifier. These are the functional building blocks into which the receiver is partitioned. In addition, these elements must be designed and fabricated to perform together, which requires an additional WBS element: *receiver integration and testing*. Each of the level 3 elements can then be partitioned

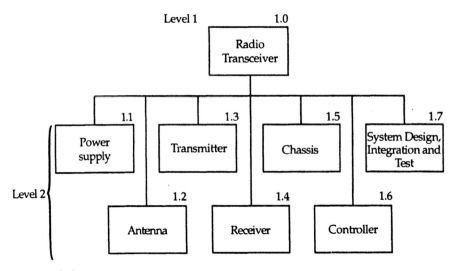

FIGURE 5.1

Work breakdown structure for a radio transceiver.

into more elements. The partitioning process ends when any element is well defined and *buildable* by a small group of engineers. For example, in Figure 5.2*a*, the low-noise preamplifier, frequency synthesizer, demodulator, and audio amplifier probably would not be partitioned further. However, one might continue to partition the tuner into a fourth level: RF amplifier, IF amplifier, mixer, and automatic gain control (AGC).

In Figure 5.2*b*, the controller can be partitioned into five software elements. Eventually, each of these elements is integrated into the hardware. Element 1.5.2, which refers to the antenna, should then be partitioned, for example, into several level 3 activities, consisting of the physical antenna, antenna hardware/software integration, and testing.

There is no right or wrong way to perform the WBS. When teams set out to develop a WBS for the same project, they may choose from among several parallel or different ways of presenting how they see the organization of the tasks and processes necessary for completing the project. Whether the WBS organizes the project by phases or functional deliverables, the exercise should help the team come to a common understanding of the project scope. Remember:

1. Always prepare a WBS.

2. Conduct an IPP session. Involve all project specialists, especially those having a diversity of expertise and experience.

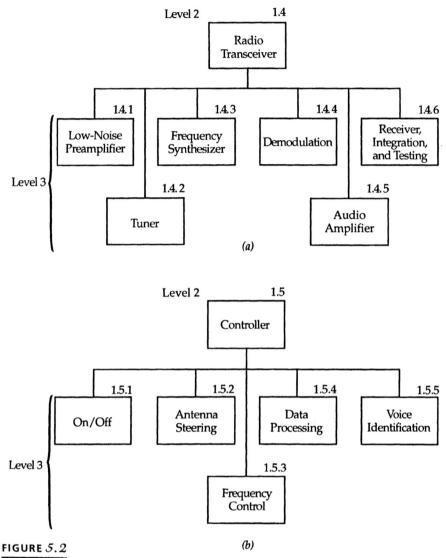

FIGURE 5.2

Level 3 WBS for (a) the receiver and (b) the controller.

3. Be complete. Any element of the project against which funds are expended must be included in the WBS. Be sure that the breakdown is a partition of functional building blocks or project activities and not a breakdown by organization or discipline.

4. The WBS is treelike. Therefore, an element in level 4, for example, must separate into two or more level 5 elements. All of the *work* performed in level 4 is performed in level 5.

5. Because the structural breakdown of project elements is hierarchical, two or more elements in level 2, 3, or 4 can never connect to the same element in a lower level.

In practice, a typical level 2 activity in a WBS generally contains the following: (1) project management, (2) documentation, (3) system design, (4) system integration, (5) testing, and (6) other functions resolved in the project and against which charges incur. Each level 2 element is then expanded as needed to illustrate the activities of which it consists.

In Figures 5.3 and 5.4, two additional WBS samples are provided. Figure 5.3 shows a WBS for a software project, while Figure 5.4 is a hardware project. Note the treelike structure and also that each branch divides into two or more branches. In Figure 5.3, for example, *coding routines* (Task 1.3.3.2) is completed when the level 5 tasks—*perform coding* (Task 1.3.3.2.1) and *unit test* (Task 1.3.3.2.2)—are completed. In practice, each coding module is delineated as task 1.3.3.X (X = 2, 3, 4, . . .).

5.4.2 The Work Package

After the WBS is constructed, the finishing touch is added. This is the *work package* or *dictionary*. The work package is a description of what must be performed, by whom, and in what time duration. The work package is always prepared for each bottom-level element of the WBS. Thus a level 2 element, if not partitioned (although an unlikely possibility), would be given a work package; and a level 4 element, if not partitioned, would also be given a work package.

A typical work package is shown in Figure 5.5.

- Item 1 gives the *project title.*
- Item 2 gives the *date* that this work package was approved.
- Item 3 gives the *WBS element* or task title and number.
- Item 3a, *revision*, lists the number of times that this work package has been revised.
- Item 4 gives the *name, telephone number,* and *e-mail address* of the *project manager.*
- Item 5 gives, for this WBS task, the *name, telephone number,* and *e-mail address* of the *functional team leader* of the group performing the work.
- Item 6 lists all the *specifications* that this task must meet. These specifications might be further enumerated in the list of product and project specifications. Or they might have to be derived from these specifications

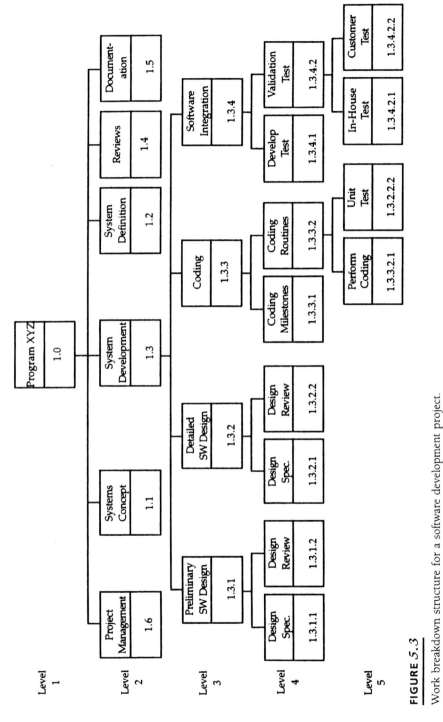

FIGURE 5.3

Work breakdown structure for a software development project.

157

158

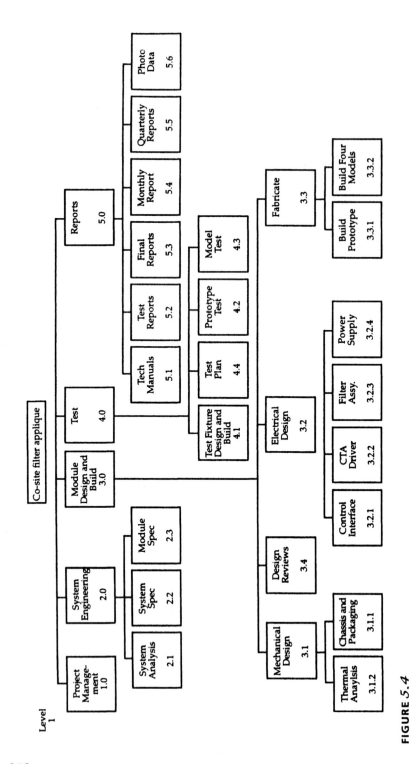

FIGURE 5.4

Work breakdown structure for a hardware development project.

WORK PACKAGE

1. Project title: 2. Date:

3. WBS Task (Element) Title/Number: 3a. Revision No:

4. Project Manager: Telephone: E-Mail:

5. Team Leader: Telephone: E-Mail:

6. Specifications:

7a. Work description:

7b. Deliverables:
       1.
       2.
       3.

8. Duration:

	to	tp	tm	Risk:	Precedence:
1.					
2.					
3.					

9. Personnel:

10. Special equipment/facilities/requirements:

11. Meeting schedule:

12. Reporting schedule:

FIGURE 5.5

A typical work package.

by the project manager, the functional team leader of the next-higher-level task (which was partitioned and thereby includes the task being specified), or the team leader of this level task.

· Item 7a *describes the work* to be performed, such as design, develop, assemble, test, and integrate for a hardware task, or design, code, test, debug, and document for a software task.

· Item 7b lists the *deliverables*. For a software project, a developer might be required to deliver the design, the source code, the object code, all algorithms, the test plan, test results, and any other documentation.

· Item 8 lists (1) the task *duration* or time needed to obtain each deliverable in item 7b, (2) associated risks, and (3) precedence relationships with other tasks, regardless of where they are located in the WBS that must be completed before work can begin. Preferably the task duration is determined in an *integrative project planning* (IPP) session by the project participants who will be responsible for performing the task. One of the many reasons that projects are delayed is that the project champion, market manager, or level 2 WBS branch manager (team leader) sets the duration of a level 4 or level 5 task. The probability is high that these managers do not know how to perform the level 4 task or performed it 10 years earlier using different development techniques or technology. What results is a highly optimistic schedule that has very little probability for success.

We recommend generating three possible durations: (1) an *optimistic* duration t_o, which is the time that the task would take if all goes well; (2) a pessimistic duration t_p, which is the time that the task would take if everything goes poorly; and (3) a *most likely* duration t_m, which is the duration that you would expect on the basis of your experience. Set t_m midway between t_o, and t_p initially, and move it toward t_o or t_p, depending on your expectation of finishing earlier or later than this midpoint. Risk is determined by the spread $(t_p - t_o)$. A low-risk task is one in which $(t_p - t_o)/t_p < 25\%$, a medium-risk task is one in which $25\% < (t_p - t_o)/t_p < 50\%$, and a high-risk task is one in which $(t_p - t_o)/t_p > 50\%$. Considered that these numbers are only guidelines, as actual numbers will depend on pervious experience performing similar tasks and some familiarity with the project's technology. Chapter 7 presents an in-depth discussion of the implications of item 8 on scheduling. The schedule presented with item 8 depends on a certain number and quality of personnel.

· Item 9, *personnel*, lists the numbers of project specialists needed and their required qualifications. Using this list, project management is able to compute the cost to complete the individual tasks, since each member of the project team is paid depending on the number of hours worked on the project.

· Item 10 helps to ascertain the project's total costs and prevent delays by ordering all *equipment* up front.

· Items 11 and 12 indicate the *meeting* and *report schedules.*

When reviewed and agreed to by the functional team leader and immediate coach, the *work package* gives both project and functional management a complete and documented description of the tasks to be performed. Once completed, the specialists assigned the work packages must sign them, and the project managers should make them available for easy reference in an on-line database in the project library. In summary, partition each element of the WBS until a meaningful work package can be prepared. Number reporting levels beginning at the total program or project level, and continue until the lowest subdivision in this progression, the end item or work package, is completed.

The integration of the project's organizational structure with the WBS occurs at the work-package level. This point of integration of work effort and organizational accountability is generally assigned a *cost accounting code* that project management can use as a reference point to track and report expenditures. Since the work packages contain information regarding cost, schedule, and technical performance requirements, functional and project managers will be able to monitor the project budget and control the overall effect of anticipated overruns on the project as the tasks are completed.

5.5 CONCLUSIONS

In this chapter we have equipped project managers with the necessary navigation aids to chart a project's course. Having brought on board a crew of technical experts and administrative specialists by communicating to them the project's objectives, the next step in executing the project plan is to build a schedule. In Chapters 6 and 7 we prepare you for this feat.

BIBLIOGRAPHY

Dor-Ner, Z., and W. Scheller, *Columbus and the Age of Discovery.* New York: William Morrow and Company, WGBH Educational Foundation, and Zvi Dor-Ner, 1991.

Project Management Institute Standards Committee, *A Guide to the Project Management Body of Knowledge.* Upper Darby, PA: Project Management Institute, 1996.

EXERCISES

5.1. Explain how your project team would use a project network to:

(a) Model the sequence of day-to-day project activities

(b) Review performance.

(c) Perform operational planning.

5.2. Your development team has just shared their detailed design estimates with the project team, and the project now faces a three-month delay. The team estimates that critical activities cannot be completed without jeopardizing the originally scheduled market introduction date.

(a) Consider and list possible options

(b) Discuss their consequences

5.3. One of the engineering groups on the project team is trying to perfect its subsystem beyond what the technical specifications require. Explain how to handle the situation if the subsystem is:

(a) On the critical path

(b) Not on the critical path

5.4. Engineers and managers often state that they cannot accurately estimate the cost of an R&D project since they have not done the work.

(a) What impact does such a statement have on the project's master budget?

(b) What techniques can be used to estimate the cost of a project more accurately?

5.5. On many projects only the project manager and a few functional managers know the true schedule and have a copy of the project plan. Discuss the disadvantages and advantages of such a procedure.

5.6. A work breakdown structure can be prepared for home projects as well as corporate projects. Consider planning a four-week trip to the American Southwest. Design a WBS having no more than five levels.

CASE STUDY: IRSS DEVELOPMENT PROJECT*

What had been perceived some nine months ago as an excellent product for launching Lutech Systems into the medical communications market was clearly in jeopardy. Ron Lewis, vice president of Lutech's newly created Medical Diagnostic Planning Division, studied closely the results of the recently commissioned project. Again and again, he asked himself: How in the world has this team of qualified designers performed so poorly? What had gone awry to cause a six-month slippage? Could I now trust their projections to resolve the identified problems? As Ron pondered the report, he realized that whatever plan he implemented to correct the situation would require the support of all his technical managers, marketing, and the respective managers from manufacturing. To keep the project viable, he'd need to improve the interfaces between these organizations, as the contribution of each was paramount to the success of Lutech's new endeavor.

What Went Wrong?

What had been conceived of in the laboratories as a single application program for a larger software system eventually became more complex in direct assessment of the market potential.

Over the past decade, medical electronics had advanced diagnostic medicine in the United States and abroad. Numbering among the modern marvels is sophisticated equipment that displays magnetic images of the human body's changing chemistry as well as ultrasonic scanners and infrared sensors that use sound waves and thermography to provide pictures of vital human organs. With these advances in medical diagnostic imaging, there was a push for innovative systems that would make possible the storage and retrieval of patient data, general medical information, and image archiving. Over the next several years it was anticipated that many private teaching hospitals and community clinics would purchase several of these anatomical machines, resulting in a growing need for more integrated and comprehensive interface systems. There would be an immediate, pent-up demand for these systems. However, given the current climate of containment of medical expenditures, at least in the United States, pricing

---

*This study was written for class discussion purposes only and was originally published in 1989. It is not intended to reflect management practices or products of any particular company organization. Any similarity to nascent companies since its original publication is purely coincidental.

would have to be sensitive to the demands of managed care institutions. So the decision to enter this new market required careful consideration. When Lutech's executives finally concurred with the market assessment, they readily set aside funds for the following fiscal year for the communication program's development. In fact, they went so far as to carve out a Medical Diagnostic Planning Division at Lutech where the R&D project was staffed. The organization had recruited talent and expertise from the country's top graduate engineering schools. All this was done in anticipation of launching several prosperous endeavours within the next 18 months.

But it soon became apparent to the information retrieval and storage system (IRSS) project auditors that the new developers were inexperienced program planners. The communication program was described to them as having the complexity of an operating system. The program would direct operations of the many hardware devices. However, concerns over the program's architecture and the security of the stored information, if it were to reside on any number of servers operating on the clinics' local and wide area networks, were still being debated.

The project's preliminary plan included only a broad definition of the program's capabilities, the primary failing being that it did not outline controlling specifications for testing software protocols, user interfaces, system performance, and recovery time in the event of a failure. A few key marketers wrote the plan. It became very apparent when reading through the preliminary plan that this group did not understand the complexities of the software system requirements or how the environments would need to be configured in the medical centers to manage the data infrastructure.

The R&D plan was also inadequate. It covered only the initial design, stopping short of the necessary system verification steps that would be required to debug the system. Also lacking were detailed plans for writing the program and installation documentation concurrently, although the preliminary project budget included funding for two technical writers for the final stages of IRSS development. Nowhere in the plan was there reference to beta test or qualification sites where the system's design would be installed and tested.

Up until now, the project had been operating in a matrix organization, where functional team leaders had retained responsibility for their own budgets and project resources. To date, there was no unified effort to staff IRSS as a single project or to dedicate full-time resources to it.

Project Organization

Although the number of specialists working on the project spanned only three departments in the new Medical Diagnostic Planning Division, project specialists also needed to interface with five or six other dispersed disciplines throughout the company. Furthermore, the manufacturing site was remote to the development and technical documentation sites. Even Lutech's long-range support strategy for the system, which was to distribute the maintenance activities for the system offshore, means that resources associated with this aspect of the project were located in several countries that operated in completely different time zones from that at company headquarters. As such, the separate work groups had proceeded independently, with no central management directing realization of the system interfaces or assignment hand-offs. Hence there was a tendency to overcommit and a great probability of error. To make matters worse, the development organizations would fail to provide what customers requested. The absence of a project management team to coordinate and integrate all the R&D and process activities now required by Lutech to maintain its international quality assurance registration resulted in conflicts regarding schedules and project deliverables. Management's expectations remained unclear in the minds of developers and turf issues dominated any attempt to communicate with them. In a very real sense, the IRSS work effort was not a "project"; nor would it ever become a true commercial venture until the lack of direction among the multidisciplined work efforts was rectified.

Questions

1. In your opinion, will Lutech's present organizational structure for the IRSS project enable the vice president to meet his market strategy?
2. What would you recommend to the new vice president as a course of action for clarifying project objectives?
3. What major hurdles face the development team?
4. Name at least four reasons why these top-notch designers might not succeed in developing the IRSS.
5. How have communication barriers among the interfacing project groups contributed to the difficulties cited in the audit report?

The Vice President's Plan

Ron considered carefully whether to salvage or abandon the project, and after much deliberation, determined that IRSS should be restructured in ways that would ensure commitment. Without a doubt, reorganizing the project and establishing new policies for meeting market demands were key actions that he needed to take. Armed with the auditor's report, Ron created a fully staffed core team of qualified team leaders who represented each of the functional disciplines necessary to realize the project. He charged these personnel with the responsibility of planning the project and overseeing the activities of the various organizations they represented. Then he authorized the direct transfer of required funds from the functional departments involved in the work to the core team. To improve communications, he named a project manager to act as the interface between the core team, the designers assigned to develop the medical information system, and each of the several technical support groups. Ron also requested that the project manager keep him informed on a regular basis of progress, problems, and the core team's examination of all alternative courses of action to correct problems. Ron pledged to reward progress and encouraged candid reporting in an attempt to ensure that the core team would attend to the quality of their problem analysis and decision making.

Directions to the Project Team

The morning kickoff meeting of the new IRSS project management team with the Medical Diagnostic Planning Division vice president was the first of many. Ron Lewis insisted that his new managers be informed, from the project outset, of the company's general thrust in the biomedical high-technology market, as well as understand the importance of coordinating tight budget and development schedules. Because of the complexity of the technical aspects of the project, they also would have to closely monitor the phased execution and testing of the product's design.

The new IRSS management team was made up of representatives from Lutech's marketing and manufacturing divisions and a core team of first-line managers who represented each of the project's technical disciplines. The project manager, Bob Anderson, was a seasoned systems engineer with expertise in digital image processing. Ron Lewis hand-selected Bob for the position after carefully reviewing several other qualified applicants. Bob had been with Lutech Systems for seven years and served as a lead systems analyst on a number of development projects under contract with the U.S. Department of Navy. Hence Bob knew how to run a project. Given full reign to organize, plan, and control a project, it was anticipated that he would succeed in doing so for the Medical Diagnostic Planning Division. In fact, the very first thing that Bob did was to organize the project structure and create the framework of an overall technical and management plan, the details of which the core team would determine together.

A Project Structure Emerges

No one had promised Bob that his job as the IRSS project manager would be easy. As he and the core team set out to determine the project's objectives, they discovered several additional market requests for functionality. It appeared that each of the new requirements was necessary to meet certified standards in key overseas markets. For nearly three weeks, they met at length to size and to estimate the extent of the requests for new functionality and customer support, which included the following:

- Turnkey solution for a professionally installed and maintained system
- Retrieval and exchange of patient information stored in other data management systems
- Special interface for manual data entry for clinical observations of patients
- Secure data links

To validate their finally agreed upon estimates, the core team invited the team leaders of the system design, engineering, and marketing groups assigned to the project to participate in an IPP session where they analyzed the requirements and aligned their schedules to a common set of project objectives. They began by breaking up the work into manageable packages that were to be assigned to individual technical specialists. Anticipating a strong early demand for the product as advised by marketing, they scheduled a phased release of the image-retrieval storage system, contracting in advance with several medical research institutions that would serve as qualification sites to test the system. Now it was necessary to acquire consensus concerning individual technical milestones.

The tracking of project-level milestones became the responsibility of Shara Jefferson, who joined the R&D team as Bob's deputy project manager. She also assumed responsibility for administering the work packages and for organizing the project office. Hence the finally agreed-on schedules were displayed, tracked, and updated. Shara also was required to report to Bob and the core team on the activities of the design organizations. To monitor progress, she would circulate among the design groups asking questions, coordinating interfaces, and determining progress on the work packages. Through her hard work she gradually gained the trust and respect of the team leaders. She also became a sounding board for several of the engineers. She would listen to their problems and advise them on all administrative project details.

Early in the process, Shara appraised Bob that the initial design specifications were inadequate. In fact, different work groups had initiated two approaches, and their developers were proceeding unaware of their duplicated efforts. Shara captured this problem in her first development report, which she and Bob sent to Ron Lewis for review. Later that week, Ron called Bob and Shara to his office to discuss their findings.

> A lot of good has been done here. But we can't have these duplicate efforts holding up progress. Even though some of the design groups may resent this, a much more formal structure must be instituted to manage design changes and control the work effort. Executing a system this complex depends entirely on how well we coordinate its design every step of the way, from its robustness in the conception phase to user acceptance when deployed in the marketplace. I'd say our job is clear. We need to institute a mechanism whereby all team members are informed of others' activities and their difficulties or problems, whether technical or administrative. Have them meet face to face and hash out their differences, if need be, on a weekly basis. I want each of these groups to learn from the others' mistakes and resolve their differences as a team solving the same problems. As project managers, you should run the meetings and keep lists of the unresolved issues. Demand weekly status reports from anyone having any responsibilities for work on this project. I want to be kept informed of any impending delays before they happen!

Questions

6. What recommendations would you make to the project management team for overcoming the many technical barriers facing them?
7. Why is it important to synchronize the various design and development activities?
8. What mechanism would you enact to provide guidelines for documenting designs and driving their execution?

Documenting Change

Shortly after their meeting with Ron Lewis, Bob and Shara learned that R&D was undecided on the appropriate hardware needed to implement the new software requirements brought to them from the core team and to act as the interface with the many diagnostic devices. By the time the project management team staged the project's first technical design review, the hardware team had, in fact, begun to address the new requirements by planning to operationalize processing intervals. To minimize signal-to-noise ratio effects on all data retrieval, these signals would need to pass through low-pass filters and amplifiers. To accomplish the interface function, an application-specific, reduced-instruction set microprocessor would be designed. Although a most feasible solution to the hardware design problem, the software designers had not yet completed the system's architecture.

Of additional concern to the hardware team was the design of the cabinetry that would house the IRSS unit. Recommendations would also need to be made regarding placement of the unit in the hospitals and medical institutions when installed. The system's circuitry would need to be shielded from the powerful magnetic fields of any magnetic-resonance imaging devices physically linked with the unit. Other technical issues surfaced concerning delays associated with synchronizing the collection of data from auxiliary storage from customer-maintained patient-data management systems. The clocks of all data-collection devices would need to be synchronized with the IRSS file servers to minimize the risk of inconveniencing physicians and media personnel relying on the data. Formats would be different for various regions of the world, and therefore localized specifications were needed to accomplish this.

The new requirements also escalated the need to resolve a previous debate about the need to include an object-oriented database at the core of the system to store patient data and retrieve medical records. The pros and cons of this type of architecture were debated at length, with arguments pointing to the need to document all the changes proposed to the original design specifications in a formal project archive. To improve communications among the various organizations, the project management team would need to maintain these records on-line.

At first, each group of designers and specialists working on the project resisted the idea of documenting every aspect of their work and making it available to team members in other organizations. But with persistent coaching, they agreed to let Bob and Shara implement the more formal structure on a trial basis. Shara and Bob would gather the core project and support teams for weekly discussions and publish the meeting agendas, minutes, and action items on-line, giving ready access to all concerned team members and functional team leaders.

To advance the project's execution, Shara concerned herself with rewriting the project's summary plan and tracking technical deliverables and milestone completions. She wrote and maintained the project's high-level requirements documentation, project description, and a detailed accounting of all the agreed-on hardware and software interfaces. The remaining sections of the project plan were to parallel the hierarchical structure of the work packages so that the various design groups were responsible for maintaining and updating the information. The actual technical specialists were free to write individual memoranda as needed regarding the specifics of their assignments. These notes might include reasons for rejecting a design, design assumptions, and special problems imposed by memory-layout or other features of the design.

Shara would continue to arrange and report on the results of the many design reviews and to update and track the associated documentation through the product's development and final testing. Also, her presence at these review meetings would help to assure that the integrity of the high-level design was not compromised in the effort to carry out these requirements. Finally, a formal sign-off of approved documentation for each project phase was instituted as a means of synchronizing the hardware and software development and testing processes as well as planning out the manufacturing intervals for the product.

Bob, meanwhile, focused the core team on broader aspects of process management of the IRSS product introduction. In particular, marketing realized the IRSS would sell more easily if a turnkey solution were offered to help customers set up and install the application. To facilitate early IRSS sales and demonstrate the system's value as a potential cost and time saver, Bob also arranged for a period of limited market introduction at a trial customer's clinic. Other systems currently offered by competitors required laborious, manual record keeping to track data stored in independently functioning units. The finally agreed-upon IRSS architecture would provide quick mass-storage retrieval and secure, rapid access to large mounts of data in these other platforms, all at optimal price points that would attract purchases from large government-sponsored research and teaching facilities as well as privately owned institutions.

Together the IRSS core management and project design and development teams proved that they could operate a project using a flexible, yet formal methodology that accommodated their different work styles. The formal database of project documentation served the team members directly, as the project information was accessible to all involved and was complete and accurate. The frequent and timely meetings helped to open a dialogue among the specialists involved. By the time the development team was ready to cut over their designs for manufacture, Shara had implemented a system whereby the project documentation was accessible on-line at the factory, along with the updated and baseline schedules of the IRSS development milestones.

Everyone associated with the project was extremely anxious to perform well. The key had been to get all parties working on the same problems and to establish a climate that stressed open and honest communication in the reconciliation of differences. The managerial actions having the greatest impact on the project's design and schedule took this into account.

Questions

9. What information did R&D learn from attending the meetings with the core and project support teams?

10. Did this information alter how the project would be brought to market?

11. Explain the impact of documenting changes to the program's technical design.

12. Describe how the project management team went about its task of facilitating the resolution of undetermined design issues.

13. Why was it essential that the many engineering interfaces be formally integrated into one project plan that was agreed on by all responsible parties?

14. Discuss the critical role played by the vice president in expediting the project schedule and mediating problems that arose from the conflicting design and marketing philosophies.

6

SCHEDULING PROJECT ACTIVITIES

*No man ever wetted clay
and left it, as if there would
be bricks by chance
and fortune.*

PLUTARCH, OF MAN'S
PROGRESS IN VIRTUE

6.1 INTRODUCTION

Looking back over the centuries, the scheduling and expediting of the engineering feats of the ancient world remain a legacy left by the master builders and architects for modern project managers to ponder. Since there remain few written records chronicling the activities surrounding the construction of these treasures, we can only surmise that their survival centuries later, against the tests of time, is due to the persistence and genius of the leaders of these civilizations to manage their fortunes. Each, indeed, was a crowning glory of its age: Olmec Mesoamerica's Teotihuacan temples erected in 1000 B.C.; Aksum, Ethiopia's granite monolith, which stood 110 feet above ground between A.D. 300 and 700.; the great Giza pyramids, built over 4000 years ago in Egypt; China's Great Wall and multistoried watch towers, constructed as barriers to marauding horsemen; the great vaulted roofs of Rome's Pantheon, completed in A.D. 120.; Celtic Britain's Maiden Castle, whose amazing labyrinths protected the fortress against sling-wielding aggressors.

The fabrication of these silent ruins was influenced by the same constraints of time and quality that projects face today. Often, quality suffered due to graft, frequent fires, and the turmoil of construction. Cost, however, was less of a concern. The ruling principals had at their disposal legions of available laborers who were forced to toil with cranes and pulleys or drag blocks of stone that exceeded 200 tons up long earthen ramps to accomplish the ambitious technological feats. In ancient Egypt, for instance, every household was required to provide either food or labor for projects' work crews. Certainly, time was of the essence, although it was easily traded off against the nonexpense of captive work gangs who received little or no wages. Moreover, each ruling monarch, pharaoh, or emperor constrained the schedule by requiring that the project be completed during his or her lifetime.

In this chapter, we describe the principal scheduling techniques that we have found to be effective for planning and later in tracking and controlling project performance. As the professional builders of the ancient construction projects learned so many years ago, project performance depends on how well *each* activity is specified in terms of time and resources, such as personnel, equipment, and facilities. We also explain how to illustrate in a project schedule the interdependent relationships of deliverables, such as test completions, reports, and equipment needed for the project to proceed. In addition, we present strategies that enable a team of programmers and system testers to prepare a schedule that accurately represents the cost of developing modern software systems, a process that many software project teams wrestle with because of the uncertainty of code development.

6.2 TYPES OF SCHEDULES

The entrepreneurial Benjamin Franklin's advice to a young tradesman 200 years ago, "time is money, time is money," is certainly an appropriate maxim for today's world. The expression captures the essence of how modern finance measures the worth of a project by specifying the time value of an investment as it corresponds to the time needed and costs required to complete specific deliverables. To accomplish this, project managers must be able to predict, with a strong sense of certainty, project performance requirements (time needed to perform the work), the cost to complete these activities (budget), and the types of specialists (project personnel) who will be assigned project tasks. In other words, basically three types of schedules are needed to manage a project: *performance, cost,* and *personnel* schedules.

Performance schedules inform project participants when each task is expected to begin and end, and therefore, when each project specialist is expected to start work on each task, how many hours per day to allocate to each task, and when a person's services are no longer needed. Personnel schedules also illustrate how the individual specialists' time is allocated, which permits project management to adjust or reassign the workloads among the specialists to shorten the project or reduce costs. *Cost schedules* inform the project team how much money has been allocated and spent for each task as a function of time. In this way, adequate money can be requested prior to starting a project. All too often, projects in the early planning phases are underfunded since a decision to continue the project must be reached before the associated expenses of bringing a product or service to market are budgeted. This means that the project team must first prove the technological or market concept before additional monies will be allocated to realize the project plan. At other times, the information used to make the original budget requests is incomplete and the work cannot be completed for the cost specified in the project plan. When no additional funds are available to finish the work, it must proceed underfunded.

In general, all three schedules are required as part of any project plan. In fact, they represent the four parameters that define and constrain projects: (1) performance or technical specifications, (2) time or schedule, (3) cost, and (4) quality. One begins preparing each of these schedules from the information contained in the project *work breakdown structure* (WBS). From the information recorded in this plan, the project management team can determine the number and levels of the tasks to be performed, the associated *milestones,* the *precedence relationships* among the tasks, and the resources required to complete them.

The level of detail reported in each schedule will depend on who will receive and act on the information it contains. For instance, is the schedule prepared for a team leader of the programming group, the cross-functional leadership team responsible for overseeing the realization of the software application project, or the programmers themselves? Although each of these individuals or groups requires a schedule, each does not require the same schedule. Typically, the various levels of functional and team management require different levels of schedule detail. A hardware engineer or software developer might describe the weekly progress report in the following way:

> I give my engineering team leader a progress report on my weekly accomplishments. The leader then summarizes the reports received from the entire development group, which are then given to our project management team. These project specialists subsequently check to see how each of our jobs is progressing by comparing our present situation to the report. They receive seven reports each week, which they summarize into a one-page report to share with the project manager and the larger project team.

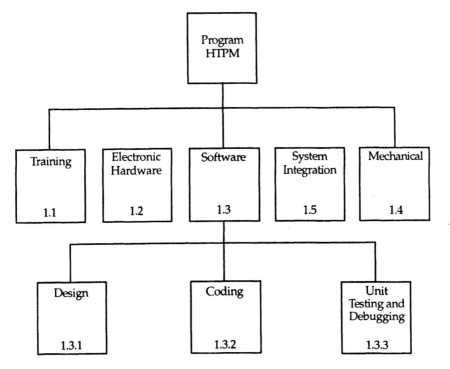

FIGURE 6.1

Partial WBS.

Each summary of a group of reports, and subsequent passing forward of the report, parallels the passage from one level of the WBS to the next. Consider the partial WBS shown in Figure 6.1. If in our case study, Ragis Kanna, the technical team supervisor, is responsible for realizing the software aspect of the HTPM project, he will be extremely interested in viewing the details of the schedules developed by Sue Kaplan (1.3.1), lead systems designer; Roger Vella (1.3.2), programming team leader; and Sandra Li (1.3.3), system test coordinator. However, Carol Nolan, project manager of the integrated product offer, is not interested in knowing all the recorded details of the hardware and software development schedules. Carol wants summary reports from Ragis (1.3) and the other supervisors (1.1 through 1.4) to ascertain the schedule and project's performance. Project teams can obtain the schedules we discuss in this chapter for any of the WBS levels. From our experience, it's best to learn the reason for the schedule and for whom it is intended before preparing one.

6.3 WHY SCHEDULE?

6.3.1 Scheduling to Meet Required Constraints

Through the centuries, successful project management has meant accomplishment of performance specifications on or before a stated deadline and within the budget allocation. However, the ability to achieve on-time delivery within budget is more difficult now than it was in the past. Today's projects are constrained by trying to accomplish what may appear to be simultaneously conflicting objectives. It is rare to encounter a team that hasn't been asked to shorten the end date of the project so that it can come to market ahead of the original performance estimate by several months, yet still cost no more to complete than originally budgeted.

Perhaps this problem stems from the project team's inability in the upfront planning phase to make accurate predictions when scoping out the work. Often, projects do not get started until a decision is finalized after a product or service concept is evaluated. By the time that market intelligence signals the need to enter a new market, several months of effort has been lost. To make up for the delay and enter the market on time, the project schedule must be accelerated. Similarly, very optimistic estimates cannot be achieved without adding additional time and resources to the project schedule. Unfortunately, when good scheduling techniques fail to be employed, the project team will realize too near the end that the planned completion date cannot be achieved. To meet the required deadline, the project specifications may be weakened, often by removing features and functionality.

When any of the project's time, cost, or technical performance specifications dimensions is altered, the other parameters must usually change as well. For example, if our project is *time limited* and we see that our initial schedules cannot meet the specified delivery date, we might conduct a series of *what-if analyses*. Their purpose is to simulate different design strategies and other precedence relationships between tasks to obtain a performance schedule consonant with the time constraint. If our project is *cost limited*, we might attempt to reallocate resources and then adjust the task durations to keep the cost within the prescribed budget.

It is very important for project managers to fully understand the impact of altering any of the various dimensions of performance specifications, cost, and time on the overall quality of the deliverables brought to market. For this reason, we consider quality, or meeting customers' expectations, the fourth element of the quadruple constraint. Today all four factors cannot be accomplished simultaneously without practicing good management, planning, and scheduling techniques.

6.3.2 Scheduling for Control, Costs, and Performance

To keep the elements of the quadruple constraint in balance, plans and schedules should be used to track and control progress starting immediately after the project team has formed and project work begins. The initial *performance schedule*, prepared in the early concept and definition project phases, becomes the project baseline and is compared biweekly to actual project performance. Slippage and the corrective actions that the team takes to put the project back on track should be noted where appropriate. The *personnel schedule* can be used to notify the project manager, who can be reassigned from one task to another to improve overall performance. The *cost schedule* provides a good indication of how much the project actually costs, compared to how much money is budgeted, or set aside, and how much money is actually being spent. This permits the project team to predict how much extra funding might be required to complete the project on time, or how long it could take were the project to continue at a given level of funding. This cost management process, known as *earned value analysis* (EVA), is mandated for all U.S. Department of Defense contracts. The EVA management technique is also used extensively by commercial project efforts and is discussed in detail in Chapter 10.

6.4 PERFORMANCE SCHEDULES

Basically there are three types of performance schedules: (1) arrow and precedence networks, (2) CPM and PERT networks, and (3) simple milestone, bar, or Gantt charts. While working with each of these performance schedules, we have discovered some inherent weaknesses and many advantages for using them, which we will share. Each technique can help a project manager to measure progress among the various functional groups that comprise the project team and control deviations across project phases.

6.4.1 Precedence and Arrow Networks

Consider a simple project ABC that consists of the activities shown in Table 6.1. The first question on every team leader's mind is: How long will this project take? The next question usually asked is: When will the project team members be needed? To answer these two questions and more, we can prepare a *precedence* or *arrow network*. A network is composed of *events* or *nodes* that represent activities interconnected by lines and arrows.

TABLE 6.1 WORK BREAKDOWN FOR PROJECT ABC

Proposed Activity	Activity Description	Proposed Activity Duration (months)	Precedence
A	Design, code, and test software	9	Start of project
B	Design, develop, and test hardware	3	Start of project
C	Integrate hardware and software	2	A, B
D	Design and fabricate chassis	5	B
E	Integrate and test system	4	C, D
F	Deliver system	2	E

The *precedence diagramming method* (PDM) is well suited for representing the uncertainly of task durations in research and development projects. In a *precedence* or *activity-on-node* (AON) network, activities are arranged in sequences that consider their relationship to one another and the project objectives. Each activity, represented by a *rectangular node*, is connected to the node that precedes (or succeeds) it in time. The first rectangle is usually considered the *start node,* and the last rectangle, the *finish node.*

Computer programs will place an event number inside the node, since events often are not labeled by name but by their start and finish numbers. The higher-event-number activity always follows a lower-numbered event. As we explain in Section 6.4.2, computer-generated networks will often include other information calculated about the activity, such as its *early start* and *late start* dates and associated *slack* or *float.*

To learn the basic drawing conventions associated with the *precedence diagram*, review Figure 6.2 which represents the precedence diagram for the project described in Table 6.1. Note that to obtain the diagram in Figure 6.2, we drew a rectangle to represent each activity A through F, as well as rectangles to represent start and finish nodes. These rectangular boxes are then connected by directional lines, shown as arrows, to indicate the precedence relationships.

The utility of the precedence diagram lies in its ability to simplify calculation of the project's duration. Because some tasks are performed in parallel, the duration is less than the sum of the times it takes to complete

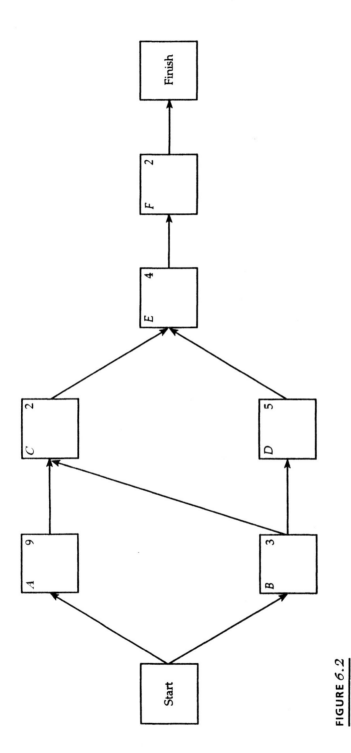

FIGURE 6.2

Precedence diagram for Table 6.1.

179

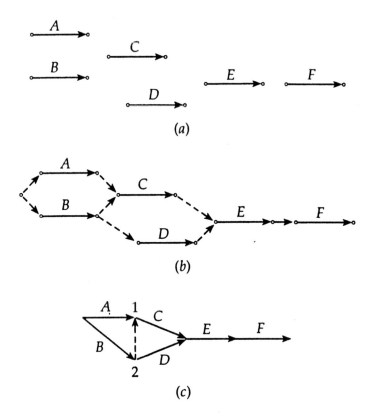

FIGURE 6.3

(a) First step in drawing the AOA; (b) introducing the dummy task; (c) AOA precedence diagram.

each activity A through F. Indeed, in this problem, the project's duration is the time it takes to finish tasks A, C, E, and F; B, C, E, and F; or B, D, E, and F. The durations of these tasks are: ACEF, 17 months; BCEF, 11 months; and BDEF, 14 months. Because the project cannot be completed until each activity is completed, the project's duration is determined by activities A, C, E, and F and is 17 months. Note that in 17 months all activities will be completed, and as a result of the precedence relationships, we cannot complete the project in less than 17 months.

An alternative to drawing a precedence diagram is the *arrow diagramming method* (ADM) or *activity-on-arrow* (AOA) technique. This approach is used more often in the construction industry, where the duration of project activities is more certain. In the AOA networking technique, the activity label is placed on the arrow connecting the nodes. Although the technique is used considerably, we caution you that it is a more complicated tool than the *activity-on-node* (AON) approach discussed earlier.

To illustrate AOA, consider the problem shown in Table 6.1. When using AOA, each activity is shown as an arrow with a node at each end. The arrow represents the activity, not the node. The first step in drawing the AOA is shown in Figure 6.3a. Note that an arrow is drawn to represent each task. Next, in Figure 6.3b, the solid arrows are connected by dashed arrows, following the precedence relationship given in Table 6.1. The dashed arrows are called *dummy tasks* and have zero duration. The final AOA diagram, shown in Figure 6.3c, contains a single dummy task. All other dummy tasks have been removed without affecting the required precedence relationship. If we remove the remaining dashed arrow, either by erasing the arrow or by combining nodes 1 and 2, the precedence relationship is altered. This first step is what complicates use of the AOA. If we stop at Figure 6.3b, the diagram is overly complicated. If we go to Figure 6.3c, the chance of error increases. We therefore recommend using AON for manual preparation of a precedence diagram.

6.4.2 Determining the Critical Path

Path *ACEF* (see Figure 6.2), which determines the project's duration, is called the *critical path*. If any activity on this path (i.e., *A, C, E,* or *F)* is delayed, the entire project will be delayed. The critical path indicates the shortest time in which the project activities can be completed. Too often, the duration of the project, as determined by the precedence diagram, does not meet the projected product entry window. To comply with this constraint, the project team must review the precedence diagram and shorten one or more of the activities on the critical path so it is finished within the 15-month duration set by marketing. Fortunately, the specialists working on the project given in Figure 6.2 have available several methods to accomplish marketing's request.

Let's look at how we might shorten the duration of a time-oriented schedule:

1. Examine whether an activity on the critical path could be started sooner. For example, let us see if activity *C* in Figure 6.2 could be started earlier. We might be able to shorten the project by integrating the hardware and software while performing the software testing and debugging phase of activity *A*. It is often possible to subdivide an activity, such as *A*, into two or more activities, for example, into A_1, A_2, and A_3, so that *C* could be started after A_1. However, *C* must start after *B* and the software coding have

both been completed. If we made these changes in the precedence diagram, Figure 6.2 could be redrawn as illustrated in Figure 6.4. In Figure 6.4, we see that the critical path would be now $A_1A_2A_3EF$ and the project duration, 15 months. We have met the required market window.

2. Determine whether a different technology could be employed that could possibly shorten the time of a given activity. For example, in project *ABC* perhaps some of the software could be replaced by hardware, for instance, programmable array logic (PAL). Because hardware development is not on the critical path, increasing *B* by a modest amount would not affect the project duration, and decreasing *A* would actually shorten the project. Of course, one must check the precedence diagram to see if the critical path has been changed.

3. Determine whether increasing the number of programmers in *A* would reduce its duration. We discuss this technique in detail in Chapter 7, but at this point, note that in this example the procedure is of limited value since doubling the number of personnel assigned to an activity rarely halves its duration. At times, adding staff late in the project actually lengthens the duration of an activity, due to the increase in the amount of communications required to bring new personnel onboard.

6.4.3 Using the Critical Path Method to Forecast Completions

In the example shown in Figure 6.2, the critical path can be found by determining every possible path and then calculating which path has the longest duration. That path is, of course, critical. However, commercial development and production projects have 100 to 1000 or more such paths, so determining the critical path using this method would be extremely difficult and time consuming. Further, because we calculate the project duration from the critical path to obtain an indication of what tasks should be shortened or what precedence relations altered, we generally recompute the critical path several times before finalizing the precedence relations, resources per task, conceptual design, and so on. Although recomputing the critical path several times in the tedious manner described above is not impossible, it is so unpleasant a chore that we have found an alternative method. This procedure employs a dynamic programming algorithm, which we illustrate in Figure 6.4.

We see from the figure that prior to starting task *C* we must complete tasks A_1 and A_2 and then complete task *B*, as tasks A_1 and A_2 take six months to complete whereas task *B* takes only three. Task *C* can start only

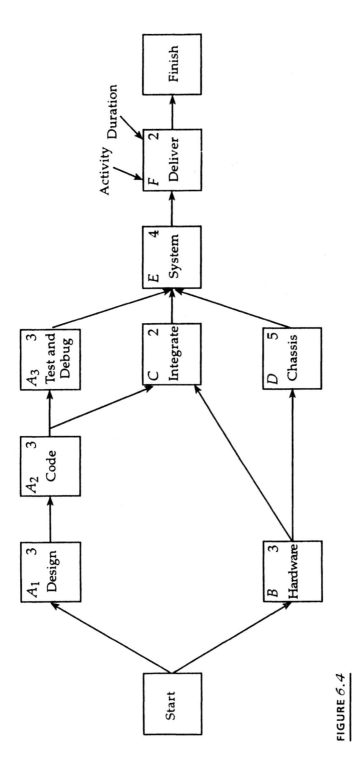

FIGURE 6.4

Precedence diagram for project ABC after splitting activity A.

183

after six months has elapsed. Tasks A_1 and A_2 are therefore critical in regard to C. For all intents and purposes, we can then ignore or cut the path from task B to task C when calculating the critical path. Next, examine the tasks in Figure 6.4 that must be completed prior to starting task E. They are tasks $A_1A_2A_3$, A_1A_2C, and BD. Note that we have omitted BC, as we have already shown that BC is not a critical path. Since $A_1A_2A_3$ is the critical (longest) path to E, we can cut paths CE and DE in determining the critical path. There is now only one path left: $A_1A_2A_3EF$.

It is important to note that we found the critical path by computing the longest path through the network. The algorithm to compute the critical path in a performance schedule was developed by the DuPont Company in the 1960s. Known as the *critical path method* (CPM), it is the most widely employed mathematical analysis technique used today by computer-based systems to calculate the critical path. As we illustrate in the following example, the CPM algorithm analyzes the early- and late-start and finish dates for each activity sequence and calculates the slack or float between predecessor activities to determine which sequences are the least flexible.

Example 6.1. A new office automation appliance intended to revolutionize productivity was designed and built using the work breakdown tasks outlined in Table 6.2. Prepare a precedence diagram and calculate the critical path.

SOLUTION: The precedence diagram is shown in Figure 6.5. To calculate the critical path, we first compute the elapsed time to complete tasks 100 through 220. The time is $T_1 = 26$ weeks. We next compute the time to complete the tasks 100 through 410. This time interval is $T_2 = 22$ weeks. Because T_1 is greater than T_2, tasks 400 and 410 *cannot* lie along the critical path. This test can also be seen by comparing the durations of tasks 210 and 220 with tasks 400 and 410. Since tasks 210 and 220 take eight weeks compared to tasks 400 and 410, which take four weeks, tasks 400 and 410 are seen not to be critical. However, task 400 can be started as early as week 9 or as late as week 18 and task 410 as early as week 11 or as late as week 20. This ability to start tasks 400 and 410 at any time over a 9-week period is called *slack* or *float*.

Next, we compute the duration of tasks 100 through 510 and compare this result to the time interval needed to complete tasks 300, 310, and 320. The result is:

TABLE 6.2 WORK BREAKDOWN OF AN OFFICE SYSTEM PROJECT

Task		Duration (weeks)	Precedence
100	Hardware design	3	Start of project
110	Breadboard construction	6	100
120	Test breadboard and correct	6	110
200	Design prototype	3	120
210	Develop prototype	4	200
220	Test prototype and correct	4	210
300	Design software	3	Start of project
310	Code all software	3	300
320	Test and debug software	3	310
400	Design mechanical system	2	200, 320
410	Fabricate mechanical system	2	400
500	Integrate electronics and mechanical systems	2	220, 410
510	Test and correct	2	500
600	Integrate software and hardware	1	510, 320
610	Test and correct	1	600
700	Prepare initial hardware documentation	1	120
710	Upgrade documentation	1	700, 320
720	Prepare final documentation	1	610, 710
800	Forward to customer for beta testing	1	720
	Finish	—	800

Tasks	Duration	Possible Critical Path
100–220 and 500–510	26 + 2 + 2 = 30	Yes
300–320	3 + 3 + 3 = 9	No

Hence tasks 100 to 220, 500, and 510 still appear to be critical. The next node having two or more inputs is task 710. Comparing the durations of the tasks involved yields:

Tasks	Duration	Possible Critical Path
100–120 and 700	16	Yes
300–320	9	No

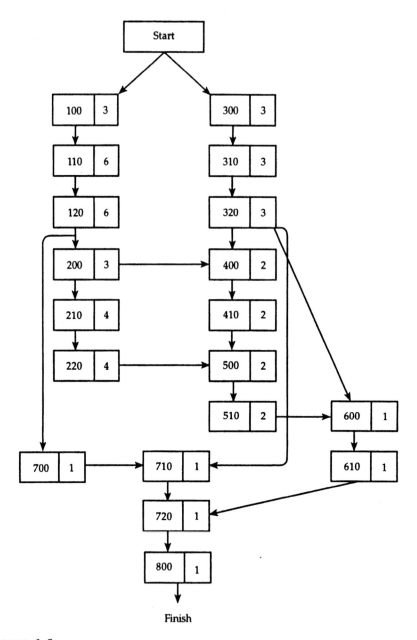

FIGURE 6.5

Precedence diagram for office system project.

Thus, since both paths leaving task 320 indicate noncritical paths, we are now sure that tasks 300, 310, and 320 are noncritical.

Entering node 720, we again have two possible critical paths:

Tasks	Duration	Possible Critical Path
100–220, 500–510, and 600–610	$26 + 4 + 2 = 32$	Yes
100–120, 700, and 710	$15 + 1 + 1 = 17$	No

The critical path is therefore comprised of tasks 100, 110, 120, 200, 210, 220, 500, 510, 600, 610, 720, and 800. The total project duration is 34 weeks.

Since management had intended to market the new office appliance within the next six months, the $8\frac{1}{2}$-month or 34-week duration is longer than desired. To comply with this request, the project team will need to investigate possible alternatives to the precedence relationships postulated. For example, the team may choose to reduce the duration by performing more of the tasks on the critical path in parallel. They might also investigate the possibility of adding personnel to tasks on the critical path to reduce the duration of these activities. Then they might attempt to determine whether any of the tasks on the critical path could be shortened, by employing a different design, better test equipment, or faster processors. As a last resort, the team may consider modifying the project's feature specifications, as it is quite possible that a minor change in the specifications could significantly shorten the total duration of the project.

Often, these procedures are sufficient to shorten a project as required to meet marketing goals or a customer's contractual needs. If, however, no modifications can be made to shorten the project adequately, this information must be communicated to the senior managers who financed the project. There's little benefit to the organization to begin a project that will be doomed to failure if not completed on time.

6.4.4 Scheduling for Uncertainty Using the Program Evaluation and Review Technique

The *program evaluation and review technique* (PERT) was introduced by the consulting firm Booz-Allen and Hamilton in the 1950s to satisfy the demands of the U.S. Navy Special Projects Office. PERT uses sequential network logic and a weighted, average-duration estimate of each activity to determine the *expected value* of a distributed mean of these activities. PERT estimations are best used to yield performance schedules for one-of-kind projects characterized by nonrepetitive tasks, such as software engineering and electronics production where there is a great deal of uncertainty associated with predicting how long it will take to carry out project tasks.

Most of today's computer-based information systems produce hybrid diagrams of performance schedules that represent both the CPM and PERT network techniques; some even use PERT-like estimates to calculate total project duration. The U.S. Navy's original specifications required the PERT (precedence) diagram to readily illustrate the project schedule of events as sequences of critical and subcritical paths. In addition, the tool was required to predict with some certainty the probability of meeting the scheduled estimates.

PERT continues to be used by project managers to include the risk and uncertainty inherent in every activity in the project schedule. PERT is a probabilistic estimate of the randomness of project events, and unlike the deterministic CPM technique, it can also be used to predict which of the scheduled tasks are likely to slip. This permits the project team to plan alternative and corrective actions to put the project back on track.

Most engineering projects are characterized by some degree of risk and uncertainty, whether they involve predicting the first fabrication yields of a new ASIC design, precisely estimating how long a novice programmer might take to code in a new and immature language, manufacturing patented hardware that advances the state of the art, or successfully compiling source code written by an internationally dispersed development team. The uncertainty associated with the time needed to complete the activities of research and development projects can be expressed mathematically if we consider the duration of each activity as a random variable. Hence the time required to complete the project's critical path, illustrated in the PERT diagram, can also be viewed as a random variable because this duration represents the sum of random variables.

Further, it is well known that the sum of random variables has a probability distribution that approaches the Gaussian probability distribution. This follows from the well-known *central limit theorem,* which requires that the number of random variables being summed be large and that each of the variables be of approximately the same order of magnitude. For project management applications, these conditions are adequately met when there are four or more activities on the critical path.

If we know the expected time t_i and the variances s_i^2 of each activity A_i on the critical path, we can calculate the *expected duration* T_E of the project:

$$T_E = \sum_{i=1}^{J} t_i \tag{6.1}$$

where there are J activities on the critical path. The *variance* S_E^2 of the project *duration time* is given by

$$S_E^2 = \sum_{i=1}^{J} s_i^2 \tag{6.2}$$

Let us define the project duration as D and define a parameter Z so that

$$Z = \frac{D - T_E}{S_E} \tag{6.3}$$

The parameter Z is the number of standard deviations by which the project duration exceeds the mean. The project duration D can then be found by using Table 6.3. For example, if the probability of completing a project is to be set at 90%, as shown in Table 6.3, $Z = 1.3$. Using Eq. (6.3), we see that the duration D_{90} corresponding to this 90% confidence level is

$$Z = 1.3 = \frac{D_{90} - T_E}{S_E} \tag{6.4}$$

Thus

$$D_{90} = T_E + 1.3S_E \tag{6.5}$$

Figure 6.6 illustrates the probability density curve, which corresponds to the cumulative distribution presented in Table 6.3. Figure 6.7 shows the probability of completing the project in time D (or less) and corresponds to the values given in Table 6.3.

We can use these figures to draw several conclusions that help estimate a project's duration:

1. If the duration of each activity in the project is given by its *expected value*, the project will be late one-half of the time. This is D_{50}, the estimated duration of the project with a 50% confidence level.

2. Since most engineers provide project management with optimistic estimates, we actually plan the schedule from the outset to be too late. When we schedule the project optimistically, it has less of a chance of succeeding than does a hand at casino blackjack!

TABLE 6.3   PROBABILITY OF MEETING THE COMPLETION DATE[a]

$$\text{Define: } Z = \frac{D - T_E}{S_E}$$

Z	Probability of Meeting Completion Date
3.0	0.999
2.8	0.997
2.6	0.995
2.4	0.992
2.2	0.986
2.0	0.977
1.8	0.964
1.6	0.945
1.4	0.919
1.2	0.885
1.0	0.841
0.8	0.788
0.6	0.726
0.4	0.655
0.2	0.579
0	0.5
−0.2	0.421
−0.4	0.345
−0.6	0.274
−0.8	0.212
−1.0	0.159
−1.2	0.115
−1.4	0.081
−1.6	0.055
−1.8	0.036
−2.0	0.023

[a]Depends on the number of standard deviations by which the proposed project duration exceeds the mean

3. When the risk and uncertainty in a project is small and the standard deviation S_E is much smaller than the expected time T_E, we can basically ignore S_E. In this case, the PERT calculations are not needed to schedule the project.

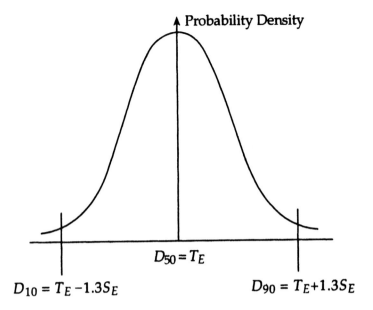

FIGURE 6.6
Gaussian probability density.

4. Once a project reaches its midpoint, the project team may stop calculating the expected time and variance of each activity, since the uncertainty inherent in these tasks will usually diminish, thereby decreasing s_i^2. When S_E has decreased sufficiently with respect to T_E, the schedule may now be represented using a standard precedence diagram rather than the PERT network.

6.4.5 Determining the Expected Time and Variance of an Activity

We can calculate the expected time t_i and variance of s_i^2 of an activity A_i if the probability density of the activity duration is known.

Uniform Density. Assuming that the probability density of the duration of an activity is uniform,

$$p(t) = \begin{cases} \dfrac{1}{t_H - t_L}; & t_L < t < t_H \\ 0; & \text{elsewhere} \end{cases} \tag{6.6}$$

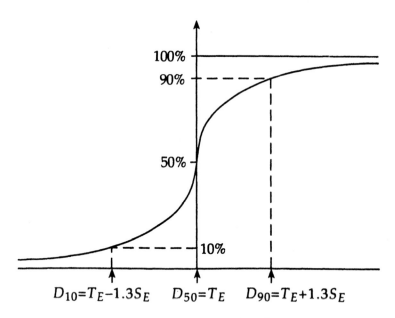

$D_{10}=T_E-1.3S_E$ $D_{50}=T_E$ $D_{90}=T_E+1.3S_E$

Project Time Duration, D

FIGURE 6.7

Gaussian distribution: probability of completing the project in a time interval, D.

then

$$t_i = \frac{t_H + t_L}{2} \tag{6.7a}$$

and

$$s_i^2 = \frac{(t_H - t_L)^2}{12} \tag{6.7b}$$

Statistics When First-Pass Duration Is T_0 with Probability P. Let's look at how the duration of an activity, such as fabricating an integrated-circuit (IC) chip, can be characterized if we note that the chip can be fabricated in a time T_0 with probability of success P. If the chip is flawed, an additional time T_0 is required. If on the third pass, the yield is also flawed, we must allow, once again, additional time T_0. We repeat this process until the chip is fabricated properly. Assuming that the probability of success remains equals to P, the expected time t_i to fabricate the chip successfully is

$$t_i = T_0P + 2T_0P(1 - P) + 3T_0P(1 - P)^2 + \cdots \tag{6.8}$$

where T_0P is the average time to complete the activity on the first try.

$2T_0 P(1 - P)$ represents the total time $2T_0$ times the probability of failures on the first try, $1 - P$ times the probability of succeeding on the second try, P; and so on.

Evaluating Eq. (6.7) yields

$$t_i = \frac{T_0}{P} \tag{6.9}$$

The second moment, $t^2 = (s_i)^2 + (t_i)^2$, is

$$t^2 = T_0^2 P + (2T_0)^2 P(1 - P) + (3T_0)^2 P(1 - P)^2 + \cdots \tag{6.10}$$

Evaluation of Eq. (6.10) yields

$$(s_i)^2 = \frac{T_0^2}{P^2}(1 - P) \tag{6.11}$$

Example 6.2. If $T_0 = 1$ month and $P = 0.5$, then

$$t_i = 2 \text{ months} \tag{6.12a}$$

and

$$s_i = \frac{T_0}{P}\sqrt{1 - P} = 2\sqrt{\tfrac{1}{2}} = 1.4 \text{ months} \tag{6.12b}$$

When Activities Approximate a Beta Distribution. Unless a database of historical events is consulted when the project team estimates the duration of project activities, the project manager must assume that the estimates lie somewhere between an optimistic value t_o and some pessimistic value t_p and that the project activity will most probably be completed at a time t_m. We *usually assume* the probability density of the activities to be similar to that of a beta distribution in which the peak of the distribution lies at $t = t_m$, and the width is chosen in an arbitrary manner. With these assumptions useful approximations for t_i and s_i are given by

$$t_i = \frac{t_p + 4t_m + t_o}{6} \tag{6.13a}$$

and

$$s_i = \frac{t_p - t_o}{6} \tag{6.13b}$$

Many authors poke fun at the estimates made by software engineers. As Fred Brooks observes in his humorous *The Mythical Man Month*, "All activities are on time until it comes time for testing," that is, until we discover that the activity does not perform as required or specified in the statement of work.[1] Brooks's remedy is to allocate sufficient time to an activity, such that one-third of the estimated duration is for design, one-sixth for coding or fabrication, and one-half of the remaining duration for preparing a test plan, testing, debugging, and making corrections to assure that the activity conforms to specifications. If we estimate the duration of this activity to be T_A, the time allocated to design is then $T_A/3$, to fabricate or code $T_A/6$, and to test $T_A/2$. This duration might be considered a *pessimistic estimate,* since it assumes that everything will not proceed as planned and that the project team will experience maximum difficulties completing the activity.

The *optimistic estimate* assumes that everything will proceed according to plan; the team will experience few or no difficulties completing the task, and that the activity, when fabricated or coded, meets the specifications. The time required for testing is merely a verification of results and might take only time $T_A/10$. In this optimistic case $t_o = 0.6T_A$. If we assume for simplicity that $t_m = 0.8T_A$, the expected time of the activity described above is $T_i = 0.8T_A$. The standard deviation of the activity is

$$s_i = \frac{t_p - t_o}{6} = \frac{0.4T_A}{6} \approx 0.07T_A \qquad (6.13c)$$

where $t_p = T_A$ and $t_o = 0.6T_A$.

Setting aside the time T_A in a pessimistic schedule for debugging and correcting errors is not inconsistent with the drive to build quality into every product that leaves the laboratory. *Doing it right the first time* means that each unit that forms the system to be built should be built, tested, and made to meet specifications before the units are integrated into the system. Then during system testing, only a few bugs are found that need to be fixed.

Unfortunately, due to today's drive to shorten the time to bring a product to market, units are submitted to integration before they are bug-free, making it very difficult to detect all the errors in system test. Quality is a continuous process, and the project team must schedule time to make each unit work according to specifications before system integration and formal testing begins. Potential or new customers will be the first to complain when the product comes to market flawed and may actually defer their purchase to a competitor once they perceive that your firm has failed to meet their expectations. *Few teams win the race to market without making quality job number one!*

---

[1] F. B. Brooks, Jr. *Mythical Man Month* (Reading, MA: Addison Wesley Longman, 1995).

TABLE 6.4 WORK BREAKDOWN OF A SAMPLE PROJECT

Task Designation	Precedence	t_o	t_p	t_m
A	Start of project	8	16	12
B	Start of project	1	5	3
C	A, B	2	5	3
D	B	4	8	6
E	C, D	2	6	4
F	E	8	12	10

Example 6.3. Table 6.4 shows the work breakdown of a quick-turnaround project in which each proposed activity duration is given by its expected value. In this example, the most probable time (t_m), the optimistic time (t_o), and the pessimistic time (t_p) specify the duration of each activity. Determine the critical path. How long will it take to complete the project with 90% confidence? With what confidence can we complete the project in 29 weeks?

SOLUTION: See Table 6.5. The precedence diagram for this example is shown in Figure 6.8. There are three paths through this network: *ACEF*, *BCEF*, and *DBEF*. Given below are the paths, their expected durations (T_E), and their variances (S_E):

Path	T_E	S_E^2	S_E
ACEF	29.2	2.91	1.71
BCEF	20.2	1.57	1.25
BDEF	23	1.76	1.33

TABLE 6.5 DATA FOR EXAMPLE 6.3

Activity	Precedence	t_o	t_p	t_m	t_i	s_i	s_i^2
A	Start	8	16	12	12	$\frac{4}{3}$	1.78
B	Start	1	5	3	3	$\frac{2}{3}$	0.44
C	A, B	2	5	3	3.2	$\frac{1}{2}$	0.25
D	B,	4	8	6	6	$\frac{2}{3}$	0.44
E	C, D	2	6	4	4	$\frac{2}{3}$	0.44
F	E	8	12	10	10	$\frac{2}{3}$	0.44

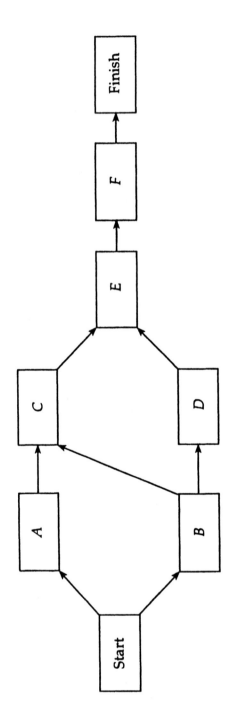

FIGURE 6.8

Precedence diagram for Table 6.4.

If the critical path were defined in terms of the expected duration, the critical path would be *ACEF*. Now, however, we see that path *BDEF* should also be investigated, since its expected duration is 23 weeks and its variance is also large.

For a 90% confidence, the duration D_{90} of each path is:

Path	D_{90}
ACEF	31.42
BCEF	21.83
BDEF	24.73

We therefore conclude that *ACEF* is critical and that there is a 90% probability of completing the project in 31.4 weeks. Because the expected project duration is $T_E = 29.2$ weeks, the probability of completing the project in 29 weeks is somewhat less than 50%.

6.4.6 Time-Based Gantt and Milestone Charts

The bar chart is another technique for displaying the project schedule. Often called a *Gantt chart* after Henry Gantt, a World War II industrialist who made the chart popular for depicting performance schedules, it can be used to portray progress graphically against a baseline plan. Project specialists can easily adapt the chart to view the *early start* (ES), *early finish* (EF) and the *late start* (LS), *late finish* (LF) times for every project activity and any subset of activities. Bar charts can also be used to highlight the project's *critical path*. Today's computer-generated bar charts easily show task interdependencies; however, these charts weren't originally designed to illustrate precedence relationships.

The *Gantt chart* shown in Figure 6.9 presents the activities given in the example in Table 6.2. Each project is represented as an open, solid bar (or rectangle) that illustrates the activity's ES and EF time. If the activity is on the *critical path*, a horizontal line is drawn through the middle of the bar, as shown in activity 100. Some software project management systems represent an activity on the critical path by using a different color or by putting a dashed box around the solid rectangle. If the activity is not on the critical path, a dashed rectangle is used to show the LF of an activity.

To determine the ES and EF times, we start at $T = 0$; to determine the LS and LF times, we start at the end of the project, which is week 34. To illustrate, consider tasks 400 and 410. As Table 6.2 indicates, task 400 starts after task 200 ends. This represents the ES time. The EF time is 2

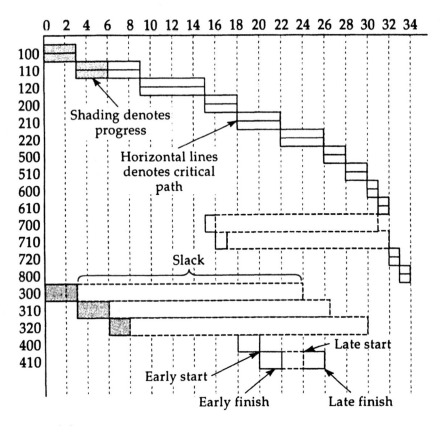

FIGURE 6.9

Gantt chart for Table 6.2 after 8 weeks has elapsed.

weeks later. To find the LF time of task 410, examine Figure 6.9. Note that task 410 must finish before task 500 begins. Since task 500 is on the critical path, 410 must end no later than week 26. Thus the LF of activity 410 is week 26 and the EF of activity 410 is 4 weeks earlier at week 22. The ES of activity 410 is then at week 20 and its EF at week 22. The slack associated with an activity is the interval between the ES and the EF.

To measure performance against a baseline plan, update and redraw the Gantt chart monthly for longer-duration projects and weekly for shorter efforts. The chart lets the project team see very easily where progress is or is not being made. Progress is usually shown by shading in a portion of each activity according to the percentage of work completed. For example, if at the end of week 8, 100% of activity 100 but only 50% of activity 110 were completed, the Gantt chart would look like Figure 6.9. After eight weeks, we are behind on our critical path, and without adjusting how project tasks are undertaken to make up for the time loss, we will delay the project by two weeks.

6.5 PERSONNEL SCHEDULES

6.5.1 Gantt Charts as Personnel Schedules

The Gantt chart is a useful tool to demonstrate how to schedule personnel. To illustrate, consider that in Figure 6.10a all the personnel listed above each activity bar are software development experts who can work on any of the activities illustrated with equal capability. Figure 6.10b shows per month the total number of the personnel used to perform each task and how many software developers are needed as a function of time when each activity starts as early as possible. Figure 6.10c shows the cumulative distribution of software developers as a function of time. This curve indicates the total number of staff months that we plan to use as a function of time and is extremely important when planning a budget. For example, if software development specialists earn a loaded salary of $150,000 per year or $12,500 per month, after six months the project is expected to spend (see Figure 6.10c) $34 \times 12,500 = \$425,000$. After completion we expect the project to spend $126 \times 12,500 = \$1,575,000$ (see Figure 6.10c).

6.5.2 Leveling

From Figure 6.10b, we see that on average, eight software development specialists are required for this project. However, between months 8 and 12, 12 specialists are needed, representing a peak period for personnel use. As is often the case, we do not have sufficient personnel available to satisfy this peak value. To determine whether we can minimize the peak number of software development specialists, we move activities D, E, and F, which are not on the critical path, from their ES positions to start times that will minimize the peak number of personnel required. Figure 6.11 shows an alternative personnel plan, obtained by starting activity F at month 12, in which the peak number of software personnel has been reduced to 10.

If the number of personnel is critical and time is not (which might be the case in a low-priority project), it is possible to decrease the number of specialists required by delaying the project. For example, if we have six software specialists available for the project shown in Figure 6.10a and b

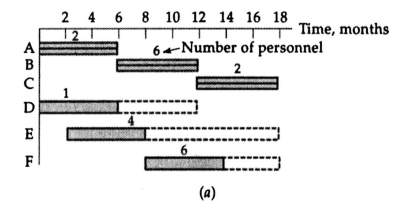

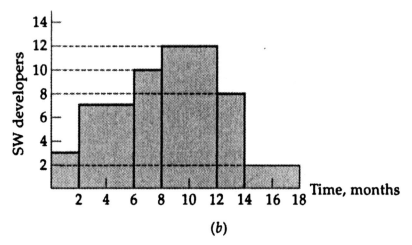

FIGURE 6.10

(a) Gantt chart showing the number of software developers for each activity; (b) software development personnel used as a function of time; (c) cumulative distribution of software development personnel as a function of time.

and additional personnel having these skills cannot be hired, then by delaying activities C, E, and F, we can reduce the personnel to six. This solution is shown in Figure 6.12. Note that it now takes 24 months to complete the project, an increase of six months from the previous total of 18.

Actual projects usually are larger than six resources and six activities and would require using an automated scheduling tool to plot the personnel schedules and to conduct these what-if scenarios (see Chapter 15).

Personnel, months employed

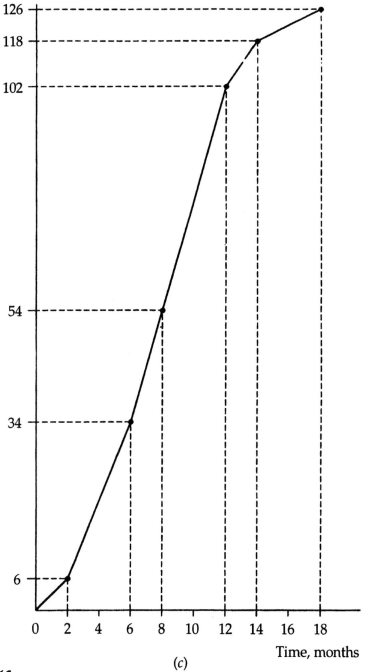

(c)

FIGURE 6.10

(Continued).

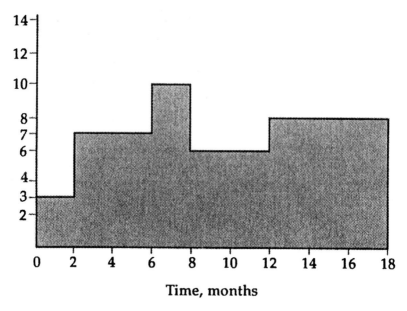

FIGURE 6.11
Late-start solution in which only 10 workers are needed.

6.6 COST SCHEDULES

6.6.1 Plotting the Cost Schedule

Project cost schedules portray the cost of scheduled project work in terms of capital outlays, resources, equipment, material supplies, and other technical requirements. Because of the risky nature of R&D activities in most industries today and the scarcity of qualified professionals available to complete project assignments, it is extremely important to monitor the performance of the project's cost schedule from its outset. To illustrate how, let's expand on the concept of the project life cycle discussed in Chapter 4.

Typically, expenditures early in the project are high compared to actual performance of the scheduled activities. Only 10% of the proposal — concept development, design feasibility, and engineering activities — are completed during the early startup phases, but as much as one-third of the allotted budget may be spent. During the middle phases, another third of the project funds are employed to procure materials and complete the engineering design. The remaining third is used during manufacturing and production to complete the final 10% of the project work.

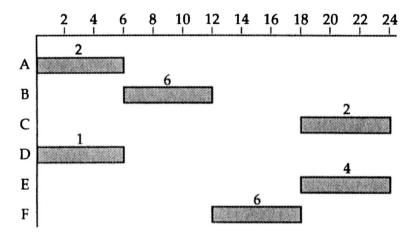

FIGURE 6.12
Gantt chart in which software development personnel is limited to six people.

If we plot the percentage of work completed versus the capital in effort spent, we observe the slope shown in Figure 6.13. Notice how the cumulative distribution of the project budget versus the percentage of work completed approximates a "lazy S" curve. If the rate of work does not progress as planned, the curve will reflect delays in the planned activity and serve to alert the project team to take corrective action. The management science associated with analyzing project cost curves such as those shown in Figure 6.14 is called *earned value analysis* and is discussed in detail in Chapter 10.

6.6.2 Integrated Schedule and Cost Control

Once the project gets under way, the project specialists look to the project management team to coordinate the control of project schedules and costs. The following are some of the measures that a project management team must track when establishing a schedule and cost control system for the project: (1) periodic reestimation of the time and cost to complete the work remaining, (2) timely measurement of physical progress and comparison to plan, (3) frequent and periodic comparison of actual progress and expenditures taken at both the activity and project completion levels, and (4) verification of authorization of expenditures.

Determining what the project costs as a function of time requires the summation of project activities in hours of work performed. Fortunately, the project management team can readily accomplish this task using

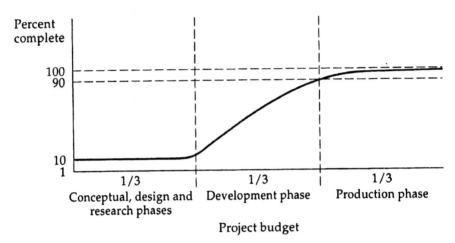

FIGURE 6.13

Plot of percent of work completed versus dollars in effort spent.

computerized project management software that plots time-phased network displays of the project's schedule using cost-coded WBS element descriptors. However, the team may discover that inputting the cost account information from a corporate financial ledger may not be automatic unless the two systems are integrated.

Establishing an integrated schedule and cost control system for the project serves to alert the project team when problems have occurred and allows the project manager to forecast new estimates of how long the project will take if the spending rate for any scheduled activities is adjusted or revised. Such control systems allow management the flexibility to reallocate resources to prevent unwanted delays or premature termination of promising projects as a result of inadequate funding.

6.7 ESTIMATING SOFTWARE DEVELOPMENT PROJECTS

In today's rapidly paced project environments, there is an unfortunately negative perception that software projects regularly get out of hand. Although there really is no simple solution to the challenge of estimating the cost and schedule of software development activities, project management can help the project team to understand the characteristics of this type of development. It is also the primary responsibility of the project managers to work more closely with the application programmers and system testers to assure them that the project environment will support quality estimates.

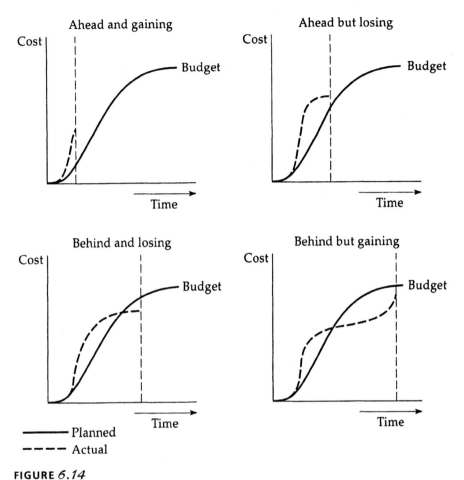

FIGURE 6.14

Typical project rate curves over time.

The problem of predicting accurate duration and cost estimates deserves special attention:

· Unlike material construction projects, software is intangible and its fabrication highly probabilistic.

· The levels of abstraction and complexity associated with software development make accurate estimates very difficult.

· The apparent bias for software developers to underestimate the time required to complete tasks by less experienced personnel often lowers the entire project's productivity.

- The existing misconception that there's a linear relationship between available time and capacity seriously jeopardizes the scheduling of software projects, particularly if it is believed that by doubling the staff to 25 people, a project originally slated for completion in 24 months can now be completed in just 12 months.
- Frequently, product management will ignore true estimates because there is an overarching need to meet a market window, make a bid more acceptable, or secure customers who otherwise would not buy.
- Often, the development team will fail to define product scope that allows customer expectations for new features and functionality to drive up the cost of the original estimate.

Considering the magnitude and complexity of many software development efforts, many managers might debate whether it's truly possible to get a handle on the uncertainty of estimating the cost of software development projects. Nevertheless, it is our intent to help the software team tackle these issues head on. For example, in Chapter 2 we discussed how the project might go about organizing to improve teamwork among the developers. In Chapters 3 and 4 we discussed specific development methodologies that foster better program designs. In Chapter 5 we emphasized planning techniques to size project work from the bottom up and address the relationship between components. In this chapter we explain how to define and quantify specific cost drivers typically experienced in software development efforts and how to use these drivers as input into one or more estimation tools to predict a schedule.

As we explained in Section 6.4.4, the typical schedule that results must also be regarded as an average solution with a 50% probability that the task durations are equally likely to over- or underrun. To improve the accuracy of the schedule, estimates should be revised continuously once the project is under way.

6.7.1 Sizing the Effort

The discipline of software engineering generally considers the following six characteristics or factors as dominant cost drivers in a development effort:

1. Size of the software application, including customer-required quality levels, requirements volatility, program complexity, level of planned reuse of previously developed components, amount of programmer and end-user documentation, and type of application

2. Computer constraints, that is, time to execute the program on the targeted computers that will run the program along with the response time and memory capacity of these machines

3. Use of modern programming techniques

4. Requirements duration, that is, whether the time taken to research and write the specifications is "compressed" to a specific phase of the project or is allowed to "stretch out" across several phases

5. Basis for project control, including the organizational structure imposed on the programmers (matrix, pure project, etc.) and the type of programming methodology chosen for the project (incremental, linear, or concurrent)

6. Participation and number of simultaneous users for the application, along with their education and familiarity using similar programs

Project teams quickly learn that they must define or objectively quantify and correlate these project drivers with important human factors studies that analyze how potential customers will react to the user interface and software complexity. If they fail to do this, they also fail to size the development effort properly. For all intents and purposes, the *teams who ignore human factors data or refuse to gather this important customer input do not know the actual cost of bringing their applications to market!* Aftermarket field support required to teach users how to interact with these programs is a hidden expense for the organization, as is the burden of retaining programming and testing staff to complete unplanned "dot" releases of the code to address customer concerns.

Project personnel can choose from among several prevalent techniques to estimate the size and cost of a software effort. This may include:

· Using experts to add up all the defined lines of code in each compilation segment of the program

· Basing predictions on a database of similar program developments

· Pricing the program to win over competitive offers, and holding firm to these parameters in the business case

· Using capacity, especially personnel, to determine how long the effort will take

· Using metrics or *function points* (inputs, outputs, inquiries, file types, interfaces) as early indicators of program size

- Conducting a bottom-up evaluative session such as an IPP session with a professional facilitator, who acts as the estimator, to brainstorm program size with prospective application developers
- Using parametric models that apply the project's global characteristics to determine its size

Corporate guidelines may specify which technique is preferred in your company; otherwise, your team may choose at least two to reduce the level of uncertainty. Following is a summary of the most prevalently used *software cost estimation* (SCE) techniques and approaches practiced today.

- SCE based on expert judgment
  - The approach is used primarily to give an indication of the effort and time needed during the early phases of a project when specifications are vague.
  - Qualitative estimation is as accurate as the estimator's experience in performing similar tasks on other project efforts.
  - Expert's rules of thumb may be applied to inapplicable situations.
- SCE using archived data from similar software projects
  - Database of similar historical projects is analyzed.
  - Estimation is based on reasoning by analogy.
  - Basing new project estimates on older ones that used a similar programming language or structured development approach requires that all projects be recorded continuously to keep the archive current.
- SCE timed to market demand
  - This approach bypasses standard estimation procedures because commercial directives constrain the amount of time available to complete the project.
  - Programming tools, structured design, level of reuse, and compression (time allotted for the research and writing) or volatility of requirements influence the project team's ability to complete the project as estimated.
  - The ability to deploy additional resources to complete the project when needed in the marketplace is essential.
- SCE based on available resources
  - The capacity of the organization to provide the personnel determines the duration and level of effort required to complete the project.

- Projects are commonly overestimated to engage all available resources that may drive up the cost of the development.
- SCE using metrics
  - The effort required to plan and complete the project is based on measurements collected and used to estimate the software's size:
    1. Number of new and modified requirements.
    2. Number of function points.
    3. Number of pages of software documentation or lines per screen of electronic documentation.
    4. Number of new and modified functions in the software design.
    5. Number of lines of base, new, and modified code in the current program release.
    6. Amount of bug-fixed and reused code.
    7. Total net developed code and bytes of executable code delivered.
    8. Required test case objectives to verify that the program complies with specific requirements and setup and operation procedures.
    9. Staff and materials needed to produce end-user documentation.
  - A staffing metric averages the number of personnel needed to complete the project. Years of experience working on earlier versions of the product and with the software technology is factored into the estimate.
  - The cost metric includes all direct (tools, workstations, and testing facilities) and indirect expenses that will be incurred to complete the development tasks.
  - Product performance metrics that have a direct impact on the development, such as quality constraints (reliability, maintainability, and usability), are considered along with memory requirements and timing constraints or throughput on the product.
  - To refine the metrics estimation process, additional measures are collected to compare the accuracy of the estimates against the project's actual performance.
- SCE by summing individual estimates
  - This is a bottom-up approach, which sums individual estimates from each contributing team member responsible for the project deliverable.
  - Automated tools are available to help developers estimate specific project activities based on questionnaires that use measures of their own productivity derived from historical data.

- SCE based on parametric models
  - Models use a number of product variables, such as staff years and length of time to build the product, which are stored in a database of completed projects to estimate development time and effort.
  - Estimates made by different models for the same project are known to vary widely.
  - Organizations should train estimation specialists to use these tools and position them to conduct this service for multiple projects to generate more reliable estimates based on consistent data input and interpretation of output.

6.7.2 Using Models

A project team might also be instructed to follow corporate directives to produce software estimates that conform to industry standards established by the *Software Engineering Institute's capability maturity model*. Procedures acceptable to the software engineering community require that the project team follow a documented process that gathers a series of estimates that reflect life-cycle opportunities. All major deliverables must be sized and a record of the estimate and associated documentation archived for future retrieval. The team must collect project metrics, audit estimates against actual data, and use these historical data in formulating future estimates. Training is required for project personnel involved in software estimation. The organization's ability to implement standard procedures when sizing the development also may be benchmarked against industry or best-in-class norms. Software estimation succeeds if it provides project management with reasonable predictions of the staff, duration, and cost to complete the development effort. The estimates target a *range of costs* and *task durations* for each work activity and are often accompanied by a *statement of likelihood (probability)* of actually achieving the targeted values, as described in Section 6.4.4.

Today, to address the uncertainty associated with software development, most corporations are establishing estimation process models that take into account the fact that estimates will improve as the project team progresses through product development. Many models establish five estimation phases. As more data are collected and fed back to the project team planning the development effort, this information is used to refine the estimates. The early estimates, known as *feasibility* or *commitment estimates*, are used by product and market management to support the business plans and fund the project. As shown in Figure 6.15, as soon as the software

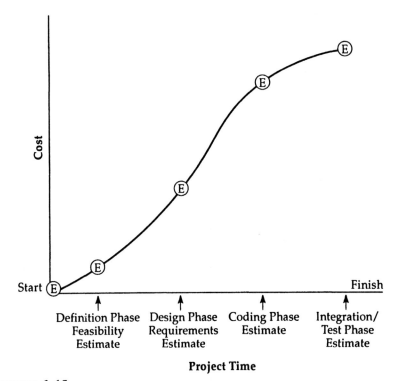

FIGURE 6.15

Main estimation phases of a software development project.

application's design is complete, the project team should generate the *scheduling* estimates to determine the costs and duration of each project activity. When the project enters into the next phases of subsystem or module design, it probably will be necessary to perform additional trade-offs and revise the estimates based on who will be doing the work. Precedence relationships between the tasks determine the critical path for the project. How frequently projects reestimate during the additional coding and testing phases will depend on the overall project duration. Facing a tight market window, some projects may need to reestimate biweekly.

When the project concludes, the team undoubtedly will want to compare each estimate calculated for size, effort, and duration with the actual values for the project. Accuracy can be defined as the percentage relative error between the detailed estimate and actual cost of completing. Projects that collect such metrics advance their standing on the software industry's maturity index and enhance their company's ability to fund future project efforts adequately.

6.8 CONCLUSIONS

Now that we have planned the project, completed the WBS, finalized the work packages, prepared the Gantt chart, identified the critical path, and performed a PERT analysis, only one additional activity remains before the project can be started: optimizing the schedule. This technique will allow the project team to meet the constraints imposed by market demands for cost containment, on-time delivery, superior quality, and customer expectations. This is the focus of Chapter 7.

BIBLIOGRAPHY

AT&T Bell Laboratories Software Quality and Productivity Cabinet, *Best Current Practices: Software Cost Estimation.* Indianapolis, IN: AT&T Bell Laboratories Technical Publications Center, 1993.

Brooks, F. B., Jr., *Mythical Man Month.* Reading, MA: Addison Wesley Longman, 1995

Builders of the Ancient World: Marvels of Engineering. Washington, DC: Special Publications Division, National Geographic Society, 1986.

Checkpoint Questionnaire. Burlington, MA: Software Productivity Research, 1997.

Heemstra, F. J., "Software Cost Estimation," in *Software Engineering.* Los Alamitos, CA: IEEE Computer Society Press, 1997, pp. 374–386.

Londeix, B., *Cost Estimation for Software Development.* London: Addison-Wesley, 1987.

McConnell, S., *Rapid Development.* Redmond, WA: Microsoft Press, 1996.

Silverman, D., *Ancient Egypt.* New York: Oxford University Press, 1997.

EXERCISES

6.1. A project's performance schedule is not used to:

   (a) Model day-to-day operation

   (b) Perform strategic planning

   (c) Prepare the WBS

   (d) Estimate project completion

6.2. Which of the following describes the PDM approach to scheduling?

   (a) Rectangular nodes, which represent activities, are connected to the nodes that succeed them in time

   (b) The technique is often used in the construction industry, where the duration of project activities is more certain

   (c) Calculation of the project's duration is simplified

(d) Dashed arrows called dummy tasks are used to represent tasks that have zero duration

(e) (a) and (c)

6.3. Which of the following describes the CPM scheduling technique?

(a) Uses a deterministic algorithm to identify which sequences of project activities are the least flexible

(b) Analyzes the early- and late-start and finish dates for each activity sequence to determine the project's duration

(c) Calculates the slack or float between dependent activities

(d) Uses sequential network logic and a weighted-average duration estimate of each activity to determine the project schedule

(e) (a), (b), and (c).

6.4. Which of the following statements are true?

(a) One can readily determine the precedence relationship between project activities by viewing a Gantt chart

(b) A Gantt chart plots progress as a function of time

(c) Gantt charts can be used to portray progress graphically against a baseline plan

(d) Gantt charts are used when preparing the WBS

(e) Gantt charts are useful briefing tools for management

6.5. The purpose of preparing a project cost schedule is to:

(a) Track the information contained in the work package levels of the project's work breakdown structure

(b) Plot the percentage of work completed versus the capital in effort spent

(c) Estimate required expenditures to support the business plan and fund the project

(d) (b) and (c)

(e) All of the above

6.6. In Figure 6.2, assume that activity *A* has been completed and that no other project activities have been started. Find the critical path and the project duration.

6.7. Refer to Table 6.2. Assume that tasks 100, 110, and 300 have been completed, task 120 has two weeks of work remaining, and that no other project tasks have been initiated.

(a) Determine the critical path

(b) Calculate the time remaining in the project

TABLE 6.6 DATA FOR EXERCISE 6.8

		Time	Precedence
1.0	*Mr. Ready*		
1.1	Awakens to alarm	Start	
1.2	Awakens wife	1	1.1
1.3	Awakens John	4	1.2
1.4	Brushes teeth	1	1.3
1.5	Showers	5	1.4, 2.2
1.6	Shaves	5	1.5
1.7	Selects clothing	2	1.6
1.8	Dresses	5	1.7
1.9	Walks dog	10	1.8
1.10	Eats breakfast	10	1.9, 2.6
1.11	Gets into car	1	1.10
1.12	Takes train to work	60	1.11, 2.9
2.0	*Mrs. Ready*		
2.1	Wakes up	Start	1.2
2.2	Showers	10	1.2
2.3	Brushes teeth and puts on makeup	20	2.2
2.4	Dresses	10	2.3
2.5	Feeds dog	2	2.4
2.6	Makes breakfast	5	2.5
2.7	Eats breakfast	10	2.6
2.8	Prepares children's lunch	10	2.6
2.9	Drives children and Mr. Ready to station	10	2.8, 1.11, 3.6
2.10	Continues on and takes John to school	15	2.9
2.11	Drives to work	20	2.10
3.0	*Ready children*		
3.1	Awaken and get out of bed	2	1.3
3.2	Shower	15	2.2, 3.1
3.3	Brush teeth	1	3.3
3.4	Dress	5	3.3
3.5	Eat breakfast	1	3.4, 2.6
3.6	Get into car	1	3.5
3.7	Arrive at school	End	3.6, 2.10

6.8. Use the information in Table 6.6 to solve the following:

(a) Draw a precedence diagram

(b) Determine the critical path

(c) Draw a bar chart using early starts

TABLE 6.7 DATA FOR EXERCISE 6.12

Task	T_o	T_m	T_p
100	3	4	6
110	6	8	10
120	6	8	16
200	2	2	2
210	4	6	8
220	6	8	10
330	3	6	9
310	2	4	8
320	5	6	15
400	2	3	4
410	3	3	3
500	2	2	2
510	2	3	6
600	1	1	1
610	2	2	2
700	1	1	1
710	1	1	1
720	1	1	1
800	1	1	1

(d) How long does it take the Ready children to arrive at school?

(e) What events lie along the critical path?

(f) How would you shorten tasks, rearrange precedence relationships, or add resources so that Mr. and Mrs. Ready can sleep later in the morning?

(g) How many bathrooms are needed if Mr., Mrs., and the Ready children do not use a bathroom simultaneously (the children can share a bathroom) and early starts are used?

(h) Is it possible to have Mr. Ready, Mrs. Ready, and the Ready children use the same bathroom at different times, with Mr. or Mrs. Ready or their children wakening at an earlier time?

(i) If it is not possible, what it the latest time that either Mr. Ready, Mrs. Ready, or their children can awaken?

6.9. Consider taking a trip to visit a relative in another city.

(a) If you were to take the train, how long would the trip take? Give an off-the-cuff estimate

(b) Now size the trip's duration more carefully.
   · Estimate the time to go from your house to the train station by taxi assuming that it happens during rush hour
   · Estimate the time to walk to the ticket counter from the taxi, the time to purchase a ticket, the time to wait for the train, the time for the train ride, the time to wait for a taxi to your destination, and finally, the time to travel to your relative's house

(c) Compare your answers. Which time estimate was longer?

(d) Repeat the exercise and calculate the travel time assuming an optimistic, pessimistic, and most likely duration for each step in the trip. With 90% confidence, how long will the trip last?

6.10. An integrated circuit is to be fabricated. The time duration of each pass is 60 days. On the first pass, the probability of success is 30%; on the second pass, 60%; and on the third and subsequent passes, 90%. Estimate the expected duration t_i and the variance $s^2 i$.

6.11. Verify Eq. (6.11). If $T_0 = 90$ days and $P = 0.9$, calculate t_i and s_i.

6.12. Refer to the durations listed in Table 6.7 for t_o, t_m, and t_p.
   (a) Find the critical path
   (b) With 90% confidence, how long will the project last?
   (c) With what confidence level (probability) will the project be completed in 49 weeks?

OPTIMIZING THE SCHEDULE

If our timelines aren't responsive to the customer's needs, we are printing yesterday's newspapers tomorrow!

HARRY CHAMBERS AND
ROBERT CRAFT

7.1 INTRODUCTION

Known as the grandest enterprise under God,[1] the construction in the American West of the transcontinental railroads over the vast plains and through the Rocky Mountains was by far the nineteenth century's largest and most favorably scheduled industrial project. The country's expansion westward was spurred on by the furor of the Manifest Destiny national doctrine, which declared that the United States had the right and duty to expand throughout the North American continent. The country's migration westward would be 10 million in 20 years and was expedited by a congressional act that granted the major trans-Mississippi and Pacific railroad companies money for each mile of transcontinental rail laid. The act, in fact, started a major race to connect the Union and Central Pacific lines in the Nevada desert just north of Salt Lake City between Ogden and Promontory Point, Utah.

At first, the point where the two companies would meet was not designated. So each company raced the other to see which could lay the

---

[1]G. C. Ward, "The Grandest Enterprise Under God," in *The West: An Illustrated History* (Boston: Little, Brown, 1996), p. 215.

most track—and therefore make the most money—$16,000 per mile on level ground, $32,000 per mile across plateaus, and $48,000 per mile in the mountains. In 1869, when the completion of the long-awaited Union Pacific line was officially celebrated from San Francisco to New York by the driving in of four gold and silver spikes, the nation was gratified by its fulfillment of a major milepost along the road to westward expansion.

Officially, the scheduled completion of the Union line was to take place within two years. First came the surveyors; then came the graders, who laid out the roadbed, cut or blasted their way through gorges, and built bridges. They worked well ahead of the builders, sometimes as much as 100 miles, so that the others would not have to wait. When the graders had finished, the road was ready for the workers to lay the track.[2] Yet despite their ability to manage terrain, it quickly become apparently that only 2 miles of track were being laid a day on flat land. So the company had to step up the pace and manage an additional $5\frac{1}{2}$ miles by timing the use of heavy equipment and by calculating to the mile the number of required "rail-lifters." Each rail weighed 700 pounds, and five men were needed to raise and place the 30-foot rails. Before the men with hammers drove the spikes, a flatcar would roll forward over the new rails, allowing the men to drop the next load in place. A great anvil chorus of driven steel rang out in triple time across the great American plains: three strokes to a spike, 400 rails to a mile, 1800 miles to California; 21 million strokes before the work was complete.[3] Techniques such as these were used by all the railroads in building many more lines from the East to the Southwest and to the Pacific after the Union was completed—all with some kind of government loan or grant.

The building of rail lines in the American West illustrates how projects of all sizes need to make the most effective use of allotted and scheduled resources to meet the expectations of their stakeholders. In Chapter 7 we discuss how performance of projects in the twenty-first century must be optimized to meet the demands of a global and information-rich economy in which stakeholder needs are varied and ever changing. We start by explaining how to baseline a project's schedule. Then, given the time constraints facing most modern projects, we introduce several techniques to expedite a schedule and deliver to market sooner than expected.

---

[2]B. Currie, *Railroads and Cowboys in the American West* (London: Longman Limited Group, 1974).

[3]G. C. Ward, "The Grandest Enterprise Under God," in *The West: An Illustrated History* (Boston: Little, Brown, 1996), p. 222.

7.2 ESTABLISHING THE PROJECT BASELINE

In Chapter 6 we introduced a number of sophisticated techniques to schedule project activities and costs effectively. Having thoroughly mastered the concepts of how to allocate project resources, budget actual costs, and develop a timeline that accounts for the uncertainty and risk inherent in the completion of project tasks, we can concentrate on predicting the end date for a project. Our focus now takes into consideration how to manage the variables that constrain a project's finish. From the project's outset, of foremost concern to the leadership team is how to schedule and utilize project activities optimally to meet, for example, constraints such as time-to-market deadlines, scarcity of skilled personnel, budget limitations, or finite availability of necessary equipment and laboratory facilities.

Although there are no silver bullets to compensate a project for the pressure imposed by such constraints, leading-edge teams have a full understanding of the potential consequences of their impact on a project's final cost and completion date. Early in the process they establish alternative plans to preserve the schedule by establishing baseline intervals needed to complete these activities. As a result, the typical 5 to 10% variance between baseline estimates and final outcomes experienced by these projects rarely gets senior management's attention.

In essence, projects following this approach are able to deal consistently with high-risk activities and still balance the costs required to produce the desired features within the time frames required to satisfy customers and other major project stakeholders. Using the approach also preserves the team's morale and the members' reputation for on-time delivery of high-performance products and services.

Essentially, there are five major steps that successful teams pursue to manage a baseline schedule and achieve the desired on-time finish at the agreed-upon cost.

Step 1: Manage the critical path. As we explained in Chapter 6, teams need to prepare a master plan for the project that (1) forecasts when each task is expected to start and finish and (2) baselines when each task must start and finish to achieve the project objectives. Consequently, the schedule becomes the warning mechanism that alerts the project team about impending troubles that must be corrected to prevent disasters and save time and money.

Since the critical path is the longest sequence of project tasks, the astute management team will see to it that all the activities on this path are completed on time to guarantee that the project finishes when estimated. Delays experienced on project tasks not on this path do not push out the project's end date. By definition, preventing overruns or the extension of time needed to complete the activities on the critical path assures the project's on-time completion. Thus, to preserve the critical path, take advantage of the available *job slack* in your project's schedule. This time, when added to noncritical activities, does not increase total project time. As we illustrated in Chapter 6, you can compute slack for any project activity by determining the difference between the earliest time and the latest time it can begin. This permits teams to reassign the scheduled start and finish times of all critical tasks to position the slack to absorb unplanned overruns of critical path tasks that would otherwise jeopardize the project's success.

The team must vigilantly monitor progress of the activities on the critical path and be keenly aware that as activities are completed ahead or behind schedule, the critical path may be recalculated. For this reason, some management theorists postulate that it is essential to identify the scarce resources needed by critical tasks on these paths and any constraints on their availability to keep activity flowing through a single chain of constrained events that compose these paths.[4] It is actually possible for project managers to deliver the project sooner than originally estimated by minimizing the time needed to complete the critical path! How else to preserve time estimated to complete the critical path and accomplish the baseline schedule is presented in steps 2 to 4.

Step 2: Manage technical and business risks. Risk affects the project team's ability to manage the uncertainty of project activities associated with occurrences that change the outcome of the project for better or worse. Often, risk can positively affect a project by introducing an opportunity to capitalize on a new success. In other words, risk does not always result in a negative series of events or occurrences that affect the project adversely. As a general rule, try to implement actions that reduce risk in the project, such as relegating high-risk activities to the research department, where schedules usually are more relaxed. A major impediment to fast development is the inaccessibility of the technology needed to develop the product. In general, projects cannot consistently achieve rapid development by doing technology innovation on the critical path.[5]

---

[4]E. M. Goldratt, *Critical Chain* (Great Barrington, MA: North River Press, 1997).
[5]P. G. Smith, and D. G. Reinertsen, *Developing Products in Half the Time: New Rules, New Tools* (New York: Van Nostrand Reinhold, 1998).

When this level of control cannot be exercised, use the project work breakdown structure to identify risks associated with project activities. First, have project specialists record in their work packages the expected risk that might occur for each of their assigned tasks. Then, when estimating how long to complete tasks on this list, have them calculate, for each high- and medium-risk activity identified, the standard deviation to complete the activity. Their estimates should reflect the uncertainty of completing the task and the probability that some unforeseen impediment will disrupt progress toward this objective. If they cannot calculate the deviation, have them at least give a range of pessimistic and optimistic task completion times.

Project teams can, in part, account for the statistical variability of schedules by using the techniques taught in Chapter 6. As we have already stated, a useful rule of thumb is to calculate the standard deviation of any task due to the uncertainty in performing it. The risk and uncertainty is related directly to the difference between the pessimistic t_p and optimistic t_o times. In general, if there is no uncertainty or risk involved in performing an activity, $t_p = t_o = t_m$.

We can illustrate the impact of risk on the ability of a factory to produce an incoming new order. If a factory were to build 1 million widgets, in addition to the typical production run of 20 million, the risk of producing the new order would be precisely the duration of performing the task, and the standard deviation; therefore, $s_i = 0$. Here, the probability of not completing on time is rather low, less than 25% (see the guideline in Table 7.1), which means that the risk is negligible. However, if the risk were high or greater than 50%, the delay would affect the scheduled order significantly.

Consequently, to account for the risk of the expected time loss, the management team should use a range of task-duration estimates when building the project network. Then, to determine the likeliness that each event might occur, they must ask the project specialists to answer the question: What is the probability of completing this project in X months,

TABLE 7.1   SIMPLE RISK GUIDELINE

Risk	$R = (t_p - t_o)/t_p$
Low	$R < 0.25$
Medium	$0.25 < R < 0.5$
High	$R > 0.5$

for Y dollars, with Z people? They also might involve the entire team in establishing the criterion for describing the technical and business risks, such as high, medium, or low, so that everyone uses consistent definitions. To get started, review the criteria suggested for classifying risks as high, medium, or low given in Appendix 7.1. With this information, project management can quantify the impact on the total project effort.

Once the team knows the cost of these impacts in terms of schedule delay or overall project performance, project management can then choose to (1) ignore the risk, (2) develop a contingency plan (see *Step 3*), or (3) add additional time and resources to complete the activity when configuring the total project time.

Following the severity definitions in Table 7.1, the team would replan high-risk events, develop a contingency plan for medium risks, and ignore anything of lesser consequence. By identifying *trigger* or predictor events that signify where risk items might be expected to occur in the project network, the team has a better chance to formulate effective contingency plans. The effort also helps the team envision particular strategies to mitigate the probable effects the occurrence of these unfavorable events would have on the overall project. Of course, the team can always deflect the risk by transferring it to someone else. A manufacturing project might take such action by purchasing an insurance policy to protect its capital investment against natural disasters such as fire or malicious activity due to political insurrection abroad.

Step 3: Reserve contingencies to avoid cost overruns when emergencies occur. The project management team should assign contingency reserves to the project budget to cushion the impact of unwanted events that take the project off track. In Chapter 3 we stated that the rule of thumb for estimating reserves for contingencies is 10 to 20% of the total project budget. Now, however, we can be more precise in estimating the needed reserves by calculating the expected value of specific risky activities and determining the gain or loss that will be incurred if the identified risk events occur.

To determine the overall positive or negative impact of the risk events on a performance schedule, first sum the expected values of the duration of the individual tasks or activities associated with the risk event. Then, using the guidelines established in Step 2, assess the likeliness of the event's occurrence. Finally, to calculate the monetary value of the risk event, multiply the event probability by the event value. This is the amount the project team should consider holding in reserve.

A software development project, for instance, may require contingency reserves to pay the salaries of its staff that lack experience programming in a new language. The time spent learning the new language may significantly increase the effort required to code and test specific features required by customers purchasing the software. If five of 10 programmers on this project have shown the ability to learn the new language quickly, the additional time required to learn the language by the other five is a medium severity risk with a probability of 0.5, or 50%. This means that the project's original staffing effort of 36 months might increase by an additional 18 months. If a staff month of effort costs $12,500, the likely increase would cost an additional $562,500. This is the amount the project might reserve in a contingency fund to cover this event.

Quite often, in a production contract, funds are needed to cover price adjustments to maintain the value of a contact when using subcontracted labor or resources whose daily and hourly usage varies. Project teams might also consider using contingency funds to *optimize* a tight schedule through the purchase of additional resources to expedite project work. The funds may be required to time the market entry of a new product or service. In Section 7.3, we discuss how to calculate the needed amount of funds to cover these events.

Step 4: Use a range of duration-based estimates to protect activities with major dependencies and shared resources. Using a range of estimates to build the baseline schedule helps the project team to focus its attention on the fact that the schedule is duration and not milestone-based. The activities associated with the project require a certain amount of time to complete and are not readily accomplished as first predicted. This is especially true in such uncertain R&D ventures as the fabrication of a new microprocessor or software development, where preliminary estimates may not be valid. As we explained in Chapter 6, the range of estimates highlights the uncertainty associated with the schedule and focuses the team's attention on the action it must take to minimize these inherent risks when communicating with senior management. No matter how hard the team tries, planning for a shorter schedule won't shorten the amount of time it will take to create the product, but only increases the risk of being late!

Also, by focusing its attention on the high-risk activities that have critical dependencies but do not lie on the critical path, projects can verify whether these activities might impede the ability to accomplish associated critical path events because they share scarce resources. In other words, these activities may become bottlenecks because there is a resource contention. For this reason, the management team may need to verify that project resources are distributed adequately among these tasks, that the tasks are not scheduled in parallel, and that estimates to complete are firmly connected to reality. The standard deviations given with each estimate can be used to create a buffer of time needed to absorb the risk associated with completing the task.

Step 5. Measure progress against the estimated project completion date and focus the project team's attention on this final end date. When managing the activities required to complete the project, project management should avoid translating the estimates into due dates. Activity start time may not necessarily be keyed off these early completion dates but be based on the actual availability of the next resource needed to start a task. Remember that the schedule is not the project! Ask your project specialists to report the amount of time remaining to completion and to notify team members responsible for accomplishing the succeeding task when a hand-off will occur. This practice will help to (1) minimize any potential downtime between predecessor and successor events. (2) serve to alert team members if a task is going to take longer, (3) notify project management if the slack time between events is adequate to absorb the overage, and (4) signal the need for additional action, such as whether to enact a contingency plan.

To understand the impact of risk and the inherent uncertainties associated with heavily loaded and dependent tasks on the project end date, project managers might also consider running a computerized *Monte Carlo simulation* of the schedule. This technique expands the PERT networking technique explained in Chapter 6 by displaying a distribution of the project's riskiest activities. This gives the team a graphical warning device to explain to management why the high probability of extending the duration of key activities will have a negative impact on the final project end date.

7.3   CRASHING THE PROJECT

If after baselining the scheduled duration of critical path activities, the planned completion of the project misses the due date set by the customer, an alternative plan must be put in place to meet the market entry requirements. One technique that can be employed to shorten or *crash* the project is to increase the number of personnel in selected activities.

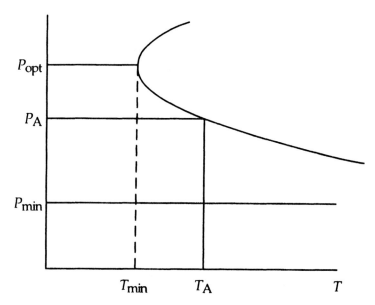

FIGURE 7.1

People trade-off for an activity, A, being performed.

Unfortunately, adding up to twice the personnel will not half the duration. Schedules simply are missed by about one-half because each activity takes twice as long as estimated, especially when personnel are shared among projects and managers are competing for their time. Also, there are an optimum number of personnel who can be assigned to complete an activity, which we illustrate in Figure 7.1.

Here the number of people needed to complete activity A in a given period of time is P_A. If we choose to decrease the duration of activity A to T_{min}, we would need to increase to its optimum level (P_{opt}) the number of people assigned. If we employ more personnel than P_{opt} to complete an activity, however, the duration will actually increase. Thus T_{min} is the minimum possible time to complete task A.

If an activity can be partitioned, project management can shorten the activity by increasing the number of appropriate personnel. However, increasing personnel often increases the need to communicate across a greater number of groups of specialists, which may increase the actual time spent accomplishing the task. If the activity cannot be partitioned, additional personnel will not decrease the time on task; rather, time will be wasted. For example, suppose that two expert chefs are engaged in creating a delicious soufflé. How would the addition of a third reduce a customer's wait time for the dessert? Might the third chef only increase conversation and conflict in the kitchen?

On the other hand, were a project team to remove specialists from an activity, the team would quickly reach a point where further reduction of personnel would result in the activity taking on an infinite amount of time. For instance, consider a capital project where a new plant is under construction and a minimum number of personnel must be engaged to move a steel beam. If all but one employee were removed from this task, which in reality requires several, the job would quickly become impossible and its duration infinite. A project team might also crash an activity by increasing the nonhuman resources required for its completion. For example, often a system test team will have access to a targeted test environment (i.e., production hardware, prototype equipment, etc.) on a limited time basis only. So the testers must wait until other projects' priority tests are completed before starting theirs. To expedite the schedule, project management could shorten the elapsed time required to complete the testing activities by purchasing and dedicating to the project additional system test equipment that duplicates the target environment.

7.3.1   Which Tasks to Crash

Now that we have seen that it is possible to shorten an activity by increasing personnel or nonhuman resources, we must determine which activities should be crashed. The answer entails a two-part decision:

1. Crash an activity only if it is on the critical path.
2. To minimize cost, crash the activity that costs the least to shorten. For instance, if two activities, A and B, are on the critical path, and if adding one person to activity A will decrease its duration by 2 months, but adding one person to activity B will decrease B by 3 months, add the one person to activity B.

Example 7.1. Development and installation of a new high-speed data networking system connecting the New York branch of BTR Investments with their London office is late. The project had already cost $92 million. The project management office has calculated that each month the system is late results in a $50,000 loss for the company, due to expenses that would not occur once the new network is complete. The new project manager's job is in jeopardy. She calls an emergency meeting of the project team to determine what strategies can be employed to shorten their

TABLE 7.2 ACTIVITY CRASH REPORT

Activity	Precedence	Time Remaining (months)	Cost Remaining	Additional Cost/Month to Crash Activity	Number of Months That Activity Can Be Crashed
A	—	3	$ 60,000	$30,000	1
B	A	4	120,000	20,000	1
C	A	5	160,000	50,000	2
D	A	4	80,000	10,000	2
E	C,D	2	60,000	15,000	1
F	B,E	3	140,000	40,000	2
			$620,000		

activities and hence the project. The report received is shown in Table 7.2. The project was originally scheduled to be completed in 7 months. Plot the cost of the project as a function of time. What is the project duration if the new project manager is to complete the project with minimum cost?

SOLUTION: Figure 7.2 represents BTR Investments' possible system's precedence diagram. As indicated, the critical path, shown by the solid arrows, contains activities ACEF. If nothing is done to crash the project, it will take an additional 13 months for completion. Here the project will be 6 months late and BTR would lose $300,000. Naturally, management wants to minimize this loss. After reviewing Table 7.2 and the precedence diagram, the project manager plans to crash an activity on the critical path. Since E is the least costly at $15,000 per month, her first selection is to crash E from 2 months to 1 month. This would result in shortening the project by 1 month, thereby saving $50,000 less the $15,000 increased cost of crashing, or $35,000.

After examining the precedence diagram, shown in Figure 7.3, the critical path is seen to be ACEF. Now A is the least costly to crash at $30,000. The project manager decides that activity A should be crashed by 1 month. The savings would then be $50,000 less $30,000 (the increased cost that the project will incur by crashing the activity), which represents a net savings of $20,000.

Figure 7.4 illustrates the new precedence diagram obtained after crashing A. The critical path remains ACEF. However, we can further reduce the project by 2 months since we can crash activity F by 2 months. The net

228

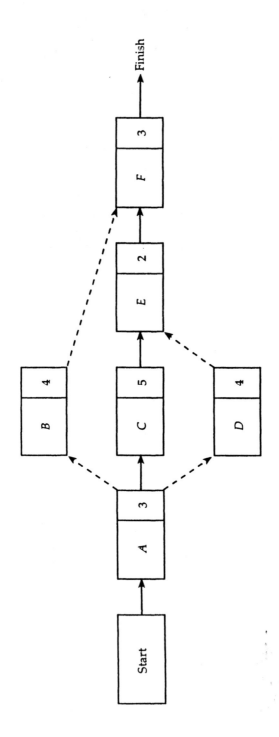

FIGURE 7.2

Precedence diagram for Example 7.1.

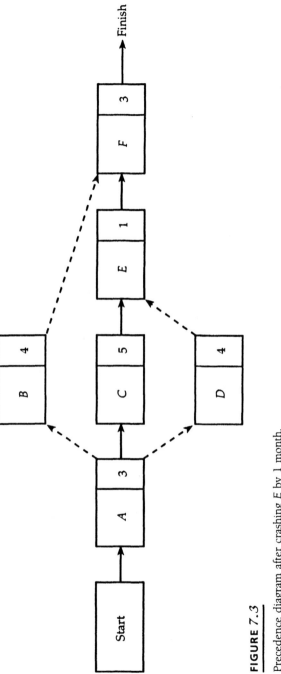

FIGURE 7.3

Precedence diagram after crashing *E* by 1 month.

229

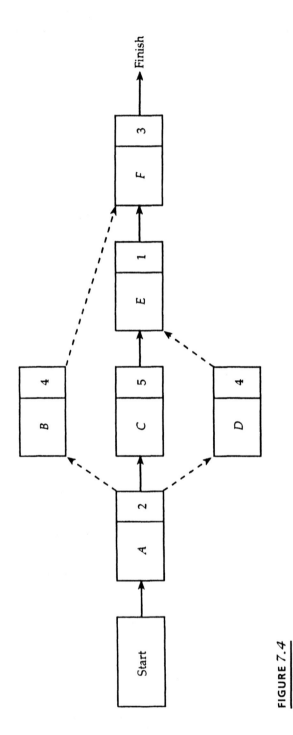

FIGURE 7.4

Precedence diagram after crashing A by 1 month.

savings is

Gross savings by shortening project by 2 months	$100,000
Cost of crashing the project by 2 months	80,000
Net savings	$20,000

The resulting precedence diagram illustrating these effects is shown in Figure 7.5. Note that *ACEF* remains the critical path. Only activity *C* is left to crash. However, crashing *C* by 1 month results in *no net savings*, since we actually incur an increased cost of $50,000 by crashing *C*. As Figure 7.6 illustrates, after crashing *C* by 1 month (i.e., from 5 to 4 months), all three paths become critical.

Keeping in line with her plans to reduce costs, the project manager decides to continue crashing the project further, although this requires that activities *B*, *C*, and *D* each be reduced by 1 month which results in a net loss.

Crashing would involve the following costs:

Activity	Added Cost
B	$20,000
C	50,000
D	10,000
Total increased cost due to crashing	$80,000

We would gain $50,000 and therefore lose $30,000 by crashing activities B, C, and D by an additional month.

The final cost-time curve is shown in Figure 7.7. As a result of crashing the project, the project manager selected to crash activities *ACE* each by 1 month and activity *F* by 2 months. In this way the project may be completed in 8 months. The total increase in cost due to project delays and crashing costs was $225,000 rather than the $300,000, which would have resulted from not crashing the project.

7.3.2 Flow Diagram for Crashing a Project

A flow diagram that can be used to crash a project is shown in Figure 7.8. Begin by listing each activity, its normal duration, the additional cost-unit time incurred by crashing an activity, and the total time that an

232

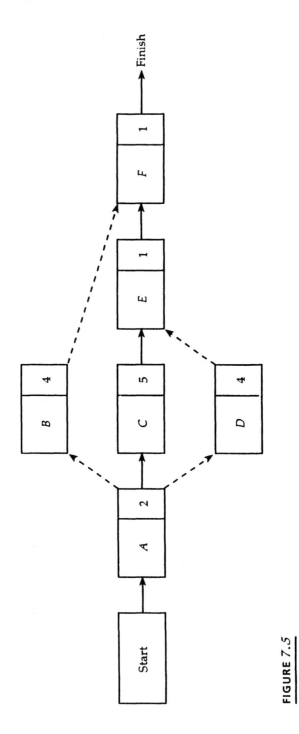

FIGURE 7.5

Precedence diagram after crashing *F* by 2 months.

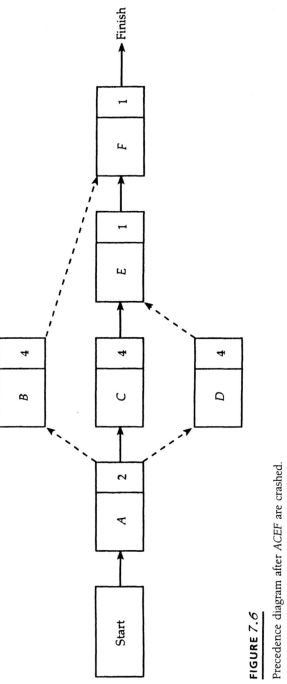

FIGURE 7.6

Precedence diagram after *ACEF* are crashed.

233

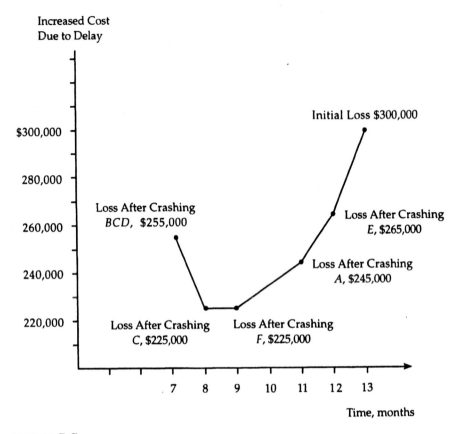

FIGURE 7.7

Cost–time curve showing that by crashing the project the increased project cost can be reduced from $300,000 to $225,000.

activity can be crashed. Next, the critical path or paths is (are) found. Then determine the activities on the critical path and locate the activities that offer the most favorable cost–time trade-off. The activity (or activities) selected is (are) then crashed by 1 unit of time, represented by the number of days, weeks, or months used in the scheduling process. Then plot a cost–time curve (similar to Figure 7.7) so that a business decision can be made regarding to what degree the project should be shortened.

Keep in mind that crashing an R&D program that has inherent uncertainty in each activity may actually extend the total project duration rather than decrease it. This is because after crashing, the project may have multiple critical paths. Therefore, the project team must proceed cautiously

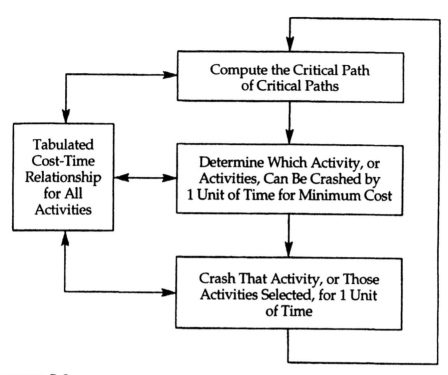

FIGURE *7.8*

Flow diagram for crashing a project.

and crash only activities that actually advance progress toward meeting or advancing the required project end date.

7.4 CONCLUSIONS

The rapid pace of changing technology and the ever-increasing market demands of the twenty-first century will require project team members collectively to baseline and optimize the project plan and schedule. This will also require the project manager to have the leadership skills to get the project participants' collective commitment to manage shifting requirements and accelerating project timelines to satisfy customer expectations. In Chapters 8 and 9 we address the role of project manager as team leader to help future and current managers develop the leadership and conflict management skills needed to meet these challenges. Then, in Chapter 10, we discuss processes to track and control actual project performance.

EXERCISES

7.1. Which of the following statements pertaining to how projects assess risk is not true?

(a) Risk and uncertainty is related directly to the difference between the pessimistic and optimistic task completion estimates generated by project specialists.

(b) Project management may chose to ignore the risk associated with not completing project activities if the impact in terms of overall schedule or performance delay is relatively low.

(c) Risk can positively affect a project by introducing new opportunities that were not considered previously.

(d) Project management can exercise little control over events that might cause slippage or affect a project schedule adversely.

7.2. Budgeted contingencies to avoid cost overruns can be determined by:

(a) Applying a standard allowance.

(b) Using past experience.

(c) Calculating the monetary value or impact of the sum total of the expected values of the most probable risk items.

(d) All of the above.

7.3. Explain how using a range of duration-based estimates, particularly in an engineering development project, helps a project team to highlight for senior management the uncertainty associated with preliminary schedules.

7.4. There are several advantages to measuring progress toward completing a baseline schedule by asking team members to report the time remaining to complete individual tasks. Explain why this practice can help optimize a team's chances for success.

7.5. Refer to Table 7.2 and change the precedence relationships for each activity to those given below. Assuming that all other information is the same, obtain a *crash* curve similar to Figure 7.7.

Activity	Precedence	Activity	Precedence
A	—	D	B
B	A	E	C,D
C	A	F	B,E

CASE STUDY: THE ÉLAN PROJECT[1]

Mary Catherine Raute's dream was about to be realized. As head of Softech's new venture department, her team of crackerjack programmers was about to embark upon an adventure she had started half a decade ago. The Élan project was one of the newest in Softech's new consumer and education division. The venture also promised to advance the software company into the home entertainment, educational reference, and corporate training markets — a leap that forecast positive returns in less than three years with improved market share in these segments for the enterprise's stable operating system.

The venture was gearing up to achieve results by the all-important, end-of-year holiday season for the consumer market. The educational on-line system would be targeted to debut six months earlier at the industry's largest training and development trade show. Reviews, it was anticipated, would favorably predispose professional multimedia designers to choosing the system over competitive, but less sophisticated, offers already in the market. Financing for the project was therefore to include enough capital for equipment and specialized staff to test an early prototype with potential customers and to finalize the user interface prior to coding the system. A core product development team coached by Mary Catherine would manage the project. Together they would present their findings to the vice president overseeing the venture.

Project Background

It was more than half a decade since Mary Catherine had managed the forerunner of the Élan Multimedia Training System. Initially, the computer-based authoring program had been developed for internal release by Softech's parent, a corporate entity with offices dispersed throughout several continents whose trainers and documentors needed a hand in the difficult task of writing on-line training lessons. Unfortunately, the full potential of this early software system had never been exploited, primarily because it lacked several critical components — graphics, sound, still pictures, and full-motion video — which could not be delivered to the desktop over the infrastructure available at that time.

Over the years, requests for system enhancements were prioritized and presented to senior management from more than half a dozen internal

---

*This study was written for class discussion purposes only. It is not intended to reflect management practices or products of any particular company organization.

organizations, which stood to benefit from financing their development. Among them were Softech's Federal Systems Group and Corporate Education and Training Center. But until the company's foray into the home entertainment and education markets, Mary Catherine's business plan received little enthusiasm.

The needs in the industrial market for on-demand interactive training had changed, fueled by the desire to provide support tools to help employees in key service positions perform their jobs. The fast-paced product delivery cycles at the turn of the century now required that employees be equipped with portable communications terminals. To perform their jobs competitively, service technicians needed immediate access to information in the field and assistance on an "as needed basis" outside the classroom setting.

The growing demand to provide educational materials over the Internet was also rapidly changing the global education market affecting how students would learn at school and at home. For the first time, children truly might be able to progress through a lesson at their own pace and be tutored in an environment designed to address unique or special needs. At home, the same technology was rapidly spanning an embedded base of networked and accessibly distributed computers that could be tapped to display educational games, play movies, and so on.

The potential opportunity to sell educational software written for these markets was tremendous. To become a major player, Softech needed to enhance the capacity of its existing training system to use the latest multimedia technology, then arrange to license it to educational writers and university consortia, paying royalties on the educational titles released in partnership. The whirlwind pace of this burgeoning industry was only hinted at when Mary Catherine conceived her first business plan so many years ago. A turning point in her own career, today she assembled the new members of the project team who would take the next step toward realizing the reborn system.

Organizing the Project

Mary Catherine's team included members from the software engineering group, marketing, user interface development, and a project management specialist who would oversee the preparation of the initial WBS for the preliminary project plan and also coordinate the efforts of the intimate group as they rallied together to accomplish it. She enthusiastically gathered her team together for an afternoon meeting in her office, hoping to inspire and motivate them as she shared her vision for the Élan project's charter.

The team was hand picked. The lead software engineer had been instrumental in advancing the earlier, in-house version of the computer-based training system, which had been used throughout the company to provide training to new hires on the flagship operating system. Tom Woo, in fact, had received a U.S. government patent for his programming routine that permitted multitasking of simultaneous users and training authors to the same software source during lesson preparation. Judy Nolan, market development manager, had past experience managing the parent company's software product offers. She had personal reasons for joining the new venture team, having assisted Mary Catherine in writing several versions of the new venture business plan for the family of interrelated software releases with which the Élan project would debut in the market. Brian Katz was enthusiastic about researching the system's user interface design. He had vast experience working with competitive on-line authoring systems in academia and joined Mary Catherine's new venture group after spending several years writing a professional standard or body of knowledge for the multimedia industry's largest instructional systems design association. Beth Everett, Élan's project manager, had recently received her certification as a professional project management specialist and was concentrating specifically on the front end of the software development process. A second specialist might be brought onto the project to manage the R&D plan if funding permitted. Otherwise, Beth would serve the team in this capacity as well.

Mary Catherine: Welcome, welcome to the team! What a great opportunity we have before us to enhance the marketability of Softech's flagship operating system by providing a much-needed application for the consumer and education segments. Over the next few weeks, we face the challenge of researching the concept of offering the education, industrial training, and home entertainment markets with an innovative solution that supports the needs of on-line, multimedia program authors. At the quarterly budget committee meeting about 10 weeks from now, the venture will be expecting us to present a high-level design of our proposed solution for these segments. We have initial funding to explore the product concept before a definite decision is made to bring the proposed solution to market. To start, I recommend that we engage R&D — Tom's design group — to produce the application's architecture. I believe we should take the concept as far as having the systems engineers team with Brian's human factors group to market test some preliminary user interface designs with potential users. Besides making the user interface almost foolproof, we'll create a strong market demand by getting future customers to participate in the testing of the product. They'll be anxious, in fact, to share their likes and dislikes with product development. Call it *relationship selling*— we'll guarantee the future, first sales next year when the product's out in the market. The feature list should be right on target, too!

Tom: I guess you'll be looking for a preliminary set of R&D estimates fairly soon, then. When, Judy, will you have available a prioritized list of what customers say they want so that the systems engineers can determine what needs to go into a release?

Judy: First, my team must review the critical assumptions about market size and take rates and learn what risks are associated with this project, so we can update the business plan. As we uncover any new information about the different segments' needs, we'll share that with your team.

Beth: To manage this project and secure the venture committee's approval, I'm going to need to put together a fairly comprehensive project plan. May I ask your developers, Tom, to firm up those estimates as soon as Judy gets you the requirements list? I've been reading that it's best to ask the programmers for a range of estimates for each proposed feature. This way, we can learn the risks associated with meeting these estimates — the technical issues as well as any related to specific resources that we anticipate needing the purchasing organization to procure. Do you think we'll need to order any equipment to get the job done?

Mary Catherine: You have my go-ahead to request funds for what you need to order now. We don't want bottlenecks impeding progress. Just remember to include any capital expenses in the cost estimates that go into Judy's business plan. I also caution you, Tom, to have your programmers give realistic estimates and a confidence indication around their proposed delivery dates. At this point, if we're successful at understanding the magnitude of their work, we can cover these risks through contingency schedules. The venture wants us to avoid committing prematurely to meet unrealistic constraints before we really understand how complex developing the new system is going to be.

Beth: So what's the next step, Mary Catherine? Would you like weekly progress reports from us until we meet again?

Mary Catherine: Excellent suggestion, Beth. Let's keep the momentum going and schedule a meeting two weeks from now. I'll want a brief update on your progress so that we can prepare for the meeting with the venture committee. Finally, the time is right to leap ahead into the multimedia information age. We're about to embark on a great adventure that will be most profitable. Good luck; Softech is counting on us!

Passing the First Quality Gate

Since funding had been allocated to Mary Catherine's concept evaluation team to explore the product feasibility and application architecture, the team immediately began to market-test some of the researcher's preliminary designs with potential users. Encouraged by their findings, the team asked R&D for some initial dates to develop the proposed list of features. Judy carefully composed the list after meeting with several customer groups. Each group represented the three targeted market segments: education, home entertainment, and industrial training. Customer interviews were also conducted in key regions around the world. It seemed that the team had gathered enough input from the market to prioritize the list of customer-requested features so that Tom's group could design an early prototype of the authoring system and begin testing it with customers. At this rate the team would be able to present to the new venture committee 10 weeks from now a positive business assessment of the concept along with a project schedule that would meet Softech's objectives.

As Mary Catherine had recommended, Tom and Beth reviewed this list of proposed features with all of the product development organizations involved. Before asking each of these groups how long it would take to complete their associated deliverables, Beth and Tom used this information to prepare a five-level work breakdown structure for the software development portion of the project. The documentation and training organization found this information extremely helpful in understanding the tasks performed by the software engineers.

Once the venture committee reviewed the team's product concept and approved additional funding for the project, R&D would need to resize the software development effort. This would let the entire project team prepare more complete work packages and a sufficiently detailed requirements document. The requirements would be accepted by all members of the team and effectively frozen to permit the test team to write an assessment plan that could be used to test the software. Each team anticipated the activities they needed to include in the plan and made their estimates as accurate as possible, adding enough detail to the draft to summarize all project activities. The finally agreed upon estimates for the project are shown in Table I.

Questions

As a member of Mary Catherine's concept evaluation team, use the estimates from each of the Softech teams given in Table I to accomplish the

TABLE I   PROJECT ESTIMATES

Task ID	Task	Planned Duration (weeks)	Staff	Uncertainty (standard deviation, weeks)	Precedence
1.	Concept definition				
A.	Business case	4	Judy Nolan	$\frac{1}{4}$	—
B.	Preliminary project plan	4	Beth Evert	$\frac{1}{4}$	—
2.	Logical design				
C.	Authoring subsystem architecture	4	Systems engineers (3)	$\frac{1}{2}$	A, B
D.	Delivery subsystem architecture	4	Systems engineers (3)	$\frac{1}{2}$	A, B
E.	Communications subsystem architecture	4	Systems engineers (3)	$\frac{1}{2}$	A, B
F.	System requirements	12	Systems engineers (3)	$\frac{1}{2}$	C, D, E
3.	Detailed design				
G.	Lesson and test construction modules	6	Software engineers (9), human factors engineers (3)	1	C, F
H.	Answer judging and administration modules	6	Software engineers (9), human factors engineers (3)	1	D, F
I.	Student registration and report-writing modules	6	Software engineers (9), human factors engineers (3)	1	D, F
J.	Hot link administration, user interface, and on-line help modules	9	Software engineers (9), human factors engineers (3)	1	E, F
4.	Coding				
K.	Lesson and test construction modules	6	Software engineers (9)	1	G
L.	Answer judging and administration, modules	6	Software engineers (9)	1	H
M.	Student registration and report writing modules	6	Software engineers (9)	1	I
N.	Hot link administration, user interface, and on-line help modules	9	Software engineers (9)	1	J
5.	Debug				
O.	Lesson and test construction modules	6	Software engineers (9)	$\frac{1}{3}$	K
P.	Answer judging and administration modules	6	Software engineers (9)	$\frac{1}{3}$	L
Q.	Student registration and report writing modules	6	Software engineers (9)	$\frac{1}{3}$	M

242

TABLE I *(Continued)*

Task ID	Task	Planned Duration (weeks)	Staff	Uncertainty (standard deviation, weeks)	Precedence
R.	Hot link administration, user interface, and on-line help modules	9	Software engineers (9)	$\frac{1}{3}$	N
6.	Unit test				
S.	Lesson and test construction modules	6	Software engineers (9)	$\frac{1}{3}$	O
T.	Answer judging and administration modules	6	Software engineers (9)	$\frac{1}{3}$	P
U.	Student registration and report writing modules	6	Software engineers (9)	$\frac{1}{3}$	Q
V.	Hot link administration, user interface, and on-line help modules	9	Software engineers (9)	$\frac{1}{3}$	R
7.	Systems integration				
W.	System test plan	3	System testers (3)	$\frac{1}{3}$	F
X.	Systems verification	9	System testers (3)	$\frac{1}{3}$	S, T, U, V, W
8.	Beta acceptance test				
Y.	Qualification site plan	2	Tom Woo, Brian Katz, Beth Everett, and Judy Nolan	$\frac{1}{4}$	B
Z.	Qualification test	4	Judy Nolan	$\frac{1}{2}$	X, Y
9.	Marketing and sales				
AA.	Marketing plan	4	Marketing (2)	$\frac{1}{4}$	A, B
BB.	Sales forecasts	4	Marketing (2)	$\frac{1}{4}$	A
CC.	Sales materials	12	Marketing (2)	$\frac{1}{4}$	K, L, M, N
DD.	Sales training available to distribution center	8	Trainer	$\frac{1}{4}$	CC, Z
EE.	Announcement materials	4	Marketing (2)	$\frac{1}{4}$	AA, BB
FF.	International promotional event	2	Marketing (2)		CC, EE
10.	Production				
GG.	Software replication	3	Replication engineer, system tester	$\frac{1}{3}$	Z
HH.	Inventory available	1	Replication engineer		GG
11.	Distribution				
II.	Stock warehouse	2	Distributor	$\frac{1}{4}$	HH
JJ.	Shipment procedures	2	Distributor		AA, BB
KK.	Ship intervals	2	Distributor		AA, BB

following.

1. Draw a precedence diagram.
2. Determine the critical path.
3. Draw a bar chart using the *most likely* and *best start* estimates.
4. With 90% confidence, how long will the project last?
5. At what confidence level will the project last no longer than 12 months?

Heading Off Unanticipated Delay

Four weeks after the end of the design phase just after baselining the system requirements, Judy learned of a new but previously unanticipated request for Softech to offer the product in at least one other language besides English. In fact, Judy was strongly convinced by the market reports that Softech's market share in all three identified market segments would increase significantly were the company to offer the multimedia training system in multiple languages.

First, Judy discussed the probable delay undertaking this new work could have on the project with Beth and Tom. They insisted, however, that she consult Brian prior to trying to estimate how long the new activity might take. It appeared that the user interface would need to be translated along with all the on-line help screens. Accommodating the different languages on the screen would probably change or even cause their current designs to be scrapped. Fortunately, the programmers had not yet begun to code this portion of the current interface. Then, depending on where the translations would be staged, Beth was aware that coordination across continents might add an element of risk to the already tight schedule and perhaps even jeopardize the anticipated user trails planned to take place only a few weeks before the product's trade show debut.

Translating the interface and end-user documentation would require the developers to attend meetings outside normal working hours so that they'd be able to communicate during different time zones. Project documentation would need to be posted electronically and special requirements would be placed on the configuration system for managing the distributed development environment. Incremental delivery of each language interface would need to be planned with system test groups collocated at the translation sites. Tom would require these organizations to oversee the testing of each interface and the on-line help messages supported in the test scenarios.

To assure meeting their initial commitment, the team met to calculate the probable impact on the schedule. They decided to ask Mary Catherine

TABLE II   COST DATA FOR MULTILINGUAL VERSIONS OF THE
TRAINING SYSTEM

Activity	Normal Time	Cost	Crash Time	Cost	Additional (Crash) Cost/Week
1. English language version					
a. Install user interface	1	$ 2,250	0.5	$ 2,250	—
b. Test user interface	2	9,000	1.5	10,125	$1,125
c. Install on-line help	1	2,250	0.5	2,250	—
d. Test on-line help	2	9,000	1.5	10,125	1,125
e. Install final system	1	2,250	0.5	2,250	—
f. Integrate final system	2	9,000	1.5	10,125	1,125
2. German language version					
g. Install user interface	1	3,375	0.5	3,375	—
h. Test user interface	4	27,000	2	33,750	6,750
i. Install on-line help	1	3,375	0.5	3,375	—
j. Test on-line help	4	27,000	2	33,750	6,750
k. Install final system	1	3,375	0.5	3,375	—
l. Integrate final system	4	27,000	2	33,750	6,750
3. Spanish language version					
m. Install user interface	1	2,700	0.5	2,700	—
n. Test user interface	4	21,600	2	24,300	2,700
o. Install on-line help	1	2,700	0.5	2,700	—
p. Test on-line help	4	21,600	2	24,300	2,700
q. Install final system	1	2,700	0.5	2,700	—
r. Integrate final system	4	21,600	2	24,300	2,700
4. French language version					
s. Install user interface	1	3,375	0.5	3,375	—
t. Test user interface	4	27,000	2	33,750	6,750
u. Install on-line help	1	3,375	0.5	3,375	—
v. Test on-line help	4	27,000	2	33,750	6,750
w. Install final system	1	3,375	0.5	3,375	—
x. Integrate final system	4	27,000	2	33,750	6,750

to secure funds from the venture committee to expedite a contingency plan to bring additional programmers on board and assign them to feature teams so that the developers, translators, and testers could proceed in parallel. Despite their careful attention to preserving the developers' initial estimates, it became clear very quickly that they'd need to crash the final testing phase of the project for the team to complete on time. Otherwise, the product could not be distributed to the market in time for authors using the system to produce titles for sale during the holiday season. Also, as the vice president had insisted, it was imperative that Softech demonstrate the multimedia training system at the targeted international trade show so widely attended that 50% of all future sales for the academic, industrial, and entertainment markets could be made there.

So Mary Catherine agreed to proceed with optimizing the project's resource assignments. She was certain that the venture committee would see the value of the time–cost trade-off exercise and absorb some extra expense to bring the multimedia offer to market on time. Her team's recommendations to proceed with the exercise would most assuredly be a contributing factor to the venture's achievement of its business plan for the year and its strategy of penetrating the home computer market.

Questions

The costs associated with testing the English, French, German, and Spanish language versions of the multimedia training system by the R&D team are listed in Table II.

6. Assume that there are multiple system test groups assigned to the project, working in parallel at various Softech facilities in the United States and at its international headquarters in Brussels. Determine the least-cost schedules for completing all the testing activities.

7. Suppose your calculations indicate that there will be additional costs for finishing the testing in less than either a 9- or 15-week interval. If Mary Catherine's team must deliver the system in time to demonstrate it at the international trade show six weeks from now, what factors would you discuss with the team before deciding to expedite the tests?

8. Marketing has stated that Softech stands to lose $20,000 for every French language version of the system not sold at the trade show. What is the minimum amount of additional funding that you should request to complete the testing activities in time to demonstrate the French language version at the trade show?

APPENDIX 7.1: CRITERIA FOR CLASSIFYING RISKS AS HIGH, MEDIUM, OR LOW

Risk	Cost	Schedule	Technical	Market
High	• Estimated software budget > $5 million[a] • Estimated hardware budget > 10 million[a] • Requires outside resources of uncertain cost • Lack of definition prevents estimating • Lack of process/tools prevents estimating • Forecast for production likely to be missed by > 50%[a]	• Requires extensive research and laboratory studies • Requires extensive testing and evaluation • Requires unfamiliar tools and new designs • Requires new production processes and manufacturing lines • Requires new facilities to be defined, located, or constructed • Requires new personnel to be recruited • Productivity of recruited personnel questionable • Contracts new vendors/suppliers not used previously • Multiple vendors/suppliers not available • Extensive dependencies on other development projects/other organizations • Lack of support from one or more key organizations • Excessive changes in specifications/user requirements	• Technology beyond state of the art • Research required • No in-house experience • Poor manufacturing technology • No field experience • Availability of property trained personnel questionable • Quality process assurance and defect removal testing questionable	• Core component of business strategy • Estimate gross margin at launch < 20%[a] • New product type to new market • Will lag behind competition by > 3 months • Competitors have stronger market position

247

APPENDIX 7.1 (Continued)

Risk	Cost	Schedule	Technical	Market
Medium	• Estimated software budget $1–5 million • Estimated hardware budget $2–10 million • Estimates are based on extrapolations • Forecasts for production likely to be missed by 20–50%[a]	• Development and associated production process currently started • Previously achieved schedule improvements planned • Reusable/adaptable software, hardware readily available • Some facilities to be modified • Vendors available but not used previously • Few dependencies on other development projects/other organizations • Few changes in specifications/user requirements	• Technology at state of the art • Research in process • In-house experience available • Manufacturing technology being implemented or planned • Some prior field experience • Properly trained personnel available • Quality process assurance and defect removal testing planned	• Moderately important component of business strategy • Estimated gross margin at launch between 20 and 25%[a] • Current product type to current market • Competitors will introduce competitive product within 3 months[a] • Considered major competitor in product category
Low	• Estimated software budget < $1 million[a] • Estimated hardware budget < $2 million[a] • Estimates are based on past history on similar jobs • Under cost on similar programs • Forecasts for production likely to be missed by < 20%[a] • Firm quotations from all suppliers • Multiple sources exist for all supplies	• Past history on similar projects • Previously performed similar tasks within or below projected schedule • Existing software and hardware readily available • Facilities/capacity committed • Previous vendor histories satisfactory • No dependencies on other development projects/other organizations • Firm specifications/user requirements	• Applicable research complete • Technology demonstrated • Extensive experience • Prototype developed • Proven manufacturing technology • Extensive field experience • Properly trained personnel available • Quality process assurance and defect removal testing in place	• Gross margin at launch > 25%[a] • Current product type to current market • Competitors will introduce competitive product > 3 months[a] • Considered market leader in product category

[a]Data from Booz Allen Hamilton, Inc.

8

LEADERSHIP IN A PROJECT ENVIRONMENT

When the best leader's work is done, the people say, "We did it ourselves!".

OLD CHINESE PROVERB

8.1 INTRODUCTION

Every thirty to forty years, significant new organizational concepts and practices emerge. Alfred P. Sloan developed the concept of decentralized operation with centralized control at General Motors in 1920. Thirty to forty years later, major defense and space programs led to highly structured project organizations and to the application of the matrix concept. Now, an additional thirty some odd years later, we have witnessed the birth of streamlined, downsized organizational models. Entire levels of managers have disappeared, to be replaced by horizontal, interdisciplinary project teams.[1]

The organizations of today face unprecedented challenges. These challenges hold important implications for managers and specialists alike. Rigorous competition, changing market environments, and dynamic technologies all contribute to shorter product life cycles. These shorter life cycles, in turn, place even greater demands on project teams and call into

---

[1]R. G. Donnelly and D. S. Kezsbom, "Overcoming the Responsibility–Authority Gap: An Investigation of Effective Project Team Leadership for a New Decade," *1993 AACE Transactions*, pp. Q2.1–Q2.12.

249

question the continued efficacy of certain long-established methods of leading and managing them. As firms respond to the new business environment by streamlining and creating flatter, more flexible organizations that emphasize quality, speed to market, and cost efficiency, managers who remain will need to enhance and recast those leadership skills that had proved successful in the more vertical organization of merely a decade ago.

Working with and through others to accomplish organizational objectives is the cornerstone of any managerial position. There is a major difference, however, between the degree of control held by the traditional line manager and that available to the project manager. The difference lies in the degree of formal authority conferred on the two managers. In more traditional organizations, manager's schedule and control work, evaluate, reward, and discipline employees, and hire and fire staff, as may be required. In many cases, traditional managers have been solely responsible for the quality of the product or service delivered to the customer.

There are fewer examples of managers exercising this degree of control in organizations that are project team-based. Project team managers typically operate in a complex, multidisciplinary setting, and possess little "right to command." Lacking the traditional boss–subordinate relationships, a project manager's success is contingent upon effectively managing and influencing the diverse personnel responsible for designing and implementing project components. Project managers must, in fact, often rely on developing informal modes of authority through a variety of influence techniques.

Effective management of project teams demands skills and qualities of leadership that differ from those of the traditional manager. Project managers frequently cross company and industry lines to achieve total project integration and successful implementation. Project leadership involves understanding and working within the physical and political boundaries not only of the organization, but of the global marketplace as well. It also requires a firm appreciation of the diverse needs of the professionals who are part of the project team.

In this chapter we examine the leadership role of the project manager in the dynamic project environment. We present the leadership characteristics or qualities that enable a person to be an effective leader in the project organizations that characterize contemporary firms. We further explore the leadership skills and techniques through which project managers may develop the support and commitment necessary from the people needed to get the job done.

8.2 THE PRACTICE OF PROJECT LEADERSHIP

Leadership has attracted much attention over the years, as evidenced by numerous research investigations. In much of the research there has been an emphasis on the in-born characteristics or *traits* that successful leaders appear to possess. Over time, however, research efforts have focused more on determining effective leader behaviors, sometimes referred to as *leadership style*. The duality foundation of leadership qualities and skills leads to the question of whether leadership can be taught. Although some astute managers contend that leaders are born and not made, many psychologists and sociologists see it quite differently. They assert that leadership emerges from the demands and needs of the group and that everyone within the group has at least the potential to lead.

We may address this issue, in part, by using an appropriate and practical definition of leadership. Especially within a project setting, we believe that this is a useful definition of *leadership*:

- Leadership is a social influence process in which the leader seeks the participation of individuals in an effort to obtain organizational objectives.[2]

To encourage such participation, effective leaders utilize a variety of practices. Some of these practices depend on innate talents; others depend on more learned behaviors. Naturally, some of these leadership practices have greater impact than others, depending on the situation.

Today's business environment calls for strong leadership skills. Corporations are undergoing fundamental changes in organizational foundations while often simultaneously experiencing rapid technological changes. Organizationally, firms are taking steps to reduce operating costs and remain lean, placing greater reliance on cross-functional teams. At the same time that these organizational changes have been occurring, firms have been attempting to gain the competitive advantage by developing and introducing new products, processes, and services. Together, these trends require increased emphasis on team leadership and on supporting the transition from traditional leader of the past to the progressive leader required by project teams now and in the near future.

---

[2]D. S. Kezsbom and R. G. Donnelly, "Managing the Project Organization of the Nineties: A Survey of Practical Qualities of Effective Project Leadership," *1992 Project Management Institute Proceedings*, pp. 415–421.

8.3 THE ROLE OF THE PROJECT MANAGER

Historically, the project manager's role grew out of the need for someone to supervise, coordinate, and engineer work related specifically to a project. The project manager's job was to plan, control, organize, and direct the work of several individuals so that the project could succeed. In some cases, teamwork might be required; in others, it may not. Frequently, team members were delegated a project-related job that required minimal interfacing with others. Like the job of a systems engineer, the project manager was required to "engineer" the pieces of the project puzzle into a coherent whole.

Several old paradigms supported this past concept of the project manager:

- The project manager is the technical expert.
- The project manager is delegated all decision-making authority and must therefore make all the decisions.
- The project manager defines the job and how it will be done.
- The organization is structured hierarchically.
- Teams are formed only when needed.
- The focus of the organization is specialization.
- The workforce is homogeneous.

Certainly, much has occurred that illustrates the limitations of these former paradigms. With the advance of high technology, project managers can no longer be the sole experts on all project-related technologies. They must rely heavily on the technical expertise and decisions of others. The strong focus on specialization, moreover, has made it difficult to resolve disagreements involving several different functions. Division loyalty clouded judgment, and a "That's not my job" attitude prevailed. The old paradigms also placed a great deal of emphasis on doing the task without addressing the need for socialization, teamwork, or commitment.

To deal effectively with change characteristic of this new era, new paradigms are rapidly working their way into today's organizations. Companies are experimenting with these new methods associated with these paradigms with varying degrees of success. These new paradigms are:

- Employees are considered experts and possess unique knowledge and skills.
- Controls are set collectively.

- Employees participate in goal setting and define how the work should be accomplished.
- Organizations are flatter and more flexible in structure.
- The organization is customer-driven.
- The workforce is heavily diverse.
- Change is a normal process.

These new paradigms result in organizations with employees who are being educated in the processes of group dynamics and team building. They also mean the creation of a more flexible and knowledgeable workforce, which expects to be a vital part of the problem-solving, decision-making processes. This greater emphasis on teams, employee participation, and empowerment requires a new style of leadership and new leadership skills. Employee involvement requires project managers to be more facilitative and less administrative. It requires leaders to empower more and direct less. The participative environment of today requires project managers to develop a team and to use group processes to derive a more quality-oriented product.

Characteristics necessary for project managers to lead project participants today are as follows:

- Better educated
- People-oriented
- Quality conscious
- Better listeners
- Facilitators
- Skilled at group dynamics
- Understand how to coach, inspire, motivate
- No longer the "expert"
- Comfortable relying on the expertise of others
- Work across the organization to acquire resources for the team (*boundary spanner*)

8.4 PROJECT MANAGER SKILLS

Often, promotion to a project manager's position results not necessarily from one's excellence in managing people or in managing projects, but may result from one's superior performance in meeting technical demands. An excellent design engineer, software specialist, or top-notch developer frequently earns his or her first opportunity in management through the promotion to the position of project manager. Unfortunately, the skills and activities required to manage a project may be considerably more demanding than even the most traditional manager's job. In the more traditional functional organization, for instance, one is granted authority largely on the basis of one's position within the chain of command. Project managers, on the other hand, frequently find themselves placed *diagonally* within the organizational structure, somewhere outside this chain of command. They must frequently work across units, departments, divisions, and companies to ensure that project objectives are achieved. These horizontal and diagonal relationships offer the project manager little traditional or delegated authority to accomplish the work. The activities and functions necessary to manage a project, therefore, require a variety of interpersonal skills and informal sources of influence and power. In many instances, the newly appointed project manager must struggle against tough odds. Faced with little legal or formal authority and a lack of managerial experience and skills, newly delegated project mangers often find themselves in what may well be considered a "sink or swim" situation.

How can project managers make it through what appears, at times, to be an overwhelming professional challenge? First, they must recognize that how well one accomplishes project work depends on their skills in fostering project integration and managing team performance. The characteristics and skills needed for working with and through others predominates. A project manager must possess a desire to get things done with and through other people rather than despite them. Concern for project team members, an ability to integrate the personal and professional needs of project specialists, and a talent for generating enthusiasm for the work create a climate that is high in involvement, motivation, communication, and performance.

A prime concern is that project managers understand that first, last, and always they are the managers of the project. There is a strong temptation for newly appointed project managers to practice their technical discipline rather than manage the process of the project. This is one of the classic reasons for project failure. Usually in this case, not only are project planning and team building functions neglected, but the project may also suffer from a narrow or myopic technical view.

A project manager's skills and background should be varied. At best, the project manager should be a technical generalist who can relate to the greatest number of team members and attend to the broader requirements of the project. Except for extremely lean organizations, or extremely small projects, project managers must *emphasize the less technical* and deal with the *more human side* of project performance. They must demonstrate and practice a keen understanding of human behavior and deal, at times, with seemingly emotionally charged issues.

8.5 AUTHORITY AND INFLUENCE: KEYS TO EFFECTIVE PROJECT LEADERSHIP

Project managers are often part of complex multidiscipline organizational environments and have little control over contributing functional specialists. Lacking the once traditional boss–subordinate relationship, project managers must frequently rely on developing the *perception* of authority and utilize the influence strategies that one may tap from a variety of interpersonal and political sources. One of the first steps project managers must take before influencing others successfully is to examine available sources of power and influence, as they operate across the multiproject, multi-disciplined, and often multichaotic environment.

To deal effectively with the plethora of reporting relationships, lines of communication, and conflicting priorities characteristic of matrix relationships, project managers must be able to communicate and model a variety of things successfully in their environments. Particularly, they must be able to communicate and display respect for the concerns and perspectives of others and welcome and encourage views and perspectives that may differ from their own. Above all, project managers must demonstrate that they are trustworthy. Project team members will share their difficulties and problems with the project manager only if a sense of trust has been established, that is, if they truly believe that the project manager understands the problems and challenges they face. Since they lack direct authority over project personnel, project managers must give and request support not only by using logic and data, but also by applying a variety of interpersonal influence bases.

8.6 AUTHORITY, RESPONSIBILITY, AND ACCOUNTABILITY: THE PROJECT MANAGEMENT CHALLENGE

Our traditional concepts of authority and the power it generates, stem from the following definitions:

- *Authority*. Traditionally, this term refers to the organizationally sanctioned right to make the necessary decisions that others are required to follow. This includes issuing orders, determining directives, outlining missions, setting or defining priorities, and scheduling target dates.

· *Responsibility*. We generally think of responsibility as consisting of tasks, obligations, or activities that result from the specific assignments, delegated in accordance with formal organizational roles or positions.

· *Accountability*. Once our integrity or professionalism is linked to the successful or unsuccessful completion of a task or assignment, we can consider ourselves to be accountable for the job. Most managers in a traditionally vertical organization are accountable for an employee's decisions, even though the task and the authority necessary to accomplish that task have been fully delegated.

Generally, our responsibilities are derived from our job descriptions and therefore from our formal role within the organizational structure. Although somewhat organizationally derived accountability relies more heavily on one's sense of professional integrity and liability. Both authority and responsibility flow down the chain of command. Accountability, on the other hand, is represented by the notion of "the buck stops here," and moves up the organization, landing in the lap of someone with a *formal role* or *legal authority* position. This person generally can make the necessary decisions for design improvements or changes, for example, or may be in a position to clarify organizational objectives. In the case of project team environments, the accountability for project success (or failure) typically flows horizontally as well as diagonally across the organization until it reaches the project manager. The traditional dilemma of the project manager, therefore, is to be able to work effectively within an organization that grants this position total responsibility, total accountability, and yet minimum authority to get the job done. The project manager must rely on a variety of informal sources of authority and power to manage a project effectively.

Following are some of the responsibilities and challenges of the project manager in the team-based project environment.

· Defining and negotiating the appropriate personnel for the project team

· Moving the group of project personnel toward becoming a team

· Integrating people with diverse skills, attitudes, and cultural backgrounds into a cohesive team with a unified purpose

· Juggling support department, customer, and management priorities

· Directing multifunctional work teams across organizational lines with little legal authority

· Maintaining project direction while fostering creativity

- Coordinating and integrating various task group activities into a total system
- Coping with changing technologies, requirements, priorities, and resources while maintaining project focus and integrity
- Fostering a professionally stimulating environment where people are motivated to work effectively within a stressful dynamic project environment
- Managing and dealing with power struggles and interpersonal and organizational conflict
- Maintaining a leadership position across internal and external organizational boundaries
- Dealing with technical, organizational, and cultural complexities
- Maintaining management's support and interest
- Sustaining project team performance
- Building lines of communication across organizations, states, and at times, nations
- Demonstrating and encouraging innovation and risk taking without jeopardizing the project
- Encouraging and rewarding individual and team commitment to project objectives
- Building skills within the project team
- Facilitating team problem solving and decision making
- Providing overall project leadership in an environment with complex and ambiguous reporting relationships

8.7 DEVELOPING PROJECT INFLUENCE AND AUTHORITY

To be effective in a multiproject environment, project managers and team leaders must possess skills that go beyond technical analysis and cost and schedule estimating. In fact, there have been several studies focusing on identifying several *influence bases* that project managers have available to generate support from functional managers and the specialists whom they employ. Competence, persuasive ability, negotiation, and reciprocal favors are among those "influence techniques" that have been found to be useful in overcoming what is frequently referred to as the project manager's *authority gap.*[3]

---

[3]R. M. Hodgettes, "Leadership Techniques in the Project Organization," *Academy of Management Journal*, Vol. 11, 1968, pp. 211–219.

Several investigations by Gemmill, Wilemon, and Thamhain[4,5] found that several sources of influence were available to project managers, contingent on the specific organizational form and project charter. These influence measures were rank ordered by support personnel from 1 (most important) to 9 (least important) according to their perception of the effectiveness of the influence source. These include:

Influence Source	Mean
1. *Authority:* the legitimate hierarchical right to issue orders	3.0
2. *Work challenge:* the project manager's ability to capitalize on a worker's task enjoyment	3.2
3. *Expertise:* special knowledge the project manager possesses that others deem necessary or important	3.3
4. *Future work assignments:* the project manager's perceived ability to influence a worker's future task assignments	4.6
5. *Salary:* the project manager's perceived ability to influence or increase monetary remuneration	4.6
6. *Promotion:* the project manager's perceived ability to improve a worker's position	4.8
7. *Friendship:* friendly personal relationships between the project manager and others	6.2

As may be expected, each of these seven sources of influence has a different effect on the morale and climate of the project team. In fact, it was discovered that the *less* project mangers emphasize organizationally derived influence bases, such as authority, salary, and coercion, and the *more* they relied on work challenge and expertise, the higher they were rated by project support personnel in their ability to manage projects effectively. The use of position power or authority (as in muscling or strong-arming techniques) by project managers as a means of influencing support personnel led to lower ratings with regard to overall project performance. Open communication and task involvement among project participants were, on

[4]H. J. Thamhain and G. R. Gemmill, "Influence Styles of Project Managers: Some Performance Correlates," *Academy of Management Journal*, Vol. 17, June 1974, pp. 216–224.
[5]G. Gemmill and H. Thamhain, "The Effectiveness of Different Power Styles of Project Management in Gaining Support," *IEEE Transactions on Engineering Management*, Vol. 20, May 1973, pp. 38–44.

the other hand, more positively associated with higher project performance scores.[6]

Although position authority is an important basis of influence and a "license to lead" that is critical to a project manager's potential for success, legal authority must be used judiciously and in accordance with the demands and characteristics of the particular project. Although perceived to be an important influence base, research indicates that project managers who are perceived to rely more heavily on their position or legal authority to accomplish project objectives are rated by peers and followers to be less effective in resolving problems and conflicts. They are further rated lower in their overall project performance.

These investigations in no way deny the importance of the need for project managers to receive as much legal authority as the organization allows. Once legal authority is available, however, a project manager is best served by developing *expertise, interpersonal skills, and work challenge.* Moreover, although many continue to regard technical expertise as an important influence base, it is an important source of organizational power and influence if it is recognized and sought after by project personnel. Technical expertise, is therefore effective only if it is associated with strong lateral and diagonal communication and a wide spectrum of well-developed personal, professional, and organizational relationships.

8.8 UNDERSTANDING LEADERSHIP IN THE TOTAL QUALITY PROJECT ENVIRONMENT

The total quality movement, with its emphasis on continuous experimentation and feedback, has been the catalyst for the transition from the adaptive or coping organization of yesterday, to the generative or learning organization of tomorrow. These contemporary organizations emphasize creating and innovating and require new ways of looking at the world.

The traditional view of leaders as unique individuals who set direction, make key decisions, and energize the troops is undergoing its own transformation. This traditional view of leadership has its roots in the individualistic, nonsystemic worldview. Leaders, especially in Western cultures, are seen as heroes, who move to the foreground especially in times of crisis. But such a short-term charismatic focus fails to reinforce the

---

[6]G. Gemmill, H. Thamhaim, and D. L. Wilemon, "The Power Spectrum in Project Management," *Sloan Management Review*, Vol. 20, May 1973, pp. 15–25.

subtler, systemic, more collective approaches required by leaders if they are to create the more innovative, generative organizations of the future.

Innovative organizations require new techniques in understanding customers and understanding the business. Building relationships with suppliers and designing business processes to be more *integrated* or *systemic* reduces costs and builds customer loyalty.[7] Increasingly, we find that creating innovative, risk-taking organizations with the creative responses these organizations require demands a more systemic view of the source of the organization's difficulties and challenges. Without such a view, we rely more on addressing symptoms than on eliminating the problems.

8.8.1   New Work for Today's Leaders

The field of management is undergoing a fundamental shift. This shift is reflected by corporate transformations away from the traditional hierarchical structure, toward the full participation by every employee in focusing on customer needs and providing products and services that meet those needs reliably. Traditional organizations were designed to manage machine-based technologies. This required the need for stable and efficient use of resources for mass production. Organizations today, however, are knowledge-based. They are designed to handle new ideas and information, with each employee becoming an expert in one or several tasks. Rather than strive for efficiency, each employee in today's information, knowledge-based organizations must learn continuously and be able to identify and solve problems in his or her specialty or domain.

In many industries today, the ability to learn and change faster than one's competitors may be the only sustainable competitive advantage. This is a primary reason why so many organizations are redesigning themselves toward what is referred to as the *learning organization*. The learning organization is an attitude or a philosophy about what an organization is and the role of employees in developing organizational capabilities.

Traditional ways of gaining the competitive advantage have been through financial, marketing, and technological capabilities. These are, in fact, the traditional competitive capabilities taught in most business schools. But in a business world shifting rapidly from machines to ideas, these traditional capabilities require a learning capability that will take the organization to a level of efficiency simply by engaging employees in the problem-solving techniques that help the organization to realize change. This learning

---

[7]G. Stalk, Jr., "Time: The Next Source of Competitive Advantage," *Harvard Business Review*, July–August 1988, pp. 41–51.

capability supplements book knowledge with that of trial-and-error problem solving and experimentation that is directed at addressing customer needs.

The traditional authoritarian image of the leader as "the boss, calling all the shots," has been recognized as oversimplified and inadequate for some time now. In today's innovative competitive organization, the critical roles of leadership include that of designer, teacher, and facilitator.

8.8.2 Need for New Leadership Skills

The leadership roles required for competitive project organizations require new leadership *skills*. Three critical skill areas include the following:[8]

1. *Building a shared vision.* When a greater number of people share a vision, the vision becomes more of a reality that people believe they may achieve. The skills involved in building shared vision include:

- *Encouraging participation to create shared vision.* People care more about the vision when they are personally involved in creating that vision! Shared visions actually emerge from personal visions.

- *Communicating and sharing.* Leaders must be willing to share their own vision and be prepared to ask: Is this vision worthy of your commitment? This is a dramatic change for one accustomed to setting goals and presuming compliance will follow.

- *Continual visioning.* Like the process of planning, building a shared vision is a never-ending process. Images evolve, and at any one point in the future, the vision may change. Many managers today want to dispense with the "vision business" by writing and solidifying the official vision (mission) statement. In a world of intense global competition, innovation and freshness are assets for any organization and may actually be the basis of their survival. Little can compete with the enthusiasm and genuine vision that comes from people by asking: What do we really want to achieve?

- *Creating positive visions.* Two fundamental sources of energy can motivate organizations: fear and aspiration. Fear is the energy source behind negative images and can produce extraordinary changes within a short period. Aspiration, however, is the energy behind positive images, and endures as a continuing source of learning and growth.

---

[8]Peter Senge, *The Fifth Discipline: The Art and Practice of the Learning Organization* (New York: Doubleday, 1990).

2. *Surfacing assumptions.* Some of the best ideas within an organization never get put into practice because they contradict the expected paradigm. One leadership task for today and the future is to challenge assumptions without invoking defensiveness. This requires a tremendous degree of reflection and inquiry skills that few leaders, unfortunately, possess. Some of the skills one can develop to help others surface and test their assumptions are:

- *Identifying leaps of abstraction.* Most of our mind and thought processes move at tremendous speed. As a result, we tend to leap to generalizations and confuse our generalizations with observable data. We project our own beliefs onto the situation, failing then to determine alternative, perhaps more workable strategies than those related directly to our assumption.

- *Balancing inquiry with advocacy.* Managers have learned to argue their case well so as to influence others to see their points of view. Advocacy skills can become counterproductive, however, as managers rise in responsibility and confront increasingly complex issues. These new complex issues require collaboration and input from others who are knowledgeable. Leaders today not only need to advocate their own point of view, but must encourage others to test their hypotheses by questioning the data on which they are based, or the holes in their reasoning. In inquiring into others' views, leaders need to seek to understand others' opinions rather than simply restate their own, and clarify how they differ from those of the others.

- *Defusing defensive behavior or routines.* One of the greatest challenges of leadership in an innovative organization is to recognize the operation of defensive behaviors. These defensive routines generally are used to protect us from the possibility of threat or embarrassment. Defusing defensiveness is accomplished by *self-disclosure.* The leader simply discusses his or her own defensiveness in a forthright manner.

3. *Encouraging systems thinking.* One of the most significant milestones in contemporary management science is the growth of managerial systems thinking as a field of study and practice. This field suggests some key skills for future leaders:

- *Seeing interrelationships and processes.* Most of us are conditioned to see the world by focusing on "things" and static images. This confines our analyses to more linear two-dimensional management systems. Each event is part of a holistic process and cannot be seen as an isolated event, to be dealt with separately.

- *Moving beyond blame.* Although we tend to blame each other or outside circumstances for our problems, it is usually poorly designed systems that cause organizational problems. Nonlinear systems thinking brings us to the realization that there is no outside; that you and the cause of your problem are part of the same system.
- *Avoiding symptomatic solutions.* The pressures to create a fix, when things go awry in a management system or process can be tremendous. Linear thinking tends to focus on systems rather than on underlying causes. The solution is too often a short-term fix, only to create more difficulties in the future. One of the most difficult leadership acts is to refrain from intervening through popular fixes, and to keep the pressure on the team to identify more enduring solutions.

8.8.3 Leader as Facilitator

In their book *Managing for Excellence*,[9] Bradford and Cohen describe how an "overresponsible" leader can diminish the effectiveness of his or her followers. According to these authors, the overcontrolling leader gets less and less from team members because the team members have not participated enough in all levels of decision making. As a result, they become less committed to the results. In turn, the overresponsible leader takes on too much ownership for what is done. In the end, group members are not fully committed to what is planned or decided. To overcome this tendency, project managers need to place high value on the knowledge and experience of all group members. They may need, at times, to behave in such a way as to be perceived as just another member of the group. At the same time, they must actively help group members to contribute and perform needed functions in the group.

For many, especially in the more scientific or technical environments, the concept of *facilitation* conjures up images of "touchy-feely" team-building sessions and the venting of frustrations. Some common *myths about* facilitation, along with how it really works, are as follows:

1. "*Facilitation takes too much time.* We won't get anything done on our tight schedule!" Many project managers harbor a fear that the team approach requires too many meetings, during which time is spent hearing everyone out and arguing every point. It is true that facilitation encourages individual expression, discussion, and disagreement. With a planned

[9]D. L. Bradford and A. R. Cohen, *Managing for Excellence: The Guide to Developing High Performance in Contemporary Organizations* (New York: Wiley, 1984).

meeting structure, however, participation can and should be highly productive. Building consensus takes more time initially, but in the long run, consensus means more support and commitment from those who make the project happen.

2. *"It won't work in this culture."* Many organizations are still not receptive to participation and involvement on the part of their employees. Determined and knowledgeable project managers can gradually introduce participation and involvement. In one-on-one meetings, project leaders can demonstrate a willingness to listen and explore the merits of an idea, without judging it. Decisions reached in group meetings can be designated "team decisions."

3. *"Facilitative leaders give up their power and control."* Actually, facilitative leaders have access to more power, because they empower others to act and therefore get power from the people they lead. Facilitative leaders do not give up control; rather, they control things (i.e., schedules, meetings) in such a way that others are enabled to produce their highest quality of work.

4. *"Facilitation encourages anarchy."* With guidance and structure provided by the facilitator, people can focus on the issues at hand and deal within the realm of their own authority. On the other hand, for facilitation to work, some decisions still need to filter down the organization from management to the workers. This is not anarchy. This is involved decision making.

5. *"Once we start a facilitative approach, we will have to do everything this way."* Even in the most participative of environments, some decisions still must be reserved for certain levels within the organization. In fact, facilitative leaders can take a strong stand about decisions that call for group participation and those that do not. When a facilitative approach works well, people generally want to use it frequently. Project managers will have to determine when participation is appropriate.

8.9 DEVELOPING A PROJECT LEADERSHIP STYLE

It is evident that project management, if practiced properly, is one of the most participative management styles in today's corporate quality environment. To be truly successful, projects must be integrated across divisional lines through group consensus-seeking activities that produce a sense of buy-in and team commitment. This can only be accomplished with the contribution of all core project disciplines to the planning, scheduling, and tracking processes. Let us not, however, confuse a *participative management philosophy* with that of a *participative leadership style*.

Quality-oriented project management strategies are built on the foundation of cross-functional organizational participation. Specific project leadership style, however, is contingent on the combination of a number of complex elements. To be participative (or democratic) at all times would be ridiculous, and certainly counterproductive. In case of a real crisis or safety hazard, for example, only a weak or ineffective leader would turn to his or her followers and ask for a vote. Project managers must develop a variety of leadership styles that allow them to respond appropriately to the changing and complex challenges of a dynamic and competitive environment. In other words, project managers need to recognize that no single leadership style will always be productive or fit the demands of all project conditions. Effective project managers must adapt their leadership style or behaviors to fit not only the life-cycle demands of the project, but the requirements and style of the client and the professional and personal needs of the project team.

The concept of developing effective leadership through the use of a variety of styles and behaviors is not a new one. In the 1950s, in his *contingency approach to leadership*, Fiedler[10] recognized that no one best leadership style existed and that different situations call for different approaches. Fielder identified variables that had tremendous affect on leader performance. They are:

1. *Position power.* This refers to the degree to which the power position enables leaders to secure compliance from their directives. This is power arising directly from organizational authority. Fielder points out that the leader with clear position power can more easily obtain followership than one who does not have such power.

2. *Task structure.* The clearer the tasks to be performed, the greater the leader's control and power, and the easier (or more favorable) the leadership task.

3. *Leader–member relations.* Fiedler regards this dimension as most important from the leader's point of view. Position power and task structure may be totally organizationally determined. The leader–member relations dimension concerns the extent to which members trust, respect, and like the leader and are therefore willing to accept his or her influence.

---

[10]F. Fiedler, "The Leadership Game: Matching Men to the Situation," *Organizational Dynamics*, Winter 1976, p. 16.

The condition of each of these three variables determines how favorable (or unfavorable) a situation is for the leader. A most favorable situation, for instance, is one in which the leader is well liked and respected by followers and directs a task that is well structured and defined. On the other hand, a somewhat unfavorable situation is one in which the leader is not yet known, has not yet established clear leader–follower relations, may not be very well liked or respected, has little position power, and is faced with directing a rather complex, unstructured task, with fuzzy specifications or requirements. Such a situation unfortunately is not far removed from the state-of-the-art project!

Fiedler concluded that when conditions are either extremely favorable or extremely unfavorable, a leader can be successful by demonstrating a highly directive or task-oriented leadership style. When there is a combination of conditions, however, as frequently is the case in the dynamic project environment, people who are more relationship-oriented prove to be more successful as leaders!

8.9.1 The Situational Leadership Model

Hershey and Blanchard[11] subsequently expanded on Fielder's basic contingency approach to leadership and developed what they refer to as *situational leadership theory*. This theory proves to be extremely useful and applicable to the dynamic and changing requirements of a project. The researchers address the importance of the leader's style (be it task- or people-oriented) in assisting the group to attain its objectives. They also recommend that the needs, goals, and task experiences of the followers be addressed if the leader is to experience success in the position.

Hershey and Blanchard's situational leadership theory considers the interaction of three variables that in their opinion determine leadership style, and the subsequent appropriateness or effects it has on followers. These are:

1. The *task behavior*, the amount of direction a leader provides in telling people what to do, where to do it, and how it should be done. This behavior is characteristic of leaders who set goals for their followers and actually outline the necessary steps to accomplish these goals.

2. The *relationship behavior*, the degree of "stroking," feedback, and support that a leader demonstrates, often characterized by two-way communication and active listening.

---

[11]P. Hershey and K. H. Blanchard, *Management of Organizational Behavior: Utilizing Group Resources*, 3rd ed. (Upper Saddle River, NJ: Prentice Hall, 1977).

3. The *maturity level* of the followers with regard to their knowledge or experience in relation to the specific task at hand. Maturity levels are dynamic in nature and can be determined in part by followers' needs for independence, readiness to assume decision-making responsibility, present level of empowerment, tolerance for ambiguity, and need for feedback.

The interaction of task behavior, relationship behavior, and maturity level of followers therefore determines the leadership style appropriate for accomplishing objectives in any particular project team, within any specific life cycle phase. Situational leadership theory supports the authors' premise that effective leadership is a symbiotic or mutually gainful relationship between the followers and the leader. To accomplish project tasks effectively, project managers must not only be aware of the characteristics of the tasks and the needs and goals of their followers, but must vary their own leadership style and provide opportunities for follower development and maturation.

Four leadership styles occur within the context of three variables: leader task orientation, leader relationship orientation, and follower maturity. Table 8.1 presents the following four leadership styles:

• *Telling (S1)*. This style of leadership is characteristic of a leader who provides a great deal of *guidance, direction,* and *input* into the decision-making process. This input may be regarding what tasks are to be accomplished, how the task will be accomplished, and where the work is to be done. Directing or guiding leaders define roles and responsibilities for their followers and provide clear and specific instructions regarding methodology and procedures. The telling style of leadership is usually quite appropriate when faced with followers who may be unable to take responsibility for task activities because of, perhaps, limited knowledge or experience.

• *Selling (S2)*. Also referred to as the *exploring, clarifying,* or *persuading style*, the selling style of leadership still provides direction, information, and knowledge to followers, but a greater attempt is made to explain why certain actions need be taken or why certain methodologies are preferred over others. The leader further attempts to get followers to accept or buy in to the decision by pointing out how their suggestions have been addressed and why this decision is beneficial to all. This style of leadership is not only appropriate for developing followers with relatively limited knowledge growing enthusiasm, but may also be successful in selling concepts, proposals, and ideas to senior management and to the customer.

· *Participating (S3)*. As the knowledge, experience, and overall organizational maturity levels reach a comfortable and relatively moderate range, the effective leader utilizes the group's ideas and expertise to formulate strategies and arrive at decisions. This style is frequently referred to as *participating, encouraging, collaborating or committing* because the leader and followers share in the problem-solving, decision-making process, with the main role of the leader being facilitator.

· *Delegating (S4)*. As followers reach optimum maturity levels, they experience the motivation and the ability to accomplish tasks with minimum supervision or support from the leader. In the advanced stages of project design, for example, participants may be so familiar with project methodology that the leader or project manager may assume a more *monitoring* or *observing* role. As such, he or she provides the opportunities necessary for project team members to *fulfill* both task and personal objectives.

Each leadership style has its appropriate place in terms of the life cycle of the project. As the level of follower maturity continues to increase over time, leaders must reduce their task directive behavior by delegating more decision-making authority to project specialists. As the bell-shaped curve in Figure 8.1 indicates, effective leaders vary their leadership style along the curvilinear functions as followers develop greater maturity.[12] Project and functional managers alike must bear in mind that to manage a project successfully they will lead a variety of followers, including not only project specialists, but vendors, the customer, and senior management. The concept of *maturity* refers not merely to overall experience, educational level, or the need for feedback and autonomy, but to the level of understanding regarding specific technical, administrative, or monetary requirements. Customers, for example, may believe that they fully understand the requirements of a system they wish to install, yet need to be "sold" on more realistic cost requirements. The CEOs of a leading R&D firm may well possess advanced technical degrees and broad-based managerial experience, but may benefit from further clarification regarding the success and feasibility of new project technologies.

One of the keys to effective project leadership therefore is to assess the maturity needs of followers, whoever they may be, and behave as the model prescribes. Project and functional managers involved in state-of-the-art projects, for example, will find that their specialists vary greatly in

---

[12]P. Hershey, *The Situational Leader... The Other 59 Minutes* (Escondido, CA: Center for Leadership Studies, 1984).

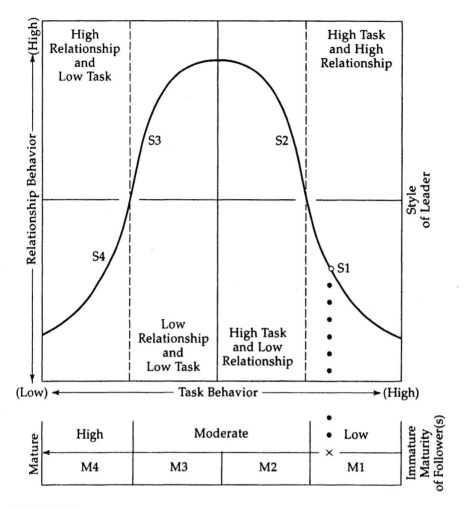

FIGURE 8.1
Determining an appropriate leadership style. (Adapted from P. Hershey and K. H. Blanchard, *Management of Organizational Behavior: Utilizing Group Resources*, Prentice Hall, Upper Saddle River, NJ, 1977.)

maturity over the duration of the project. An experienced software developer, for example, may master the technology well enough to require minimum input and direction when coding software, but may not demonstrate the same degree of maturity in documenting specifications, or in reporting to management. Thus, it may be appropriate for the project manager to offer little technical direction during the design phase, yet provide a great deal of supervision, coaching, and support when it comes to activities involved with reporting to management.

8.10 POWER: THE LEADERSHIP POTENTIAL

If we consider leadership to be the ability to influence others toward the accomplishment of organizational goals or objectives, power may be considered a leader's influence potential. It is that special something through which one achieves influence. It is, indeed, difficult to separate the concepts of leadership and power. It would not be possible to influence another without using some degree of power. Power is a variable or factor that comes to play within most organizational interactions. It is used by all employees, at some time or another, to control scare resources, negotiate agreements and reach professional and organizational goals.

In theory, it is possible to possess substantial power without holding any formal leadership position. As defined previously, legal authority is conceived of as the *right* to make final decisions that others are *required* to follow." Authority is therefore a formally sanctioned privilege that may or may not be available to all employees. Power, on the other hand, implies an *ability to* get results. As illustrated in Figure 8.2, one may conceivably possess authority, yet demonstrate no power, possess no authority, and demonstrate power, or, ideally, possess both.

8.10.1 Identifying and Developing Sources of Power

Of particular interest to project and team managers working in a hybrid or matrix environment are the various sources of power that help to increase their influence both inside and outside the organization, when dealing with external customers, vendors, or alliance organizations. The more that project managers understand power and are able to identify situations in which it arises, the more effective they will be in obtaining and tracking resources and in establishing their leadership position within the project team.

Hershey and Blanchard[13] expand on five traditional[14] bases of power to include the following:

---

[13]Hershey and Blanchard, *Management of Organizational Behavior*.
[14]J. R. P. French and B. Raven, "The Bases of Social Power," in D. Cartwright (ed.), *Studies in Social Power* (Ann Arbor, MI: Institute for Social Research, University of Michigan, 1959), pp. 150–167.

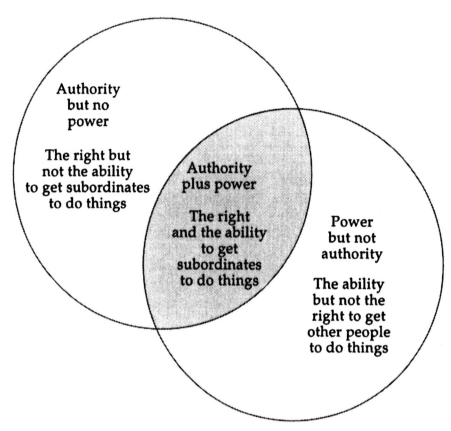

FIGURE 8.2

Relationship between authority and power.

1. *Coercive power.* This power base is, ostensibly, derived from the use of fear tactics and some kind of subtle force. It may be the use of threats of punishment or humiliation associated with, for example, demotion, an undesirable job assignment, or poor performance reviews. In more subtle terms, coercion is the power associated with voting, majority rule, or the responsibility of an assignment that may "make us or break us." Excessive use of coercive power creates a situation in which employees meet deadlines, but lack innovation, drive, or initiative. It may lead further to breakdowns in whatever influence or authority a project manager possesses. Strange illnesses that threaten the survival of the project have been known to inflict project teams working under unduly coercive project managers!

2. *Connection power.* Building alliances with people within a variety of positions both internal and external to the project is a very important source of power. Developing a variety of informal contacts can help project managers recognize project pitfalls early. People will share their project-related problems with colleagues whom they trust. These relationships may be forged through lunches, dinner, social events, sight visits, and perhaps through sports.

3. *Expert power.* Knowledge and experience rank high among influence and power bases. If project managers have prior experience working in contributing technologies, functional specialists may perceive them to be competent and respect them because of the extra knowledge they bring to project efforts. Expertise, however, is never limited to technical knowledge alone. Effective project managers are those with the ability to handle both technical and organizational challenges. Power comes from an expert understanding of administrative techniques and controls as well as a firm grasp of the customer's needs. Finding a project manager with technical expertise, a broad knowledge of the organization, and an understanding of customer requirements and their needs contributes greatly to project success.

4. *Legitimate power.* Legitimate power comes from a person's formal position or location within the organizational hierarchy. The higher the position, the greater the power or legitimate authority associated with it. Legitimate power is described as the right to command. It may vary in the degree of influence it has, depending on the values and expectations of the project team. In recent years, we have witnessed erosion of the effectiveness of legitimate power perhaps through misuse, changes in societal values, demographics, and increased levels of education among the workforce. Formal position held within the hierarchy of the organization no longer is sufficient as a sole basis of power or influence. Project managers operating within today's team-based matrix organization, moreover, lack such formal position authority and are wise to develop some of the other more informal bases of power.

5. *Referent power.* Charisma or personal characteristics have been known to influence the behavior of others simply because of the strong desire for followers to identify with their leader. Referent power is strongly associated with feelings of respect, admiration, or liking, which may result from a leader's past accomplishments or simply be a function of some magnetism or personality. We have frequently heard the comments that charisma, like leadership, cannot be developed. Our stance on the development of leadership, like other skills, should now be quite clear. Like any other skill, leadership not only can be developed but also needs continual fine-tuning.

Charisma, similarly, may be thought of as a function of effective interpersonal communication and listening skills. Developing such abilities assists the project manager in building referent power.

6. *Reward power*. One source of power on which a leader may capitalize is the ability to provide something of value to those on the project team. For such a reward to be effective, however, it must be aligned properly with the values of the person on whom they are bestowed. In other words, project managers must be familiar with the needs and values of the specialists working on the project. Since monetary rewards may not be available, project managers must consider a variety of rewards that have the potential of satisfying diverse employee needs. Research indicates that while traditional rewards, such as salary and promotion, are important, work challenge and recognition are equally important.[15] Project managers must go to great lengths to ensure that the work of team members is recognized throughout the organization. Such actions and rewards may include a letter of appreciation, a recommendation for a promotion, or a merit increase. Project managers must be perceived to have the ability to grant or bestow rewards, be they monetary or otherwise. Reward power is clearly associated with both position power and connection power.

Rewards may also be used to blend project teams effectively. Team rewards should be provided in celebration of accomplishing hard milestones and successful project performance. Rewards help to reinforce the concept of *unity of mission*. They provide an important sense of accountability as a functioning unit. Unfortunately, many organizations continue to reward employee performance solely on the basis of individual contribution; that is, "What have you done this year to contribute to the bottom line of the unit [department, etc.]?" rather than "What have you done that contributed to the success of the project team?" Failure to review performance on the basis both of unit and team performance undermines the accountability of the team and contributes to poor team performance. Peer reviews, for instance, serve as a means of emphasizing the interdependence of team players and the importance of team accountability.

7. *Information power*. Information concerning the project, its related market or its relationship to other corporate priorities perceived to be valuable to project participants are excellent sources of influence and may place you in a position of power. Like other sources of power, however, for it to be effective, one must share information with others. Many people

---

[15]B. Baher and D. Wilemon, "A Summary of Major Research Findings Regarding the Human Element in Project Management," *Project Management Journal*, Vol. 8, No. 1, March 1977, pp. 34–40.

believe erroneously that hoarding information strengthens power. Nothing could be farther from the truth! Project managers who hoard information will only discourage others from seeking them out and, therefore, will eventually erode their own power base. Information helps to influence others because people want to be kept informed and feel that something of importance is being shared with them. Project managers have many internal and external sources of information. They are in a central information-processing position and should use that position to influence their team.

Table 8.1 provides an overview of the power bases available to project managers, as presented by Hershey and Blanchard. Additional power tactics are as follows:

- *Controlling the agenda:* determining beforehand the issues, subjects, or concerns that require group action or decision

- *Using ambiguity:* keeping communication unclear and subject to several interpretations

- *Forming coalitions:* securing allies, with superiors, employees, peers, customers, vendors, etc., associated with the project or the organizations

- *Co-opting opposition members:* bringing a representative of the opposition onto a task force or team so that others opposed will sense a new alignment

- *Controlling decision criteria:* selecting criteria by which decisions are made so that desired outcomes result, regardless of who acts as decisionmaker

- *Developing others:* increasing the capabilities of others in ways designed to help them become more independent

- *Using outside experts:* involving experts expected to recommend a certain strategy or course of action, so as to avoid a unilateral decision

- *Presenting a favourable image:* displaying skills, capabilities, values, or attitudes that others respect

- *Incurring obligation:* securing indebtedness in others

- *Rationalizing:* consciously modifying facts when giving information to others

- *Selectively dispensing rewards and punishment:* using rewards and punishment to win support from others

TABLE 8.1 PROJECT MANAGER POWER BASES

Power Base	Source	Comments
Coercion	Fear and the avoidance of punishment and threats	Use of coercive power is linked to organizational position; tends to inhibit creativity and affect project team morale negatively
Connection	Alliances with influential or important people	Highly effective for developing trust and recognizing project pitfalls early
Expert	Knowledge and experience, especially in a contributing functional area	To be effective, must be perceived by project participants to be vital to project success; limited if not expanded to the administrative and/or customer knowledge arenas
Legitimate	Formal position: "right to command"	May vary in degree of potential influence, depending on the values of the group; less effective due to frequent misuse, changes in societal values, and increased education of workforce
Referent	Charisma or personal characteristics	Highly effective; may be conceived as a function of interpersonal communication and listening skills
Reward	Ability to provide positive sanctions for performance	To be effective, must correspond to participant values and/or expectations; since money is not available, project managers must consider a variety of potentially satisfying sanctions, especially those related to work challenge and recognition
Information	Knowledge or "tidbits" concerning project activities and related occurrences	Effective only if perceived valuable and shared appropriately with functional managers and project participants; erodes trust and creates resentment if hoarded

- *Selectively allocating resources:* distributing resources in ways that will produce desired outcomes
- *Using a surrogate:* having an intermediary intervene to secure compliance from others
- *Using rituals:* introducing institutionalized patterns of behavior in the group or organization
- *Training and orienting others:* instilling skills, values, beliefs, and specific behaviors in others in order for them to understand and realize certain goals or outlooks
- *Using symbols:* reinforcing control through the use of objects or ceremonies

8.10.2   Foundations of Effective Empowerment

In today's competitive environment, organizations with a commitment to continuous improvement and quality are those that have a better chance of survival. *Employee empowerment*, autonomy, and increased decision making, characteristic of more team-based approaches, are critical prerequisites for continuous improvement. The team movement of the 1980s concentrated on improving the quality of work life, increasing employee involvement and opportunities for providing suggestions for improving operations. These quality circles were the forerunners of the more self-directed, empowered teams that organizations today strive to create. In moving toward a more empowered project team approach, however, changes must be made in the typical manner of leading and conducting business.

Direction from leaders and from management helps to get the project and the project team started. Unfortunately, in numerous organizations unsuccessful empowerment and team implementation are due primarily to one major factor: erroneously, poorly, or vaguely defined performance requirements and expectations. Bob Waterman and Tom Peters refer to this phenomenon as *solution space*.[16] It involves the definition of boundaries and the scope of authority clearly enough to provide the team with direction, yet allow sufficient flexibility so that the team may provide any necessary modifications and thereby establish commitment. It provides clarity as to the vision management has, the rationale for the team's existence, establishes performance challenges, yet leaves plenty of solution space for the team to set their own goals, schedule, and overall approach to accomplishing their mission.

---

[16]T. J. Peters and R. H. Waterman, Jr., *In Search of Excellence: Lessons from America's Best-Run Companies* (New York: Warner Books, 1982).

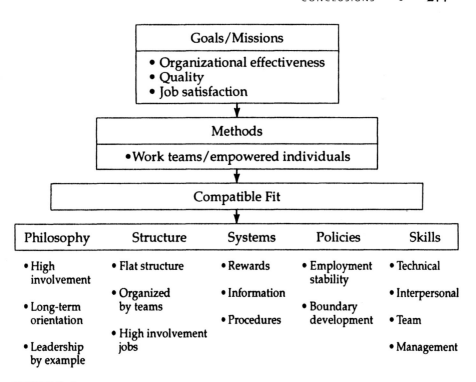

FIGURE 8.3

Foundations of effective empowerment. (Adapted from James H. Shunk, *Team-Based Organizations: Developing a Successful Team Environment,* Business One, Irwin, IL., 1992.)

It is, therefore, critical to the success of the project team, to the credibility of management, and to employee morale to set expectations clearly. Misconceptions concerning empowerment will lead to frustration for all parties involved. Clear expectations should start from the top and be consistently supported by the senior levels of the organization. When estimating boundaries for empowerment, there are several situational factors to consider. These are (1) the present level of empowerment, (2) quality programs and processes, (3) the general management style or the culture of the organization, and (4) the organization's ability to handle change. Figure 8.3 provides a graphical portrayal of the foundations of effective empowerment.

8.11 CONCLUSIONS

As the focal point of the project, the project manager has total accountability and responsibility for making it all happen. This requires outstanding selling skills, artful negotiation ability, and people prowess. Especially in the more technical communities, a project manager's credibility relies not only on others' perception of their technical competency or lack of it, but on their ability to negotiate, persuade, and communicate effectively. The relationships required to lead a project effectively are endless and supersede the boundaries of the project and the organization.

Appointing professionals because of their technical expertise is obviously insufficient in accomplishing a project manager's role. Achieving a unity of effort within a complicated and competitive project environment requires a firm appreciation of the challenges and diverse needs facing the stakeholders in any project. Understanding the nature and requirements of the *project leader* is a first step toward accomplishing this objective.

CASE STUDY: NEW WAVE II COMMUNICATIONS SYSTEM*

The executive steering committee was to meet in 15 minutes to decide the fate of the New Wave II Communications System. "I don't know what the outcome of this meeting will be," commented Tim Leary, manager of systems production. "But I know the hell we've been through and the distance the team has come.... These people really care about this project!"

Background History

TelPlus Communications provides customized development, manufacturing, and deployment of a variety of communications systems, both digital and wireless in design. They provide domestic and international network services to both commercial customers and private residences. Their goal, especially within the last five years, has been to provide their customers with ongoing upgrades in network capabilities, to handle the increasing demand in voice and data transmission. The New Wave II switching system was, perhaps, the key to TelPlus's network upgrade over the last decade.

The New Wave II project environment was characterized by a highly demanding technical effort, an unusually high level of integration required between project entities (especially between the design and production

---

*This study was written for class discussion purposes only. It is not intended to reflect management practices or products of any particular company or organization.

departments), and a high level of design changes, due to the rapidly changing technologies. Several items added to the complexity of the project. Not only did transition from the customer's previous system to the new state-of-the-art system need to occur flawlessly, without any interruption in service, but the growing importance of cellular technologies presented unprecedented competition for manufacturing resources and added to Tim's growing concerns about eventual reassignment of his manufacturing staff.

The status of the project in its earlier development was one of organizationally strained relations, characterized by mutual mistrust and insufficient communications across the organizational entities. Historically, there was little interaction between TelPlus's manufacturing and design divisions, each believing they possessed sufficient expertise and talent to make decisions without consulting other divisions. With a lack of precedence, the new project management procedures being implemented across the organization were met with either resistance or polite disdain. Relations between production and design personnel were tenuous and getting worse. Production had been accustomed to a more stable product environment than that in the New Wave II production effort. With the frequent changes being made by design, production took the point of view that design was consuming the time necessary for manufacturing process development, while design believed that production was not responding adequately to the necessity and reality of the continuing design changes. Production frequently requested more "design presence" on the factory floor, only to be ignored or told that resources were just not available. It was clear that manufacturing would have to transform its operations to be more change responsive.

Developing Organizational Commitment

Being a switching manufacturing organization within a future of wireless communication made Tim's job of developing adequate incentive for all-out commitment from his staff quite a complex task. Management's support was clearly in the future of wireless technology, which manifested itself in the levels of inadequate staff for the New Wave II production operations. Although dedicated and adequately experienced, Tim wondered how he could sustain his people's motivation in the face of the constant design changes and scheduling pressures. As "full production" loomed in the near future, the strong support of development by management was in stark contrast to Tim's urgent need for increased management staff.

Tim knew that many improvements were necessary if production was to fulfill its role on the New Wave II project. Production would now have to take much more of a proactive approach than in the past. New and more appropriate project management approaches and manufacturing processes would have to be made to obtain management's support. Additions to staff were desperately needed and had to be lobbied vigorously with management.

With the aid of Ricky Brentwood and Ken Nelson, Tim outlined a plan for the New Wave production organization that would make it more consistent with the transient nature of the project. To capture the attention and respect of management and its sister organizations, production began to outline a plan to restructure its operations. Roles were more clearly defined, which seemed to highlight the urgent need for additional department managers and for professional and manufacturing staff. Cross-functional *process review teams* (PRTs) were established to address key manufacturing issues. Tim also decided to take the bull by the horns and fully embrace a project management approach within the production domain. To create a stronger internal identity for the New Wave II production organization, Tim requested the formation of the new position of production program manager to serve within his own manufacturing structure.

It was a difficult decision for the New Wave II steering committee to consider, but through his comprehensive plan, Tim had made a very convincing argument for hiring his old friend Hank Galaway to the position of production program manager. The number of shop floor personnel increased dramatically to 200, and the request for additional supervisory personnel, although not fully accepted, had been recognized with the addition of both the production program and quality process managers.

On Hank Galaway's first official day on the job, Tim called a meeting of all his management staff. This included not only Ken Nelson and Rick Brentwood, but all the manufacturing supervisors and shop foremen. Tim knew that he had to make Hank's role within the organization perfectly clear to production staff, and that meant from the top down. Hank would need a "license to lead" without overstepping his boundaries. Staff would need to see that Hank not only had Tim's blessings to improve production's project and process management structure, but the blessing and support of senior management as well. Hank would have to make not only structural changes within the organization but *cultural* changes as well.

"Ladies and gentlemen, it gives me great pleasure to introduce Hank Galaway to all of you," Tim Leary began. "Hank and I have worked together before in many other projects within TelPlus and we're lucky to have him as our new production program manager. Hank has worked on both the development and the production sides of the organization, so he brings a lot to the game. He will be working with all of you, trying to get you what you each need to get this job done. He'll also be working closely with the process review teams and will be coordinating all of our production management processes."

"Great to have you on board, Hank" commented Rick Brentwood. "I mean *really* great! We all know you'll have your work cut out for you, and I for one, am behind you all the way."

"Yeah," responded the others, "good luck!"

"Boy, am I going to need it!" Hank reflected silently.

Smoothing Out Relations

With the new and visible commitment from management, Hank and Tim now believed that production was truly "energized." To improve relationships between design and production, Hank was determined to build upon the generally comfortable working relationships between the development staff that visited the shop floor and the production staff. Hank encouraged production staff to work more closely with the visiting developers rather than turn over responsibility to a development rep, as was previously the case. Now production staff would actively be involved with development on the shop floor and were more readily able to illustrate the circuit pack and testing issues, which were at the heart of many of the delays.

Prior to Hank's appointment, the cross-function process review teams (PRTs) limped along, with no formal procedures or project reporting mechanisms. Meetings were low in attendance, if held at all. Little documentation from the meetings that were held existed, and communication between functions were strained and lacked documentation.

Adversarial feelings existed to some degree at the project specialist and task leader levels. On the design/development side, there was the view that manufacturing had not done enough to accommodate the reality of design changes, which appeared to continue to the last minute. On the manufacturing side, there had been the view that design/development displayed a rather insular attitude, failing to appreciate how the continuing changes in design significantly affected production's work. As far as production staff was concerned, the changes from design meant little more than a lack of self-discipline in seeking a stable configuration. The date for rollout of the product was not changing, but the point at which full production could proceed based on a stable product design has remained the same. Production felt it was being required to make up the time to keep product rollout on schedule.

Design/development, on the other hand, took the point of view that design changes had been forecast to proceed into the period at which production needed to begin developing processes. It was their contention that production was not responding to the situation. Design/development believed that the changes were inevitable and due to the complexity of the hardware and the criticality of achieving stable operation of the product once it was deployed and installed. A more prototype-oriented approach was needed to make production responsive to the high degree of configuration changes.

Hank knew that such adversarial feelings had to be rooted out proactively. Hank concentrated on using the PRTs to close the gap between design and production and to implement more project-oriented mechanisms into manufacturing procedures. Although project management strategies had a strong presence among the development staff, it was used infrequently in manufacturing. Believing that they had their own metrics for tracking and control, most people in the production area regarded project management reports as redundant and unnecessary.

Changeover Begins

Hank focused on directing and coordinating the PRTs and made it his business to make frequent informal visits to the development area. Since he never possessed official authority over all the cross-functional resources that comprised the PRTs, Hank had to rely on his strong working relationships with both development and production and their common goal of the success of the project.

Hank viewed the PRTs as a way to harness the combined efforts of production, design, and the customer. Hank began to structure process review team meetings to encourage frequent interactions and communication between the design and production teams. Agendas were distributed prior to each regularly scheduled meeting and briefings held if problems existed with the agenda items. After each of these meetings, Hank would debrief Tim, outlining what was accomplished and what still needed to be done.

With the new structure and increased volume of process review team meetings, collaboration began to improve dramatically between the organizations. A higher level of respect and design responsiveness emerged from the issues raised in the PRT sessions. Brainstorming sessions, for example, were the key to reducing circuit pack cycle time. Whenever possible, sessions were held on the production shop floor, using any method possible to get design personnel physically present at the site. If travel was not possible, audio- and videoconferencing were often used.

The PRTs established the manufacturing organization as a "sound" contributor through more detailed reporting and a larger share of podium time at monthly project meetings. The PRTs also helped to establish a stronger identity not just for members of the New Wave II production operations but also for the new support staff and support teams external to the project. The PRT approach also helped people within the manufacturing organization to be recognized, and to recognize themselves, for the good work that they had done as a team and in their individual contributions as well.

Conclusions

Many improvements in the ability of production personnel to fulfill their role on the New Wave II project had been made in the 10 months that had passed since the steering committee meeting. Production had been permitted to add significant additional professional resources and had marshaled their efforts effectively. The PRT approach and other advances in process analysis and design had been accomplished. Most important, production and design personnel have undertaken extensive efforts at collaboration, adopting the point of view that design changes will continue into formal production launch and that efforts must be organized accordingly. Tension between production and design eased considerably, as mutual understanding of their relative positions grew. A stronger internal identity for the New Wave II manufacturing organization had been developed. The manufacturing organization was now viewed as a sound contributor through more detailed reporting and a larger share of podium time at monthly project review meetings.

Questions

1. What was Tim's style of leadership? Discuss why it was or was not appropriate for this project.

2. What was Hank's style of leadership? Discuss whether or not it was appropriate for the environment in which it occurred.

3. Discuss how the terms *authority, power,* and *influence* relate to this case.

4. Refer to the list in Section 8.6 of the responsibilities and challenges of project managers. Which of these refer more appropriately to Tim's demonstrated role? To Hank's?

5. The leadership roles required for today's competitive environment require new leadership skills. As discussed in this chapter, which of these new skills were demonstrated in the case study? By whom and how were they demonstrated?

9

PLANNING FOR AND UTILIZING CONFLICT

When two people in business always agree, one of them is unnecessary!

ANONYMOUS

9.1 INTRODUCTION

Conflict! It's one of those fascinating but frequently misunderstood concepts. More often, conflict generates ambivalence and confusion because of its potential to do great injury. Think of some of the typical feelings or behaviors you associate immediately with the word *conflict*: anger, animosity, hostility, battle, perhaps even war! Yet if harnessed and managed properly, conflict has the capacity or potential for doing great good! It is a vehicle for problem solving and a catalyst for change. If approached positively, conflict has the potential to help synergize *diverse* ideas and ultimately lead to improved relations.

No one can deny the negative and frightening potential that conflict holds for a project. It may create schedule slippage, work slowdowns, inferior quality, and even sabotage. The present trend of organizing specialists into aggressive, quality-conscious, schedule-driven teams often creates hybrid organizations with somewhat confusing reporting relationships. The members of the team, moreover, carry with them an assortment of performance measurements and ambiguous and sometimes divergent divisional loyalties.

Because of the nature of project work, conflict is inevitable. Shared and limited resources, interdependent assignments, diverse technical and cultural backgrounds, stressful life cycles, and narrowing market windows all increase the potential for conflict to occur on the best of projects and in the most "blended" of teams. An important aspect of managing projects and achieving the requisite quality processes and standards therefore is the ability to manage effectively the conflict typically associated with time restrictions and resource limitations.

9.2   BASIS OF CONFLICT

When it came to minimizing the *potential* for conflict, organizations of the past were somewhat at an advantage. Manufacturing or production facilities typically produce standard goods through highly structured procedures. Organizational roles and relationships were more clearly defined and decision-making authority placed in the hands of a few. Highly centralized operations defined the organization's goals, which employees accepted willfully for the good of the organization and the continued security of their employment.

But times and markets have changed. To survive and compete in today's global market environment, organizations have undergone significant transformations. These organizational changes have not only been in structure or form, but in the *technologies* that enable them and the *ideologies* that drive them. Decision making, for example, a key to empowerment, is now more frequently delegated to cross-functional teams, who share a diversity of talent and a potfull of conflicting priorities. The changing composition of the workforce, fragmented markets, and competitive global environments all require organizations to reexamine typical business approaches. Most will discover that they are in need of creating newer, more fluid, systemic processes, capable of responding rapidly to a variety of new internal pressures and external demands.

9.3   CONTEMPORARY CONFLICT: A NEW POINT OF VIEW

People working within organizations tend to ascribe to one of two opposing views or attitudes toward conflict. The *traditional view* sees conflict as dysfunctional or harmful to the organization: as something that is best avoided, squelched, ignored, or suppressed. According to this viewpoint, conflict is seen as the result of poor leadership or personality differences and is best resolved by either physical separation or through intervention by higher levels of management. In the more traditional organization, managers and the people they manage are encouraged to ignore or avoid conflict or to squelch or suppress it whenever it occurs.

The more *contemporary view* of conflict is humanistic in nature. According to this view, conflict is an inevitable consequence of the complex organizational systems and relationships. It is perceived as the predecessor of change and may actually be beneficial to the project if handled properly. Rather than suppress conflict immediately, contemporary managers view conflict as an opportunity to confront and resolve the issues that may hinder project momentum and eventual success. It is inevitable, especially in a hybrid matrix organization, where cost, quality, time, and technical issues are often divergent. The value of conflict, however, depends on how project personnel regard and manage it.

Since conflict is a natural and inevitable occurrence on most any project venture, our goal is not necessarily to eliminate conflict but to see it as a healthy and essential characteristic of project work. Managed properly, conflict can enhance team productivity, stimulate innovative thinking, and ensure a higher-quality product. Through the constructive resolution of conflict, it is possible to gain a broader perspective and understanding of the problem by addressing a wider array of issues. By encouraging rather than suppressing the expression of divergent opinions, managers create a reservoir of alternatives from which a solution may eventually evolve. The excitement and energy associated with the expression of divergent ideas and the efforts made to resolve them may actually help to blend a work group into a cohesive *team*.

9.4 ANTECEDENTS OF CONFLICT

Although divergent ideas and disagreements may arise in just about any situation, there exist certain conditions within the hybrid project organization that predispose it toward conflict.[1,2] These conditions include:

- Ambiguous roles and overlapping responsibilities
- Inconsistent or incompatible goals
- Communication barriers

---

[1]A. C. Filey, *Interpersonal Conflict Resolution* (Glenview, IL: Scott Foresman, 1975).

[2]K. S. Kirchof and J. R. Adams, *Conflict Management for Project Managers* (Upper Darby, PA: Project Management Institute, 1982).

- Diversity of talents
- Interdependent tasks or activities
- Dynamic, stressful life cycles
- Unclear goals or objectives
- Divisional loyalties
- Need for consensus
- Global competition
- Narrow market windows
- Unresolved or prior conflict
- Differentiation or specialization

In many team-based project organizations, one may find two or more individuals, sections, or departments that have related and even overlapping roles, responsibilities, and assignments. Conflict is often generated by the need to define and establish unit goals and objectives, as well as a mission or sense of purpose. Incompatible goals often arise when departments or individuals must work interdependently, yet perceive the other as having an opposing purpose. This is common, for example, between sales and manufacturing, or engineering and marketing. On one hand, sales frequently promise customization to obtain a small but lucrative order. Manufacturing, on the other hand, prefers high-volume orders and often finds the small, customized orders to be a nuisance, having to modify tooling mechanisms to produce limited quantities.

Communication difficulties and barriers create additional misunderstandings, as efforts to explain and negotiate the needs of the parties involved become distorted or blocked. Conflict frequently arises when one discipline or function is dependent on another for hand-offs or resources. Employees who possess special talents and are responsible for accomplishing unique tasks characterize today's modern technology-driven organizations. These specialized individuals and groups possess their own jargon, perspectives, and goals—conditions that create or stimulate additional conflicts. This situation intensifies, as project teams work across the organization to accomplish project goals. When groups of divergent talents must further reach consensus or agree among themselves, disagreements naturally occur and may be difficult to manage. As rules, procedures, policies, and regulations, natural components of the project environment, restrict team member actions, team members may feel that they are in opposition (or in conflict) with the very organization they attempt to serve. Finally, unresolved prior conflicts tend not to dissipate but intensify and

TABLE 9.1 ANTECEDENTS OF CONFLICT

Factors Encouraging Cooperation	Factors Associated with Conflict
Common goals	Incompatible goals
Common values	Lack of common values
Similar backgrounds and perceptions	Disparate backgrounds and perceptions
Centralized authority	Decentralized authority
Consistent supervisor–subordinate expectations	Inconsistent supervisor–subordinate expectations
Well-defined responsibilities and roles	Ambiguous responsibilities and roles
Proximity in the workplace	Physical and time separation
Consensus not necessary	Consensus necessary
Independent tasks	Interdependent tasks
Equitable workloads and work systems	Inequitable workloads and work systems
Adequate resources	Limited resources from a common pool
Noncompetitive relationships among participants	Win–lose relationships among members

create a strained environment that precipitates conflict of an even more intense and destructive nature.

Table 9.1 lists a variety of factors that serve as antecedents of conflict and cooperation.

9.5 DESTRUCTIVE VERSUS CONSTRUCTIVE ELEMENTS OF CONFLICT

Conflict is destructive, disruptive, or distributive when project team members fail to understand the value of conflict as a source of diverse ideas and alternatives or do not have or use constructive means, skills, or behaviors to channel the conflict into problem-solving sessions and deliberation. *Destructive conflict* is frequently associated with competitive win–lose situations. It is characterized by a lack of team spirit and a "not invented here" (NIH), "get my own way" attitude. Group members frequently "position" themselves as they stubbornly adhere to their own narrow viewpoints and fail to consider the possible value of another approach. Group decisions are frequently stalemated, while members avoid the critical issues and engage in personal attacks. Negative feelings and attitudes grow more entrenched each day. Even if a decision is reached eventually, it is likely that few people will be pleased or satisfied. Destructive conflict therefore produces defensive and disruptive behavior. It is the type of conflict that gives conflict its bad name.

Constructive or *integrative conflict* arises from a climate characterized by open communication and mutual understanding. It develops only when project team members have acquired skills in effective communication and possess a variety of methods with which to handle conflict. Members demonstrate a keen sense of team spirit and recognize that disagreements evolve from a sincere commitment to the successful realization of project goals. Project managers who model integrative conflict management strategies tend to be less apprehensive of disagreement and more willing to approach conflict rather than avoid it.[3]

Integrative situations are characterized by supportive rather than defensive communication. Active empathetic listening (as discussed further in Chapter 10) prevails as team members grow to appreciate the merits of considering others' opinions and integrating a variety of viewpoints into the best possible solution. Group cohesiveness and trust evolve from the support and openness that team members display toward one another. Team decisions are reached primarily by consensus and represent not merely a compilation but a synergy of the most positive aspects of the group's total solutions.

9.5.1   How Differences Escalate

As Figure 9.1 illustrates, even what appears to be a simple difference can escalate all too rapidly into destructive conflict. Conflict may start as a simple difference of opinion over, perhaps, roles, procedures, processes, or values, which may seem to be a simple misunderstanding of "I see it my way and you see it yours." Unfortunately, because so many of us find opposition distasteful and are uncomfortable with conflict, we frequently react quite negatively and, at times, defensively in such sensitive situations. Feeling uncomfortable with and perhaps even a little fearful of losing control of the situation, we may actually push to eliminate the conflict by forcing a solution through unilateral escalation or by establishing a virtually ineffective, watered-down approach. We do not allow ourselves the time to determine sufficiently where the actual differences lie and, thus, permit the real sources of the conflict to grow stronger and more harmful. Participants

---

[3]R. Hill, "Managing Interpersonal Conflict in Project Teams," *Slogan Management Review*, Vol. 19, No. 2, 1977, pp. 45–61.

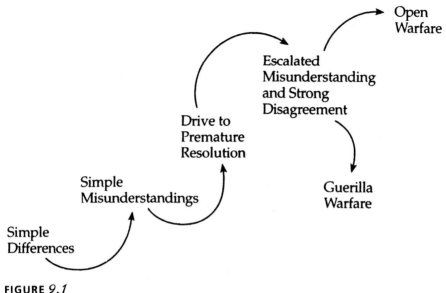

FIGURE 9.1

How differences escalate.

grow entrenched in their positions, as they become increasingly concerned over supporting their respective procedures or viewpoints, rather than about the end result or common goal. This, in turn, leads to a state of open warfare, making it almost impossible to perceive or agree on a mutually beneficial result. Guerrilla warfare may arise further in the form of excessive project turnover, increased absences from meetings, lack of commitment to project timetables, or even sabotage.

9.6 A PROBLEM SITUATION

Taking the time to examine the many sources of conflict is the first step toward dealing effectively in such a situation. Unfortunately, finding the time to accomplish this objective is itself often a problem. Let us examine a short and simple case situation.

Ms. Sharp, supervisor of product sales, receives a rush order from one of her company's most important customers. The customer requests a slight modification in a stock item. Ms. Sharp is particularly concerned about this request since she is responsible for satisfying this influential customer. She receives the call at 9:00 A.M.; the revised items have to be shipped by 5:00 P.M.

To save time, Ms. Sharp goes directly to the manufacturing supervisor, Leslie Payne. Payne wastes no words. Because he has "his own emergencies," he is not going to interrupt his busy work schedule for this special order. Further, he complains about the continuous interruptions and requests from Sharp's area.

Ms. Sharp then goes to her manager, who, in turn, approaches Payne's manager. Payne's manager orders him to put the rush job through. The special order is complete early that afternoon; however, quality control rejects the entire lot because of an extremely high error rate. The job is redone the next day — 12 hours too late! This event accentuated the existing differences between the sales and manufacturing areas. What are the various sources of conflict? How could the conflict have been avoided or utilized?

For but a few moments we have taken you out of the complicated world of high-tech projects and into the simpler world of widgets. Yet many of us will easily identify with the laundry list of conflict sources that this simple case generates. It appears that one source of the conflict may indeed be *personality differences*; that is, Sharp and Payne just do not like each other. Let us bear in mind, however, that what may appear to be personality conflict may at times actually represent *organizational role conflict*. This occurs when we ascribe and generalize negative characteristics to individuals who represent the various divisions within the organization and thus personify the organization itself. This may be represented by such statements as "People from marketing never understand engineering's problems," or "Engineers make terrible marketers!" It also appears that *scheduling conflicts* exist and that Sharp and Payne have *different priorities*. While Sharp may be rewarded for her small but lucrative special orders, Payne has his huge volume of standard widgets to produce. In addition, there appear to be no procedures, or perhaps, different processes for special orders. Or, as the phrase "to save time" may indicate, such procedures, if they exist, may have been bypassed. *Lack of communication* or consultation with manufacturing certainly contributed to the ill feelings and may have led to premature commitment to the customer. We advocate that saying "no" to a customer is rarely heard. One should rarely, if ever, say no to a customer, but "Indeed!" and then *carefully inform the customer of the impact of that special request on schedules and on cost*. A baseline plan and schedule allows you to view customer changes as an opportunity for increased scope and, hopefully, increased profits. It is easy to see that the lack of negotiation on Sharp's part, and the lack of "buy-in" from manufacturing (and quality control), did the job in, not to mention the hundreds of unresolved conflicts that probably occurred prior to this incident.

We can go on and on. The point is that in today's modern organizations, there are many sources of conflict. The first step toward working with conflict is to recognize its inevitable existence, and to identify and understand its wide variety of sources.

9.7 SOURCES OF CONFLICT IN THE PROJECT ENVIRONMENT

Among the several forms of project organizations, the one that breeds the greatest amount of conflict is the matrix or hybrid organization. As discussed in Chapter 6, the matrix project manager may perform the functions of planning, organizing, controlling, and directing the project, but has little direct authority over the people contributing to project objectives. Situations that are characteristic of the matrix environment, such as overlapping responsibilities, the tendency toward ambiguous roles or reporting relationships, shifting personnel assignments, and constant changes, all increase the potential for conflict.

Recognizing the primary sources of conflict is the first step toward managing conflict effectively. In fact, several studies[4–6] indicate that across many major technology-oriented businesses, there appear to be several common sources of conflict that vary in intensity over the life cycle of the project.

Exercise. For the next few moments, think of your present or past project situation(s). List what you believe are the seven top sources of conflict and rank-order these sources according to their negative impact on project performance and morale. Assume that rank 1 has the greatest impact, rank 7 the least. Compare your sources and rankings according to the research discussed on subsequent pages.

Rank	Source
1	_____
2	_____
3	_____
4	_____
5	_____
6	_____
7	_____

---

[4]H. J. Thamhain and D. L. Wilemon, "Conflict Management in Project Life Cycles," *Sloan Management Review*, Vol. 17, No. 3, Summer 1975, pp. 31–50.
[5]H. J. Thamhain and D. L. Wilemon, "Leadership, Conflict, and Program Management Effectiveness," *Sloan Management Review*, Vol. 19, No. 1, Fall 1977, pp. 69–89.
[6]Barry Posner, "What's All the Fighting About? Conflicts in Project Management," *IEEE Transactions on Engineering Management*, Vol. 33, No. 4, November 1986, pp. 207–211.

By now the message should be increasingly clear. Conflict is a natural and inevitable phenomenon that results from organizing work by interdisciplinary teams across some form of matrix or hybrid organization. The unique conditions characteristic of a matrix environment, such as dual-boss relationships, shared resources, and perceived incompatible goals or differing technical outlooks, all contribute to the potential for and the intensity of conflict.

A number of investigators have addressed the inevitability of conflict in project environments[7,8] and have attempted to pinpoint some causes of typical disagreements that have a negative impact not only on the quality of technical performance, but also have such an impact on the quality of work life. For example, Wilemon[9] found that the greater the diversity of expertise among project teams specialists, the greater the potential for conflict to develop. The less the specific objectives of a project are understood, moreover, the more likely conflict is to develop. Baker[10] found further that competition over resources, pressure for consensus, lack of professional incentives, and the overall stress that results from the project environment, are also primary causes of conflict.

Thamhain and Wilemon[11] further categorized these conflict issues into seven fundamental sources: schedules, priorities, technical opinion, personality, cost, administrative procedures, and resources. Their research, moreover, attempted to illustrate the relative intensity of these conflict sources to one another. In a survey of 100 project managers, Thamhain and Wilemon measured the rank order of these disagreement-generating variables. When presented with these seven predetermined sources of conflict, the following rank order was revealed:

1. Schedules
2. Project priorities
3. Resources (staff)
4. Technical opinion
5. Administrative procedures

---

[7]G. R. Gemmill and D. L. Wilemon, "The Power Spectrum in Project Management," *Sloan Management Review*, Fall 1970, pp. 15–25.

[8]Posner, "What's All the Fighting About?"

[9]B. N. Baker and D. L. Wilemon, "A Summary of Major Research Findings Regarding the Human Elements in Project Management," *Project Management Quarterly*, March 1977, Vol. 8, No. 1.

[10]Ibid.

[11]Thamhain and Wilemon, "Conflict Management in Project Life Cycles."

6. Cost objectives and estimates

7. Personality

Table 9.2 provides detailed explanations of each of these classic conflict sources.

Following this earlier research, Posner[12] replicated the classic study and examined the pattern of conflict intensity. His study occurred over a decade after Thamhain and Wilemon's classic study. Posner used survey instruments developed and validated in the earlier studies; respondents indicated the issues that were most likely to create conflict during a project and how the intensity of these sources varied over the project life cycle. Posner's 1986 study indicates the following changes in rank:

1976	1986
Schedules	Schedules
Priorities	Costs
Staffing	Priorities
Technical opinions	Staffing
Procedures	Technical opinions
Costs	Personality
Personality	Procedures

A variety of circumstances may explain the differences between the classic study and Posner's work. Differences over cost conflict may reflect the change from a U.S.-dominated market to that of an intensely competitive global economy. It may further be confounded by changes in government contract pricing strategies from a flexible cost-plus basis to the more rigorous price-fixed approach. The diminished intensity of procedural conflict may further indicate the increased acceptance of project management methodology and related organizational forms.

9.7.1 Conflict in a New Perspective

The classic literature and research in conflict occurred at a time when technology, markets, organizational structure, and even the needs of employees differed dramatically from the contemporary project environment. The theoretical categories of conflict selected and used by past investigators reflected their times. But times and organizations have changed, and so have the categories that reflect the sources of project conflict.

---

[12]Posner, "What's All the Fighting About?"

TABLE 9.2 SEVEN CLASSIC CAUSES OF CONFLICT IN PROJECT MANAGEMENT

Potential Cause	Characteristics
Schedules	Disagreements that develop around the timing, sequencing, and scheduling of project-related tasks
Project priorities	Differing views by project participants over the importance of activities, tasks, and trade-offs that should be undertaken to achieve successful project completion
Personnel (staff)	Conflicts that arise around the staffing of the project team personnel from other functional and staff support areas from the desire to use another department's personnel for project support
Technical opinions and performance trade-offs	Disagreements that arise, particularly in technology-oriented projects, over technical issues, performance specifications, technical trade-offs, and the means to achieve performance
Administrative procedures	Managerial and administrative-oriented conflicts that develop over how the project will be managed (i.e., the definition of the project manager's reporting relationship, definition of responsibilities, interface relationships, project scope, operational requirements, plans of execution, negotiated work agreements with other groups, and procedures for administrative support)
Cost	Conflict that develops over cost estimates from support areas regarding various project work breakdown packages; for example, the funds allocated to a functional support group might be perceived as insufficient for the support requested
Personality	Disagreements that tend to center on interpersonal differences rather than on technical issues; conflicts that are ego-centered

Source: Adapted from H. J. Thamhain and D. L. Wilemon, "Leadership, Conflict, and Program Management Effectiveness," Sloan Management Review, Vol. 19, No. 1, Fall 1977, pp. 69–89.

In a 1991 investigation, Kezsbom[13] not only sought to investigate the sources of conflict experienced by managers and specialists in the 1990s, but also added to the literature additional categories reflective of some of the more current trends and issues. Table 9.3 presents Kezsbom's 13 conflict categories.

Kezsbom investigated the sources of conflict experienced by a sample of project specialists and managers in Fortune 500 corporations. The 13 theoretical categories or sources of conflict were extracted from the literature and from an analysis of the respondents' open-ended responses. The 13 categories used in this study contained Thamhain and Wilemon's original seven sources and six additional areas particularly conducive to disagreements. These six additional categories are:

1. Unresolved prior conflict
2. Politics or turf battles
3. Leadership
4. Communication or the flow of information
5. Ambiguous roles/organizational structure
6. Reward systems

The original category of "staffing" was expanded to include all resources other than money (resource allocation).

To assess the overall distribution and relatively intensity of the 13 conflict sources, points were allocated to each source according to its individual rank and then assigned a ranked weight. By summing the total of each ranked response, final ranking of the 13 overall sources was obtained. Table 9.4 presents the descending order of the 13 conflict categories by total scores. What we found in this investigation differed somewhat from earlier studies. The leading conflict issue, *found consistently across organizational positions and technology*, was the lack of or poor definition of goals and priorities. The intensity of this source of conflict appears to be consistent with the nature of today's competitive and at times complex multiproject environment. Employees typically find themselves working in a variety of teams and on a multitude of projects, which create, at times, somewhat complex reporting relationships. It is also consistent with the relatively high ranking of the third conflict source, communication. When organizational or project priorities are unclear, fail to be

---

[13]Deborah S. Kezsbom, "Bringing Order to Chaos: Pinpointing Sources of Conflict in the Nineties," *Cost Engineering*, Vol. 34, No. 11, November 1992, pp. 9–16.

TABLE 9.3 DEFINITIONS OF THE 13 CONFLICT CATEGORIES

1. Scheduling	Disagreements that develop around the timing, sequencing, duration of projects, and the feasibility of project schedules or tasks
2. Managerial/ administration	Disagreements that develop over how the project will be managed; the definition of reporting relationships, responsibilities, project scope, plans of execution, negotiated work agreements, and procedures with other groups
3. Communication	Disagreements resulting from poor or lack of information flow among staff or between management and technical staff, including such topics as misunderstanding project goals, strategic mission of the organization, and the general flow of information from management to staff and staff to management
4. Goals/priority definition	Disagreements arising from the lack of specified goals or poorly defined goals; disagreements regarding the project mission and related tasks, the importance of activities or tasks, and/or frequent shifting of priorities
5. Resource allocation	Disagreements resulting from competition over resources (e.g., people, materials, facilities, and equipment) among project team members or across teams; perceived insufficient, inadequate, or inappropriate resources assigned to project
6. Reward structure/ performance appraisal/ measurements	Disagreements that originate over differences in reward structure; insufficient match between the project team approach and the performance appraisal system
7. Personality/ interpersonal relations	Disagreements that focus on personal and interpersonal differences rather than technical issues; includes differences arising from ego, values, prejudices, stereotyping, sexism, etc.
8. Costs	Disagreements that arise from the lack of cost control or authority within either project or functional group; disagreements over the allocation of funds
9. Technical opinion	Disagreements that arise particularly in technology-oriented projects over technical issues, performance, specifications, requirements, technical trade-offs, etc.
10. Politics	Disagreements that center on issues related to territorial/organizational power, NIH (not-invented-here) attitudes, or hidden agendas
11. Leadership/input/ direction	Disagreements that rise from a need for clarification of project-related goals or the strategic mission of the organization; perceived lack of decision making regarding goals
12. Roles/structure	Disagreements over perceived, related or overlapping assignments, tasks, or missions
13. Unresolved prior conflict	Disagreements that are perceived to stem from prior unresolved opposition

298

TABLE 9.4   TOTAL SCORES ALLOCATED FOR 13 CONFLICT CATEGORIES, IN DESCENDING ORDER

Category/Source	Total Score
Goals/priority definition	1877
Personality	1201
Communication	1183
Politics	655
Administrative procedures	633
Resource allocation	571
Scheduling	527
Leadership	453
Ambiguous roles/Organizational structure	185
Costs	174
Reward structure	151
Technical opinions	131
Unresolved prior conflicts	96

communicated effectively, or are frequently ignored or changed by management, disagreements will occur. Such an environment produces additional stress and frustration on the very teams trying to complete their objectives; and when stress and frustration run high, patience and tolerance run low. Diagreements that in the "proper" environment may be worked out, turn to heated debates and poorly utilized conflict.

The category of personality and interpersonal relations presented another dramatic change from previous studies. When examining the overall distribution of conflict scores, personality and interpersonal relations was the second greatest source of conflict.[14] This dramatic rise in interpersonal conflict may be related to the increased use of cross-functional teams. The team approach may be especially challenging to those who are more technically oriented and who may prefer working with tangibles (such as numbers) rather than the intangibles one must deal with when working with people. This dramatic change may have tremendous implications to human resource specialists and those with responsibility for human relations training. With the increased importance placed on flatter organizational structures, every effort must be made to provide cross-functional team specialists with a keen understanding of what teamwork truly entails, and a deeper appreciation of the value of differences.

---

[14]Ibid., p. 14.

9.8 PLANNING FOR CONFLICT

One way to manage conflict is to wait for it to happen, and then smother it with a barrage of tactical skills and interpersonal charm. For this approach to be successful, project team members must be well versed in conflict resolution strategies and constantly on the alert for eruptions or dissonance. Another route to managing conflict is through preventive planning. If project managers are aware of the impact or potential of each conflict source, they may be in a better position to encourage synergy and change. Mapping out sound conflict moves and making them at the right time will substantially reduce major crises and keep conflict at manageable levels.

Project priorities, schedules, resource allocation, and communication are but some of the key issues that have strong potential for conflict. If project planning and scheduling functions are performed properly, the odds of meeting project deadlines and control parameters are increased, and conflict levels tend to diminish. Poor communication, conflicts of interest, and personality differences are also primary sources of project conflict. If blended properly, project teams can be better prepared to deal with disagreements on a routine basis.

In managing conflict, the process of project planning may prove to be as important as the plan itself. By involving participants in the planning process, project managers may generate personal commitment among their team members. Some strategies for minimizing the destructive effects of some key conflict areas and maximizing their potential for more effective project communication and control follow:

· *Priorities*

1. Jointly define and establish a team mission statement.
2. Jointly define a master project plan.
3. Jointly develop a work breakdown structure.
4. Define customer needs and solicit input.
5. Provide feedback as plans are implemented.
6. Review work packages periodically.
7. Establish contingency plans.
8. Obtain early buy-in from all program participants.
9. Establish a change control process.

· *Schedules*

1. Jointly establish a preliminary project schedule, with major hard milestones.

2. Solicit input from all organizations involved.

3. Identify risk areas and define contingency plans.

4. Establish regularly scheduled status review meetings.

5. Utilize the critical path method to track and control the project.

6. Track progress and update schedules regularly. Provide feedback to team members.

7. Reward accomplishment of milestones.

· *Procedures*

1. Establish a project focal point and clearly delineate project administrative procedures.

2. Define roles and reporting relationships.

3. Jointly develop a project responsibility chart that defines roles and responsibilities of the project team.

· *Personality issues*

1. Provide training in team dynamics, listening techniques, and conflict management strategies.

2. Develop and maintain harmonious working relationships through team-building approaches and interventions.

3. Identify new project resource needs early.

4. Use the talents of an outside facilitator to mediate in the resolution of interpersonal and team conflicts.

9.9 CONFLICT RESOLUTION STRATEGIES

The potential effects of conflict, whether detrimental or beneficial, will depend on the environment created by project managers as they manage the numerous conflict situations that arise. There is no single right way to handle conflict. Managing conflict situations requires flexibility, an awareness of the most appropriate resolution strategies available, and a knowledge of the impact that such strategies may have on the project team.

Basically, there are five common methods available for handling conflict:[15] withdrawal, smoothing, compromise, forcing, and collaboration. As

---

[15]R. Blake and J. S. Mouton, *The Managerial Grid* (Houston, TX: Gulf Publishing, 1964).

is the case with any leadership function, the specific method of handling conflict will depend on a number of situational variables. The best approach, however, will be the one that removes the obstacles that block or prevent project completion and works to build a better project team. Analyzing the alternatives and their possible effects places project managers and team leaders in a better position to choose their strategies wisely.

1. *Withdrawal (denial/retreating)*. The withdrawal approach is frequently associated with low levels of conflict, or situations in which the conflict appears to have little impact on project activities. Here the project manager, or project team, chooses to ignore the conflict or deny that it even exists. As a cooling-off period, or a method of buying time, the strategy can be quite effective. Withdrawal usually is a short-term strategy that neither deals directly with the conflict at hand nor serves to blend a better or cohesive team. Conflict tends not to evaporate, but to build in intensity. Withdrawal is therefore the least appropriate conflict resolution strategy when the issues are critical to project success.

2. *Smoothing (suppression)*. Sometimes, when people fail to recognize the constructive aspects of handling conflict openly, they may tend to suppress it by playing down the differences in viewpoints or opinions and emphasizing the commonalties or strong points. The result may be superficial harmony. Smoothing may, however, serve as a clarifying tactic and may be used effectively to emphasize certain points to a colleague (or opponent) prior to entering negotiations. In this sense, the typical pep talk may be regarded as a smoothing tactic. Smoothing over the conflict does not, however, help to eliminate it and is, of course, an inappropriate approach to use when others are ready and willing to deal with the issue. If the issues are relatively minor, however, and time for problem solving is limited, smoothing or suppression may be the most appropriate action.

3. *Forcing (power or dominance)*. The use of power position, perhaps even majority rule, is frequently used to force or strongly persuade participants to reach decisions. It usually involves exerting one's point of view at the expense of another's. In many instances, the use of forcing results in antagonism and resentment, which can increase conflicts. Forcing is a method that is most appropriate when used as a last resort.

4. *Compromise (negotiation)*. Compromise ("You give a little and I'll give a little") involves considering various issues and determining a middle-of-the-road solution. Although valued within our society as a conflict resolution strategy, compromise often leads to watered down and ineffective solutions, in which commitment by all parties may be dubious. Compromise is a situation in which neither party can win but that generates some

acceptable form of resolution. It becomes a somewhat risky resolution strategy when considering disagreements over quality or technical performance.

5. *Collaboration (integration).* The collaborative conflict resolution strategy involves recognition of the positive aspects of the conflict in terms of generating a variety of alternatives and solutions with regard to the specific problem at hand. Unlike compromise, where parties enter into negotiation prepared to experience some degree of loss, those involved in a collaborative effort fully expect to modify their original view as the group's work progresses. Emphasis is placed on confronting the conflict, determining common goals, and generating a group solution. Often referred to as a *confronting strategy*, the assumption underlying this integrative approach is that the value of the group's effort exceeds the sum of each person's contribution. Collaboration becomes even more difficult, however, if the time to apply consensus-seeking strategies is perceived not to be available, or if commitment to group goals is not present.

Choosing an appropriate conflict strategy is not an easy task. Conditions in which conflict occurs vary and therefore demand a sensitive, yet objective eye. Effective conflict management requires much more than selecting one's best personal style and eradicating the differences among opposing parties. It involves recognizing and investigating the various sources of conflict that exist and then determining and predicting the effects of a resolution strategy on the project team's performance. This may involve using a variety of conflict resolution strategies.

9.10 THE BEST CONFLICT RESOLUTION STRATEGY

Of the five basic resolution strategies, the collaborative– integrative approach appears to be the ideal method for resolving conflict. The collaborative strategy is preferred because it assumes that all parties involved will come out of the negotiation as winners. In fact, a variety of investigations indicate that although project managers use the full spectrum of conflict-handling modes in managing diverse conflict situations, they most frequently rely on collaborative techniques. Compromise is ranked second preferred, with smoothing ranked third, followed by forcing and withdrawal.[16,17]

---

[16]Thamhain and Wilemon, "Conflict Management in Project Life Cycles," and "Leadership, Conflict, and Program Management Effectiveness."

[17]Posner, "What's All the Fighting About?"

From a management perspective, it is impossible to use collaborative strategies at all times for all conflict situations. The choice of an appropriate strategy relies on a number of variables, which include (1) power positions of the conflicting parties, (2) managerial and personal philosophies, (3) impact of the conflict situation on the project's schedule, (4) bottom-line monetary impact, and (5) effect on the project team.

Project managers who hold positions of power within the organization favor both collaborative and forcing resolution strategies when the negative consequences of the conflict are high. These same project managers tend, however, to prefer and use smoothing, compromise, and withdrawal when the stakes are low. Project managers with lower organizational power positions, such as those frequently found in matrix organizations, tend to use collaboration and compromise for high-risk conflicts, while substituting compromise for withdrawal when the effects of the conflict were considered less detrimental.[18]

It appears, therefore, that there is no one best conflict resolution strategy. Smoothing works well to clarify certain issues and to stress common goals, while withdrawal may be appropriate when one is missing all the facts and needs to delay making a decision. Compromise is used best when both parties can afford to give something up. Forcing, in turn, although a win–lose situation, may be necessary when the conflict has reached a crisis phase. Integrative collaboration, although lengthy and requiring skill at consensus seeking, actually synthesizes all approaches and should be used as the preferred mode when time and skill are available.

Exercise: Using Conflict Resolution Strategies. Like effective leadership, conflict resolution strategies depend on a number of variables or conditions that help determine the most appropriate course of action. Power position, the impact of the strategy on project team members, and where you are in the project's life cycle are among the determining factors. On the basis of your knowledge of conflict resolution strategies and their appropriateness for certain situations, complete the sentences below. Try to be as specific as possible regarding the characteristics of the situation that make this conflict management style most suitable.

1. Smoothing or suppressing makes sense when/if _____.
2. Avoiding or withdrawing makes sense when/if_____.
3. Forcing or dominating makes sense when/if _____.

---

[18]Thamhain and Wilemon, "Conflict Management in Project Life Cycles," and "Leadership, Conflict, and Program Management Effectiveness."

4. Compromising makes sense when/if _____.

5. Collaborating or integrating makes sense when/if _____.

9.11 KEY STEPS TO MANAGING CONFLICT

Successful project managers realize that conflict is inevitable and therefore develop procedures and techniques for minimizing its negative effects and maximizing its constructive potential. Such procedures may include (1) analyzing the problem in terms of the variety of situations that lead to conflict, (2) assessing the effect of a particular approach or methodology on the conflict and on the project team, and (3) developing the appropriate environment or conditions that encourage negotiation and resolution. Should a conflict occur within a project team, a confrontation meeting is necessary between the conflicting parties, units, or departments. The project manager should then be aware of the recommended actions and sequence of events as listed in Table 9.5.

Developing expertise in project management strategies will help project managers to minimize conflict throughout the project life cycle. Recommendations for improving project manager effectiveness and minimizing the negative effects of conflict include:

- Communicating key decisions in a timely fashion to project specialists
- Adapting leadership styles to the status of the project and the needs of project personnel
- Recognizing the primary sources of conflict and the effectiveness of each of the conflict-handling approaches
- Experimenting with alternative conflict-handling modes
- Providing work challenge to motivate support staff
- Developing and maintaining technical expertise
- Planning early and collectively in the life cycle of the project
- Demonstrating concern for project team members

Every effort should be made, right from the start, to integrate the various functional groups affected by the project. Product development core teams, for example, should include at least one person from each of the disciplines or functions contributing to the project. Letting people know what life will be like after the project is complete helps to keep them on track and headed toward the goal. High-stress environments should be loosened up by "hokey," but effective team-blending activities, such as prizes, functions,

TABLE 9.5   APPLICATIONS OF CONFLICT MANAGEMENT

Key Step	Application Ideas
1. Ensure that key people affected by the conflict are involved.	
2. State the desired outcome of the situation.	• State the status quo of the conflict in terms of available facts, and point out the impact on the organization and the team.
3. Have members of the group describe their view of the situation	• Hidden issues or agendas may surface. • Avoid discussion of solutions at this time. • Avoid arguing. • Probe for what facts have led to a particular point of view, but do not criticize any member of the group for taking a position. • Seek additional viewpoints, if necessary.
4. State the problem as understood by members of the group.	• Most problems are a product of misunderstandings; often restating the problems clarifies the issues.
5. Speculate on possible solutions.	• Be creative! Encourage others to come up with unique ideas. • Allow people to contribute freely and openly. • Build on points of agreement and reconciliation. • Focus on the overall goals of the organization. • Allow ideas to develop without prejudgment. • Do not let the absence of data lead to reject concepts that may be valuable. • Develop a list of possible solutions.
6. Evaluate ideas in terms of objectives.	• Decide how ideas would contribute to solutions and to subordinate goals. • Follow ideas through to their consequences and ultimate impact on people, costs, and systems. • Seek additional data to validate ideas.
7. Negotiate a resolution; test for clarity and agreement.	• Conflicts are resolved through flexibility and creativity. Each group member should see clearly how the solution would benefit the individual, the team, and the organization.
8. If the problem cannot be resolved in the group, seek mediation from a third party.	• Sometimes, conflicts are reduced to contests because of negativity and inflexibility; in that case, a third party should review the merits of each side's point of view and help make a final determination.

and parties. One top 500 corporation, during a major product development effort, celebrated the accomplishment of hard milestones by renting a football stadium for family picnics. Each team player had the opportunity to run through the goal posts as the scoreboard flashed his or her name to the jubilant crowd. Finally, since research consistently predicts what we can expect, resolving conflicts early in the project life cycle through collective project management planning, scheduling, and tracking strategies will help to prevent slippage and ill feelings as the project progresses.

9.12 CONCLUSIONS

Conflict is an inevitable and necessary part of the project environment. Given the proper atmosphere, attitude, and training, however, disagreements can actually broaden perspectives and stimulate innovative, productive and cohesive interactions. In their efforts to deal with the uncertainty of situations such as changing technologies, dynamic competitive markets, and global economic conditions, managers have found that resulting conflicts can actually become predecessors to change. Project managers who realize that preventing conflicts is as important as solving them are likely to be effective.

CASE STUDY: INTERNATIONAL ELECTRONICS—A CONFLICT SIMULATION EXERCISE*

International Electronics (IE) is a medium-sized corporation specializing in designing and manufacturing of communications systems and components. Located within the northeastern United States, IE has experienced a relatively high growth rate in the last decade. Recent market changes, however, have had a negative impact on IE's bottom line. In the last 16 months, profits have dropped by 10%, as IE products enter into a highly fragmented and competitive market.

Last year, a project team was established to coordinate the introduction of the new PX 1090 communications system. The PX 1090 is a sophisticated and highly functional system to be used in the mobile communications industry. According to detailed market analysis, the venture has tremendous potential of capturing a lucrative market for International

*This study was written for class discussion purposes only. It is not intended to reflect management practices or products of any particular company or organization.

Electronics. You are pleased to accept the position of project manager for this important mission. You report to the vice president of engineering systems, who, in turn, reports to the divisional vice president.

This morning, 12 months into the 26 months of the project, you receive a memo from the newly appointed manager of development. As part of his first official duties, he informs you that he has changed the central switching module of the system to an all-new digital design. The memo also contains sketchy but supportive documentation as to the increased reliability of product. Changing the system will require major revisions and will set back total system implementation by at least one year. You are also aware that considerable time and research went into the design of the product, and how important timely delivery of the technology will be to future market opportunities and company growth.

Upset over the development manager's memo, you decide to walk in on him and discuss the situation. You hope that you can get him to hold off on his change request. The discussion becomes increasingly entrenched as he stresses the benefits of his plan and the errors and lack of information on which the previous plan was based. You point out the benefits of placing the current product in the hands of the user as soon as possible. Each of you becomes increasingly angry, and the meeting ends on a less than positive note.

At this point, you're extremely upset and decide that something must be done to ensure the success of the project. You see your alternatives as follows:

A. You can send a memo to the development manager, explaining your position and how releasing the present design will allow the company to maintain its current market share while positioning it for future growth.

B. You can ask the development manager to meet with you for a full day next week in order to work out your differences and come up with alternative solution.

C. Let it go for now; he will probably cool off soon enough and the crisis will be over.

D. You can go to the divisional vice president and request that former requirements be adhered to.

E. You can invite the divisional vice president to the next project review meeting to stress the importance of the project to the company's current sales.

F. You can march right into the development manager's office and demand that he justify his position.

G. You can send a letter to the division vice president, resigning from your position.

H. You can invite a representative from the divisional vice president's office to arbitrate all team problems and issues.

I. You can send a letter to the development manager (with copies to all team members, the divisional vice president, and the president) indicating that his or her opposition is holding up a potentially profitable project.

J. You can ask all the other department managers, who will realize a schedule delay, to convince the development manager to ease his request or incorporate it in the spin-off series.

Rank-order your alternatives and identify the resolution strategy you believe that each alternative represents (i.e., smoothing, withdrawal, forcing, compromising, collaborating).

Discussion

As we discussed earlier, there is never any "correct" conflict resolution strategy. In this exercise, the various alternatives represent strategies that must be viewed from the perspective of the specific life-cycle phase, the intensity of the conflict, the nature of the relationships of the individuals or departments involved, and the long- as well as short-term consequences of implementing the alternative.

Let's examine each of the alternatives presented in terms of their effect on the project team and how they address or resolve the conflict at hand:

A. This conflict resolution strategy represents a *smoothing* or clarifying strategy. By sending a memo and stating your position, you are explaining your motivations and objectives and how they match the project's needs. The purpose of this strategy would be to provide the development manager with a greater perspective and understanding of the problem at hand and an opportunity to reevaluate any previous decision. It may be wise to follow a smoothing or clarifying approach with subsequent collaboration. The danger of this strategy is that it does not deal directly with all the conflict issues at hand: namely, the development manager's objections.

B. The conflict resolution strategy here is *collaboration* or *integration*. By suggesting to meet for a full day, the project manager has indicated commitment to integrating his or her ideas with those of the development manager in the hope of generating a brand new solution. The emphasis here is on generating a solution to the problem through sharing expertise and ability. The outcome, therefore, is most likely a *win–win* solution to the conflict and will create the most positive long-term solution for the project and project team. Unfortunately, such an approach also takes the greatest amount of time. This solution, naturally, is the *most favored*, provided that time and skills are available.

C. "Letting go" represents a form of *withdrawal* or *retreating* and may be extremely appropriate as a short-term cooling-off solution. Ignoring or denying the conflict, however, will not make it go away. It is best to follow through on such a cooling-off period with more constructive, integrative, problem-solving approaches. Conflict, as you recall, intensifies over time.

D. This is a *forcing* strategy, as it is a direct request to use power and position to resolve the conflict unilaterally. If both managers, however, had jointly agreed to approach the divisional vice president for clarification of organizational objectives, we would consider the strategy to be one of escalation and in the long run more desirable than forcing. Forcing is frequently used by project managers in the face of scheduling crises; however, if used to excess, the strategy may fail to blend team members and lead to long-run disruptive feelings of resentment.

E. As indicated previously, like leadership, conflict resolution strategies are situational and depend on our attitudes, power position, project life-cycle phase, time, and other factors. This alternative can actually be viewed as two strategies. On one hand, the pep-talk that the divisional vice president delivers may be seen as a way of clarifying the importance of the project to team players and the need to reconsider the issues that are creating conflict. Such a *smoothing* strategy is represented by the vice president's attempt to play down the differences among factions and stress the common goal. Another way to view this strategy is as a subtle form of *forcing*. Since it was the project manager who invited the vice president to the meeting, it may be the project manager's desire to create an opportunity to be the one to discuss the issues and present the dangers of the conflict situation *unilaterally* to the divisional vice president. This strategy, of course, could result in resentment and hostility on the part of the other team members should they feel manipulated into submission.

F. This *confrontational* approach in some cases may place the development manager in a somewhat defensive position and may entrench the conflict further. However, confrontation of this nature may be successful depending on the previous relations of the involved parties. We all know individuals (and couples) who are capable of thrashing it out until a solution is finally generated. If the project and development managers have had a strong relationship in the past and have had positive experiences in which conflict was resolved in this manner, the approach may lead to the beneficial win–win solution. Immediately confronting a heated situation, however, should be considered with care.

G. Resigning from the project team, ostensibly, represents *withdrawal* and is the least preferred method as viewed by most project managers. However, depending on the power position of the project manager within the organization, this alternative may be viewed as a *forcing* mechanism. If the project manager's responsibilities and duties cannot at this time be adequately accomplished by someone else, and if the project is in a critical life-cycle situation, this strategy may be a powerful, although dangerous tool to accomplish unilateral objectives.

H. Again, alternative H can be viewed as a combination of conflict-resolution strategies, depending on the rationale behind the approach. Arbitration, for one, is a form of *compromise*. The arbitrator with the assistance of the opposing parties will seek to find some middle-of-the-road solution and generate some resolution to the conflict at hand. However, by inviting a representative from the divisional vice president's staff to arbitrate *all* issues and problems, you have unilaterally relinquished all decision-making authority and, in effect, have *withdrawn* from a position of independence. Since this decision was made without the negotiation and acceptance of all concerned, lack of respect may develop for the project manager.

I. Obviously, this strategy is *forcing*. What we usually recommend in our seminars is that if you implement alternative I, you really have to resign. This alternative is certainly a dangerous strategy, depending on where the project is in the life cycle, for it will serve to polarize and destroy a project team. Writing such a letter with carbon copies to all is an aggressive strategy because of the degree of forcing that is practiced. Although sending a letter might serve to clarify certain issues, in this instance it is not the content of the letter that is significant but the fact that pressure will be exerted on the development manager from *all* organizational directions. Politically, it is a very dangerous strategy and should be avoided at all costs.

J.   What a difference a well-thought-out plan makes! Although a *forcing* mechanism, this strategy is one of the preferred techniques of *using power peacefully* to one's advantage. By lobbying with the other functional managers on the team, the project manager has additional power to exert on the opposing manager. At the same time, incorporating the development manager's plan in a spin-off series provides that person with the recognition needed and the corporation with a new product line.

Now review your ranking in terms of the categories of conflict resolution. If you are like most project managers in a matrix environment, your conflict resolution strategies will be, from most preferred to least preferred:

- Collaboration
- Compromise
- Smoothing
- Forcing
- Withdrawal

10

TRACKING AND CONTROLLING THE PROJECT

*Consider the little mouse,
how sagacious an animal it
is which never entrusts its
life to one hole only.*

PLAUTUS, TRUCULENTUS,
ACT IV, SC. 4, 15

10.1 INTRODUCTION

On November 7, 1940, eyewitness accounts of easily the worst project disaster in the history of modern civil engineering spawned a controversial debate that has continued among physicists and engineers since the beginning of the twentieth century concerning the controls that must be exercised when designing and building a modern suspension bridge. At approximately 10:00 A.M., a 42-mile an hour wind jolted the mile-long structure spanning a narrow arm of Puget Sound between Gig Harbor peninsula and Tacoma, Washington. Already infamous for its springiness, since its opening on July 1 of the same year, the center span would rise and fall as much as four feet even in winds of only three or four miles an hour, earning the Tacoma Narrows Bridge the nickname of *Galloping Gerty*. Drivers would either go out of their way to avoid crossing the roadway or purposely seek out Gerty for the roller coaster thrill. It was said that the lights of the cars could be seen ahead disappearing and reappearing as they bounced up and down.

Valiant attempts by project engineers to monitor the vibrations led to the conclusion that the motions were predictable and tolerable, so only minor remedial efforts were made to reduce them before the bridge was open to traffic. Apparently, Tacoma officials were so confident in these reports that they even planned on reducing their insurance before the November 7 storm, which caused some bracing connecting the center span to give way, abruptly stopping the predictable up-and-down oscillations. Instead of raising and falling together as usual, the main span cables began to pull and wrench in opposite directions, creating a violent twisting motion that changed the bridge deck's vertical oscillation. When the roadway suddenly tilted side to side, as much as 45° from the horizon, Gerty became too much even for the thrill seekers, and everyone got off. Finally, at 11:00, the center span buckled and a 600-foot-long piece of the first section broke in two and plummeted 90 feet into the water.

An observer's recording on film of the collapse of the Tacoma Narrows Bridge presents scientific evidence for the much-debated corroborating mathematical analysis of the project disaster. The bridge, which was rebuilt in 1951, is one of the world's longest. Since the collapse of the first Tacoma Narrows Bridge, bridge construction project managers have used wind tunnels, computer simulations, and mathematics to create sophisticated models of a bridge's vibrations to test their designs in an attempt to guarantee safety.

From this infamous example, we can easily understand how not establishing control over project events can be devastating for an organization, causing injury, inconvenience, delay, and loss of life, not to mention loss of an assumed profit or even economic ruin. As managers of modern development projects, it is our responsibility to assure that our teams comply with the defined performance standards and established fiscal targets set for our projects by monitoring daily operations through formal and informal reporting procedures. These procedures help determine how a project is progressing and whether the initial goals and objectives will be met.

In this chapter we discuss a variety of appropriate tracking and controlling techniques used by commercial and government-sponsored program efforts to manage ongoing project performance. Using these tracking and controlling techniques to exercise control over the cost, schedule, and quality of program and project work efforts helps guarantee the successful achievement of sought-after objectives. Today, the bridge that fell four months and seven days after its opening remains a tourist site. One of the world's longest artificial reefs, thrill seekers now dive to look at the chucks of Galloping Gerty in Puget Sound.

10.2 COMPONENTS OF PROJECT CONTROL

10.2.1 When Planning Isn't Enough

Projects often experience too much or too little control, perhaps due to the naive optimism that everything will proceed as planned. The former demotivates personnel, typically consumes too much time, and costs too much money. Insufficient control, on the other hand, can lead to disaster.

The following memorandum was written when the program management team of a major software release reached what seemed to be an insurmountable impasse. Admittedly, this team has committed several classic mistakes. But could the project have avoided the problems by the regular monitoring and reporting of progress against a baseline plan? What controls should team leaders have exercised to detect deviations from plan and allow the decision makers time to redirect, or possibly terminate, the project without severe cost overruns and schedule delays?

The new client server application promised customers scalability without the added cost of new hardware upgrades. Its eagerly anticipated user interface was designed to reduce service dispatch rates by making the software simple enough to allow customers to install and administer the application themselves. Access to program updates would be deployed across the Internet and supplemented by a richly embedded, on-line help system.

The application was planned to support more than one server platform, making it possible to target customers in all three ranges of the business computing environment (low, middle, and high end). These customers valued performance and price over abundant features as important attributes of the total business solution.

1. Nine months into the endeavor, program management concluded that to realize the aggressive schedule, the number of deployed resources should be doubled.

2. The budget for the development had been allocated as planned, but the resources were shared across other mature project efforts, which also happened to be late to market.

3. Market management hadn't defined key features to position it against competitive entries.

4. The technologies being used were too old. Although the programmers had discussed using the most modern development language and a distributed architecture to improve transaction response time, they were unfamiliar with these techniques, and the schedule did not allow for ramp-up time to develop an experimental prototype.

5. The original budget assumed half the number of function points required to complete the new application. This would require an increase in technical resources equivalent to four to five staff years to code and test the additional software.

6. The current budget required that the management team keep development cost essentially flat so that it will be another five months before the application is ready to release to the market.

7. The project has yet to implement a configuration management control system to support the planned beta trails.

8. Additional funding hasn't been set aside for contingencies that might result from unanticipated requirements.

9. Market intelligence has learned a considerable amount of information about a competitor's product which one of your major accounts will test within the year.

10. To continue, the project would need to redesign the application architecture, costing some 800 hours of rework.

11. One proposed way to put the application development back on course and not overspend the development budget is to relocate the project overseas to a sister operation in newly acquired state-of-the-art facilitates.

12. If the project were to move overseas, there would be an increase in the hours for project administration, due to the longer duration and cost of travel.

This software application project illustrates that few entrepreneurial efforts proceed as originally conceived without the management team interceding to control deviations from plan. The team accomplishes control by monitoring, tracking, and guaranteeing the project's fiscal and technical performance. Much more today than in the past, this guarantee is also a promise that the cost of fabricating a product or service design will not exceed the price points set for the project by the customer's willingness to purchase. In addition, it is a pledge that the product or service the team brings to market will perform according to its technical specifications. We quickly recognize these objectives as familiar components of the quadruple constraint that all projects face and discuss methods for controlling them in Sections 10.4 and 10.5.

10.2.2 Process of Project Control

For many decades, the process of project control was exercised as an *evaluative process*, whereby line managers reported deviations from planned events and assessed probable causes. Project control was also conducted as a *performance measurement process*, with functional management sharing the responsibility for initiating any corrective action to alleviate the impact of unforeseen deviations or unfavorable trends on project schedules, budgets, resources, or staffing levels. Today, stimulated by the worldwide quality revolution that began in the mid-1980s, project control has become a *quality assurance process*. As such, organizations must demonstrate that their designs meet technical performance standards and assure that a quality product or service is delivered to customers at a price they are willing to pay.

As a result of this mandate and the drive to keep pace with incessantly changing technology and accelerating product development cycles, organizations have operationalized these controls at the project level. Where the step-by-step process of verifying whether project work is done on time and within budget was once the duty of technical supervisors and department managers, the entire project team now shares this responsibility. In many organizations, a core team of cross-functional project leaders and managers keep top management appraised of project status, especially when slippage or budget overruns jeopardize contractual agreements and market commitments. Senior management typically reviews project control documentation with team leaders. This is to help clarify new directives or share in the decision to take corrective action because of a change in the organization's priorities and strategic plans. The team also must update chief executives who need information about project status to evaluate the firm's overall financial performance. Teams may even recommend to executive management whether to terminate a nondiscernible project effort and redeploy the resources to a more profitable venture. In a participative management environment, the project's destiny is most assuredly in the hands of the project management leaders.

There are five fundamental steps in establishing a project control process as illustrated in Figure 10.1.

Step 1: Establish a measurement system. Step 1 begins in the project's planning phase and incorporates all the technical specifications, cost parameters, and market requirements that must be achieved. This is where project personnel set their first hard milestones and develop preliminary test plans that will serve to advance the project toward its established end date. Test plans often are written by a separate quality assurance or verification team, as is the case in software program testing. But no matter who writes the plan, it should be thoroughly reviewed by the appropriate members of the project team. At a minimum a good plan should (1) explain the specifications that the test is to verify, (2) specify how the test is to be performed, (3) list what equipment will be required, and (4) describe the expected results and the criteria for successfully completion.

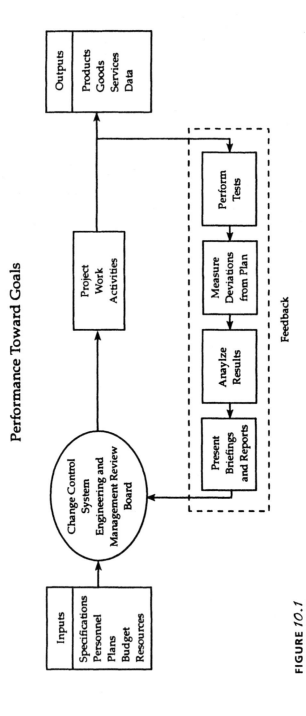

Performance Toward Goals

Inputs
Specifications
Personnel
Plans
Budget
Resources

Change Control System
Engineering and
Management Review
Board

Project
Work
Activities

Outputs
Products
Goods
Services
Data

Perform
Tests

Measure
Deviations
from Plan

Anaylze
Results

Present
Briefings
and Reports

Feedback

FIGURE *10.1*

Process of controlling project work.

To control the project effort properly, the project team will need to perform additional testing that can help determine deviations from schedule, overages in budget, and changes in the numbers of personnel assigned to complete specific tasks. Too often, the project management team is asked to "trust" a status report handed to them two months into a six-month effort. While the report may state that nothing is wrong, four months later a minor bug delays the final system integration and customer acceptance testing. Unfortunately, project management might have alleviated this several-month delay by scheduling the reporting intervals to occur more frequently and incorporating a major test at the end of a specific life-cycle phase to gauge progress. In some projects, biweekly tests may be in order so that the report recording the status is believable and accurate.

As shown in Figure 10.1, process control is a feedback system in which deviations from the plan are reduced. In any control system, increased delay results in an increased tendency for the control system to become unstable. For the control process to operate properly, the team must incorporate carefully designed tests into the control process and conduct them in a timely manner. We can accentuate their importance by calling them *milestones*.

Step 2: Measure results and assess deviations from plan. Wherever possible, it is very important that the tests conducted by the project team be very specific and carefully designed to meet specifications and assess progress. There is little point in designing and performing a test that incorporates specifications not contracted for or needed by the customer to satisfy market requirements. To ensure agreement that the tasks are completed, both the individual team members performing the work on the activity and its successor should take the measurements and conduct the assessment. In addition, no activity being tested can be judged to be complete unless all the specifications relating to that activity have been met. This is the only way that quality can be assured.

Step 3: Report results orally and in writing to the appropriate personnel. Deviations from plan can be readily assessed after measurements are taken and reported in a one-page status report. Such a report is designed to present the extent of any deviation in a clear, concise manner. While there may be a tendency to "fix" the problem before submitting the report to make the report look better, delays in reporting could be catastrophic, causing instability in the control process. Long reports also delay reporting. Too often we hear the comment, "Of course I'm late; I had to write a 20-page report."

Step 4: Evaluate the results of the report. All appropriate personnel — including representatives from engineering, system test, quality assurance, project management, market management, and customer support — evaluate the results of the report. Trends are noted and reviewed; appropriate trade-offs are studied. Of course, be cautious to avoid responding to each deviation by taking some corrective action. Where there is no trend, no corrective action should be taken. We use the physical control system as an analogy — if sufficient friction damping or resistance is not applied, large overshoots in performance may occur that may well extend the project duration rather that shorten it.

Step 5: Apply corrective action. As we introduced in Chapter 5, we recommend establishing a *change control board* made up of selective members of the testing, quality assurance, engineering, marketing, and project management teams. The change control board considers any deviation carefully and assesses the impact imposed by the proposed change on the quality, schedule, cost, and customer's acceptance of each component or project deliverable in question.

10.3   MODERN PROJECT CONTROL SYSTEMS

10.3.1   Establishing Systematic Project Control

For a control system to be meaningful, it must measure project information economically. Modern control systems must be *discernible* and *appropriate* for the complexity of the tasks being addressed and the size of the project effort. They also are *timely, simple* to employ, and *congruent* with the events being measured.[1]

1. In general, it isn't necessary to collect data the project team doesn't intend to analyze. Similarly, the project control system's effectiveness doesn't increase by using computerized reporting programs to generate huge amounts of statistical information. Unread reports do not improve by their sheer weight or volume. The project team must be careful to determine the minimum information needed to analyze project situations to redirect or apportion work efforts prior to setting up a project control system. The system's capacity need only match these minimum requirements.

---

[1]Peter Drucker, "Characteristics of Controls in Business Enterprise," in *Management: Tasks, Responsibilities, Practices* (New York: Harper & Row, 1974), pp. 496–505.

2. Trivia should never be measured. Control is maintained by assessing primary activities that affect overall performance and results.

3. Control any changes in the scope of project activity. Measure and compare results to predetermined standards; report only significant deviations and trends away from established or baseline plans. Just because a phenomenon can be measured, there may be no relevance in reporting it.

4. Controls must also be appropriate to the size and complexity of the activities or tasks being measured. In general, the more visible and larger the commitment of resources, the more managerial attention associated with the project and the greater the amount of control expended. The size of the project, however, doesn't determine if the project will go adrift. Technical, cost, and schedules performance are easily deterred in small as well as large and complex projects.[2]

5. Measurement systems should be congruent with the events being measured. A measurement isn't more accurate if calculated to the *nth* decimal place when it is, at best, a trend perceived in a diagram, chart, or graph.

6. Controls should be simple to employ. Costly systems may be too complex and may not satisfy the objectives of providing the minimum amount of information to the project management team when needed most for interpretation and decision making. A system that isn't *useful* or *usable* will not be employed.

7. The control system must provide reports and generate data on a timely basis so that adequate corrective action can be taken before deviations from plan baselines become serious. Two dangers exist: measuring too frequently so that reports are meaningless and extraordinarily expensive, and not measuring enough to monitor the "pulse" of project activity effectively. Managers must think through when and in what dimension (cost, schedule, technical performance, and customer satisfaction) they need to measure successful achievement of project objectives. This is especially complicated in R&D projects that may culminate years later as prosperous commercial ventures.

8. Operationalize control systems. Fit them into the project organization. Give access to those members of the project team who need to interpret the data and who are capable of taking corrective action regarding the problem situation. Tailor the information to these team member's needs.

---

[2]R. J. Might and W. A. Fisher, "The Role of Structural Factors in Determining Project Management Success," *IEEE Transactions on Engineering Management*, Vol. EM-32, No. 2, May 1985, pp. 71–77.

Reduce prose to graphics, and summarize project status in brief one- and two-page reports that are easy to read, yet comprehensive. Remember, establishing an effective control system also requires paying attention to the interpersonal communication subsystems that exist within the project, as well as to the roles and authority of the various project leaders and managers engaged in the exchange of project information. Chapter 11 is devoted to the topic of communication, as its importance to the establishment of meaningful and effective project control systems cannot be emphasized enough.

10.3.2 Environmental Nature of Project Control

In Table 10.1 we describe several useful techniques that project teams can use to control project costs, schedules, and technical work properly. However, the effectiveness of each technique for measuring performance deviations from initial estimates or baseline plans depends on several variables, many of which vary according to situational conditions present in the project environment. It may therefore be prudent for teams to choose to use more than one technique for a control system.

Since most project budgets today include some expenditure for software development, a new software quality assurance industry has emerged to ensure that the control techniques taught a decade ago become the accepted procedures or standards against which projects are measured. In the software industry, documentation gives the product visibility and allows the project management team to track progress against a plan much as, in a hardware project, the output of planned laboratory tests give the management team confidence that progress is under way. Tracking the completion and acceptance by the software engineering team of the system's architecture document, for example, provides a visible deliverable for the management team to access progress.

Today, it is not unusual for project managers of software development efforts to face inordinate challenges as they attempt to control the overly optimistic development and cost schedules established by teams in what is an immature industry. As Capers Jones, founder of the firm Software Productivity Research, Inc., has discovered, 20% of large software systems fail to deliver due to managerial, social, and technical reasons.[3] Since 1985, Jones's firm has published the findings, from annually administered questionnaires, in a research model that synthesizes the quality assurance

---

[3]C. Jones, *Software Quality Analysis and Guidelines for Success* (London: International Thomson Computer Press, 1998), pp. 303–308.

activities of more than 500 organizations. His findings are independent of the development methodology or practices followed in the reporting organizations, as well as the subindustries and countries where the development took place. Nearly every project team that ever produced software as a deliverable has some experience with these obstacles. These problems typically include (1) failure to assign a project manager or to use project management tools, (2) ineffective quality control, (3) ineffective software processes, and (4) ergonomic factors.

Although all contribute to the mishaps that teams experience, generally poor project control is regarded as the primary root cause of failure across six software subindustries: systems software, military software, information systems software, contract or outsourced software, commercial software, and end-user software.

Teams may learn from these empirical data what practices are followed by other organizations to control their projects. Of the management practices that cause variations in software quality and productivity, the following seven yield the best results.[4-6]

1. *Project planning:* using CPM, Gantt charts, and WBS techniques to plan, schedule, and track performance
2. *Project estimation:* using automated tools and models to:
   (a) Conduct side-by-side "what if" comparisons of different development scenarios
   (b) Demonstrate how various quality assurance approaches, such as formal design and code inspections or testing specialists, affect the project
3. *Risk analysis:* using custom-built or add-on features to project management estimating and tracking tools that predict:
   (a) Technical risks associated with inadequate defect removal or requirements volatility
   (b) Sociological risks from excessive pressure and staff burnout
   (c) Whether a project will run late, exceed the budget, or have a high probability of litigation

[4]C. Jones, *Applied Software Measurement: Assuring Productivity and Quality*, 2nd ed. (New York: McGraw-Hill, 1997).
[5]C. Jones, *Patterns of Software System Failure and Success* (London: International Thomson Computer Press, 1995).
[6]C. Jones, *Assessment and Control of Software Risks* (Upper Saddle River, NJ: Yourdon Press, 1994).

TABLE 10.1 USES AND CHARACTERISTICS OF MODERN CONTROL SYSTEMS

System	Characteristics	Action Items Reporting	Budget and Resource	Breadboard Development	Change Control	Data Analysis	Document Tracking	Performance Measurement	Prototype Development	Unit and System Test	Defect Tracking
Bar graph	Simple-format, broad-based planning and tracking tool that is used for briefing and reporting	X	X	X		X	X	X		X	X
Configuration management	Manual or computerized tracking system used to detect problems, modify requirements, and control project documentation				X	X	X				X
Change control management	Manual or computerized tracking system used to control project change or modification requests				X	X	X				X
C/SCSC-related cost reporting	DOD cost/schedule performance reports required on all major U.S. government contracts		X			X		X			
Design reviews, code inspections, and walkthroughs	Technical assessments of product's integrity and compliance to design specs held by technical experts at key decision points			X		X			X	X	X
Demonstrations	Quality assurance or end-user review of the working features of the software system to control product's reliability					X			X	X	X
Development statistics	Technical performance measurement statistics used to track progress, control quality, and predict completion date			X		X		X	X	X	X

Technique	Description									
Earned value analysis	Compares the budgeted cost of work scheduled, the budgeted cost of work performed, and the actual cost of work performed	×	×	×	×	×		×		
Gantt chart	Simple format, similar to bar graph, time-based planning and tracking tool that is used for briefing and reporting; good predictor of project completion date	×	×		×		×	×		
PERT	Each network activity is figured as three separate time estimates; used where time uncertainty is a factor; a good predictor of project completion date	×	×		×			×		
Project review meetings	Periodically scheduled business meetings of all project team members to review progress, problems, and solicit recommendations for improvements	×		×	×		×		×	×
Summary reports	Simple-format, broad-based account of project status prepared by the project manager to brief senior management, document changes, and report progress	×		×	×	×				
Automated project estimation methods	Mathematical modeling that uses project characteristics and historical data to compute the total cost	×			×					
Risk analysis methodology	Management process that concentrates on identifying, assessing, planning for, and dealing with events causing unwanted change	×	×	×	×	×		×	×	×

325

4. *Milestone tracking:* using add-ons to planning or estimating tools to track the quality assurance guidelines that drive the software development process and become the milestones around which the rest of the design, code inspections, and testing activities are scheduled

5. *Resource tracking:* using tools that apportion costs to specific accounts as well as track all the unpaid overtime to learn the real effort expended to complete the programming or testing

6. *Methodology management:* using tools that are keyed to one or more common development or conventional software structured analysis and design methodologies to plan, estimate quality, and assess risk

7. *Assessment support:* using tools that are keyed to an industrywide forum for evaluating software to:

   (a) Highlight current organizational weaknesses or ill-directed practices that need repair prior to moving projects into full development

   (b) Gather data that may be used to help control cost, schedule, and quality estimates over the course of the development

It is not surprising that this research reveals that initially fairly healthy organizations improve more rapidly than lagging ones lacking the infrastructure necessary to make improvements. Therefore, at least for software development projects, we can conclude that the management team's success implementing chosen control techniques depends to a very great extent on the organizational environment in which the project operates.

10.4   PROJECT COST CONTROL

There are two aspects to modern project cost control that management teams must take responsibility for controlling. These are (1) the cost of the scheduled work items of the product or service development in terms of capital outlays, resources, equipment, materials supplies, and other technical requirements; and (2) the cost of the product's or service's actual design, which may be specified in a contract issued by the customer or set as a targeted price point by a market assessment of the customer's willingness to purchase. Today, as more and more engineering efforts concentrate on how to fabricate products designed to meet specific target costs, the functions of the project management team will grow to include oversight of this activity. Project managers will take on this new responsibility alongside their traditional role monitoring the accrued costs for the development effort — actual project expenditures incurred to realize the scheduled work.

10.4.1 Target Costing

Since very few project teams enjoy the luxury of working on pure research projects where the output of the effort does not need to be immediately profitable, we have chosen to introduce in this chapter the cost control concept of target costing. In today's highly competitive markets, target costing is a cost control methodology that is gathering increasing attention. The technique is widely used by project teams to manage and control the costs of the physical components that make up the products and services they are developing. The practice permits an organization to determine the baseline costs at which it must produce a product with specifically proposed functionality and quality levels to generate a desired profit. Both the project management and design teams are involved in this exercise of guaranteeing that the cost to produce the product or service does not exceed the price points or selling price that marketing has verified to be within the customer's willingness to purchase.

Target costing is essentially a value engineering exercise (see Chapter 4) directed at analyzing the functional components of products and the processes used to achieve this functionality. To guarantee the lowest overall costs with no reduction in required performance or quality, the project team must engage in a interdisciplinary examination of all the factors that affect the cost of the product's development, including the purchase price of all externally acquired components and raw materials. Recommended process steps that a team may follow to conduct target costing are as follows:

Step 1. Align new product development with cost control strategy.

Step 2. Quantity customer needs.

Step 3. Determine allowable costs.

Step 4. Subdivide allowable costs into subsystems linked to customer needs.

Step 5. Further subdivide subsystems into components and determine allowable costs.

Step 6. Calculate financial return on identified path to achieve target costs.

Target costing is usually performed during the design phase of product development. Here, the team's attention is focused on reducing overall product costs by making trade-offs and changes in product design. The practice is distinctly different than another familiar cost reduction technique, *kaizen costing*, which engages project teams in exercises to drive out costs by identifying and redesigning inefficient fabrication or production processes. To proceed, the project teams needs (1) quantitative information about customer needs and requirements, (2) thorough information about competitive products that are already in the market, and (3) savvy projections concerning how a competitor will react to the release and market introduction of the team's product. These key inputs direct the team's attention on bringing to market a product with a sustainable competitive advantage because the product's functionality reflects the price points set by marketing for the specific customer base.

Target costing can also help project teams who develop products under contract to establish the cost parameters for the fixed-priced elements of the product or service development. Whether the goal is to break even or sustain a profit, the charge to both projects is not to exceed the cost targets. In each instance, the objective is *not* to minimize the cost of the products, but to *control* the design to meet a specified low level of cost.

Organizations that achieve designs that maximize the target costs of their product and service offers are able to position themselves very favorably with their customers and with other major market contenders. However, since competitors will never cease to challenge market leaders, we recommend that organizations implement continuous cost control programs in which their cost targets are adjusted to meet new thresholds over time. To guarantee repeated wins with customers and maintain their cost leadership position, organizations must further reduce production and fabrication costs by performing additional value analysis exercises.

10.4.2   Overcoming Cost Problems

In development and production projects, there are inordinate problems related to cost containment that the project management team needs to control. In the following sections we enumerate the most frequently occurring causes of cost problems and introduce several useful management techniques practiced for decades by managers on noncommerical projects that have relevance to commercial enterprises.

When a project budget begins to overrun, the first steps the project team should take are to (1) measure the overrun and (2) understand the probable causes. This action is necessary to determine the actual root cause before attempting to remedy the problem and serves as input to the collective decision making that must occur to bring the project back on track. Factors that cause cost escalation on small and large projects include:

- The tendency to base cost estimates for project resources and/or capital outlays on vague feasibility studies taken from similar project efforts rather than detailed design specifications or system requirements of the technology under development
- Failure to build slack into the project schedule to realize the economies sought from cutting project work or product features in an effort to alleviate cost overruns
- Delay in the flow of information on which to base early corrective actions
- Lack of corporate infrastructures or unforgiving management styles that do not support open and honest disclosure of information that affects or jeopardizes the project's financial success
- Failure to modify, redesign, or reengineer the project when design flaws are first discovered or technical performance capacity falls below design objectives
- Resistance to operationalize quality improvement efforts that are at odds with initial project cost goals, technical objectives, or performance specifications
- Failure to recost the entire project objectively after modification or reengineering is authorized
- Incorporating changes to the design scope of the project contract without client or management approval, or the appropriate allocation of funds for the changes
- Failure to measure the cost of authorized rework and overtime (paid and unpaid) to complete the design and development of new features that change the scope of the product
- Failure to account for the added operational costs of extending the life of late software development efforts by adding staff to write, test, and debug new code
- Indiscriminate distribution of the project's contingency budget to functional management personnel who have authorized design changes that surpass expected limitations
- Failure to estimate adequately the effects of inflation on total project costs: R&D, materials, component parts containing precious metals, labor, pilot plants, and so on

- Failure to estimate adequately the impact of new taxes or hold in reserve funds to cover the expense of unforeseen labor actions, political strife, or acts of nature

- Failure to protect the project from currency fluctuations through maintenance of value clauses in international contracts

- Failure to account for exposure to foreign currency transaction costs when converting funds between currencies

- Failure to set aside cash reserves to relocate project work should the firm not be able to repatriate or convert financial resources from a local currency

10.4.3 Cost Management Using Earned Value Analysis

The concept of *earned value analysis* (EVA) is a considerably useful approach for managing and controlling project costs. In fact, this management technique is considered so important that since 1967 the U.S. Department of Defense (DOD) has required its implementation on all contracted cost and incentive projects of major significance. In the commercial marketplace, the earned value concept permits project management teams and corporate accountants to track costs deviations and progress by comparing prepared cost estimates — the budget — with the expenses that actually occur throughout the project. These comparisons usually are made monthly or biweekly and are fundamentally different from the traditional practice of measuring variance from a spent plan. The EVA project management technique compares the cumulative actual costs, incurred up to a given date, with the estimated costs and schedule deviations of the project. By measuring the value of work performed as compared to the funds expended to accomplish that work, project personnel are able to estimate the final cost of the project and thus to forecast its final completion date. EVA also will highlight negative cost trends that project management must correct to meet original projections.

Cost Estimation (Budgeting). Budgeting of R&D and production projects is a complex process requiring input from corporate financial planners and technical managers alike. Good plans link the annual, near-term planning of the technical organizations to that of the business's overall, long-range strategic plans. No matter which management cost and control system is implemented to allocate scarce resources (incremental-, zero-, or variable-based), the objective is to develop a plan against which a time-phased budget can be established.

Budget preparation begins with the WBS. As we saw in Chapter 5, the work package associated with each lowest-level activity must describe the personnel needed to perform a given activity and the work schedule. Figure 10.2*a* shows a section of a WBS containing activities 111, 112, and 113 that together make up activity 110. Figure 10.2*b* shows the personnel

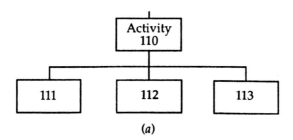

(*a*)

Activity ID	Skill time (person-months)	Loaded Cost (dollars/month/person)
111	1 (p-m) Technician 1 (p-m) Programmer	$4,500 $5,000
112	2 (p-m) Programmers 1 (p-m) Engineer	$10,000 $6,000
113	2 (p-m) Programmers 3 (p-m) Engineers	$10,000 $18,000

(*b*)

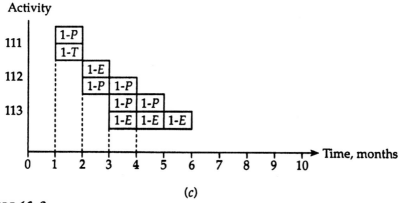

(*c*)

FIGURE *10.2*

(*a*) Part of WBS showing bottom-line activities 111, 112, and 113; (*b*) work breakdown structure activity data; (*c*) chart indicating number and type of personnel required as a function of time.

needed on each activity, the duration of time that each is needed, and the loaded cost (i.e., direct labor cost plus overhead and general and administration expenses). In a fixed-price contract environment, profit is often added later. In the commercial environment, profit would result from subtracting expenses from the revenues received after selling the final project deliverables. Figure 10.2c shows each activity, when it occurs, its duration, and the number and type of personnel required. Using Figure 10.2 we can plot the estimated cost of activity 110 as a function of time. (Note that activity 110 is the sum of activities 111, 112, and 113). We show this in Figure 10.3 along with the cost of the materials for activity 110. From this calculation we arrive at the total *budgeted cost of the work scheduled* (BCWS). Since this cost curve was prepared in the planning phase, it represents the estimate against which actual expenses are compared. The cost of all other activities is also calculated, and one final BCWS is plotted for the project. Figure 10.4 gives a typical resulting BCWS. We can also note in this figure the total planned project cost or *budget at completion* (BAC). Similarly, the planned *time at completion* is denoted TAC.

Actual Cost of Work Performed (ACWP). The actual budgeting and tracking of the performance of task-level project expenses comprise what is known as a *cost account*. Each month, as work is being performed, each employee should complete a time sheet indicating how much time has been allocated to each of the many project tasks. Time spent marketing and demonstrating the product — time not related specifically to product development — should be charged to another charging number and not directly to the project. The breakdown in some companies may be by hourly increments. Individual hours performed by all the personnel working on tasks 111, 112, and 113 would be summed to arrive at one final ACWP, as illustrated in Figure 10.4. The ACWP given in the figure shows the actual spending on the project, which currently is less money than expected. Yet how much work has actually been accomplished? If, for example, no work was accomplished at the time this report was prepared, even though money was spent on labor and material it would be feasible to argue that the project is over budget. If, on the other hand, work was completed without spending all the allocated funds, we might argue that the project is under budget, rather an envious position for any project team to be in!

Budgeted Cost of Work Performed (BCWP). The BCWP is a technique employed to indicate the budgeted or planned cost of the work actually performed. The curve shown in Figure 10.4 is often difficult to obtain since the project team will often tell management what they think the team

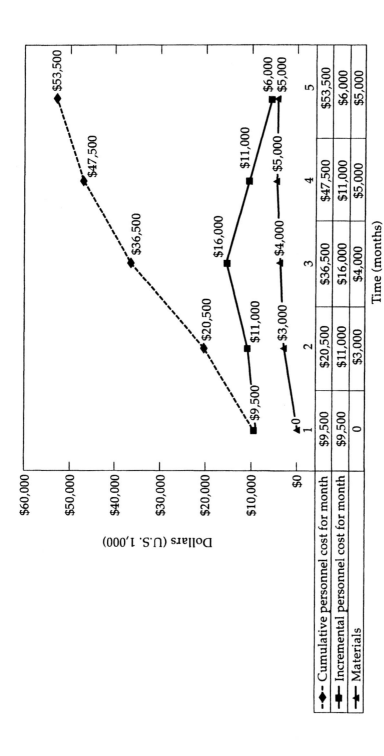

	1	2	3	4	5
Cumulative personnel cost for month	$9,500	$20,500	$36,500	$47,500	$53,500
Incremental personnel cost for month	$9,500	$11,000	$16,000	$11,000	$6,000
Materials	0	$3,000	$4,000	$5,000	$5,000

Time (months)

FIGURE 10.3

Budgeted cost of work scheduled (planned) for activity 110.

333

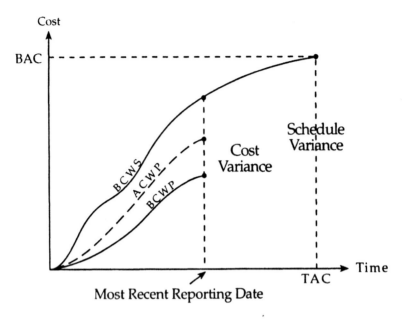

FIGURE *10.4*

Earned value.

leaders want to hear rather than what really is happening on the project. For example, suppose that at the end of the reporting period, the team planned to have completed tasks A, B, C, D, and E, and the planned cost in U.S. dollars was as follows: A, $10,000; B, $35,000; C, $47,000; D, $28,000; and E, $22,000. However, only tasks A, B, and C were completed, and D and E were not started. Note that the actual cost to complete A, B, and C (or ACWP) turned out to be $100,000, while the planned cost to complete A, B, C, D, and E (or BCWP) is $142,000. Here the calculation is easy to perform. By using a test to measure the completion of tasks A, B, and C, we determined that tasks D and E were not yet started.

Measuring work still in progress is much more difficult. Often, the percentage of completion estimates are collected from the development team by a lead engineer, programmer, or functional coach and shared with project management. But how can the project managers really know whether the engineers or programmers working on the uncompleted tasks are properly estimating the remaining completion time or hedging to tell them what they believe they want to hear. To minimize these problems, the project management team should ask the developers to perform full-scale functional tests of the integrated system on a periodic basis and report the results in the periodic status report.

Variance Analysis. *Variance* is defined as any deviation from plan. By determining a variance in the project budget, the project management team can determine if the loaded salaries of the assigned personnel and direct charges for materials are driving the project out of control. When a variance occurs, it is important first to determine its cause. In this instance, the actual expenditures on the project either are significantly different from the initially planned estimates or personnel outside the team have inaccurately accrued charges against the project. Whichever is the case, tracing the source of the variance will not help the management team to understand how the project is performing if the cumulative costs and milestones achieved are not tracked.

To forecast whether a project is over budget, the management team is encouraged to calculate the project's *cost variance* (CV), which is derived by comparing the project's earned value to actual cost using the formula

$$CV = BCWP - ACWP \tag{10.1}$$

As shown in Figure 10.4 the CV is the cost difference between the planned cost to complete tasks *A, B,* and *C* and the actual cost to complete the tasks. In Eq. (10.1), CV = \$92,000 − \$100,000 = − \$8,000, indicating that the project is \$8,000 over budget.

Figure 10.4 also shows the *schedule variance* (SV), which is denoted in a currency denomination (i.e., U.S. dollars) and not months. The SV is the comparison of the project's earned value to planned value and is the difference between the BCWP and the BCWS:

$$SV = BCWP - BCWS \tag{10.2}$$

The SV is given in terms of a currency denomination because if the project is behind, it will cost SV dollars, for example, to catch up. For instance, if a project specialist were to report two months of slippage, the project management team still might not know much money it would cost to catch up and put the project back on plan. In the preceding example, SV = \$92,000–\$142,000 = − \$50,000, which means that it will take approximately \$50,000 to put the project back on track. Thus, the project is \$50,000 behind schedule.

To understand the concept of *variance analysis*, refer to Figure 10.5 and assume that at the last report the BCWP was \$200,000. In Figure 10.5 we see that the project is on schedule with peak deviations of less than \$4,000, or 2% of the BCWP. However, the CV is approximately \$20,000. If not corrected, this high overrun, which may be due to material and component

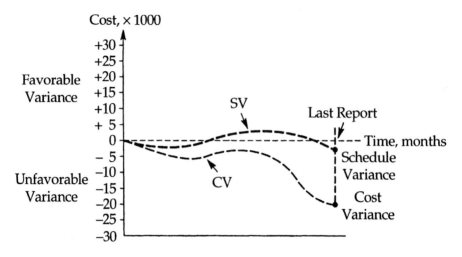

FIGURE *10.5*

Variance analysis.

charges, could greatly affect the targeted cost set for the product and hence its profitability. Each month, the project management team, along with the appropriate functional managers, should review the BCWS, BCWP, and ACWP values for the project's level 2 WBS tasks reported in the U.S. federal government mandated *cost schedule status report* (CSSR). The information in the report is also useful for commercial project team leaders, who may discover that their organizations can use the calculations to justify partial payment on time and materials contracts. A sample CSSR report is shown in Figure 10.6, where there are seven level 2 activities (tasks 1.1 to 1.7). The BCWS, BCWP, ACWP, SV, and CV are given, as are the *budget at completion* (BAC) and the *estimated cost at completion* (ECAC), along with the *variance* (difference) between them. In the next section we explain how to obtain both the ECAC and *estimated time at completion* (ETAC).

On reviewing the cost schedule status report shown in Figure 10.6 we see that system design is overspent and well behind schedule. The slippage is also beginning to affect the rest of the program. As a result, the project management team needs to speak with the manager of the systems design activity to ascertain the difficulty and to help remove roadblocks that may enable the designers working on task 1.2 to get back on schedule.

Estimation of Cost and Time at Completion. It is important not only to track actual versus planned costs but also to predict what the final cost of the project will be based on the performance to date if no unforeseen

Project Title _Alpha Beta_

Date _1/1/2000_

Project Manager _T. Williamson_

Performance Data

WBS Elements	BCWS	BCWP	ACWP	Variances		At Completion		
				Schedule (SV)	Cost (CV)	Budget (BAC)	Est (ECAC)	Var
1.1 Management	193,898	193,898	196,645	— 0 —	(2,747)	537,482	545,097	(7,615)
1.2 System Design	868,666	784,948	968,434	(83,718)	(183,486)	990,762	1,222,358	(231,596)
1.3 Training and Evaluation	166,676	134,846	149,854	(31,830)	(15,008)	504,275	560,399	(56,124)
1.4 Training	7,909	7,909	2,225	— 0 —	5,684	124,475	35,018	89,457
1.5 SW Development	61,639	9,128	5,646	(52,511)	3,482	697,844	431,642	266,202
1.6 Support	7,908	7,908	2,224	— 0 —	5,684	23,882	6,716	17,166
1.7 HW Development	72,448	72,448	72,448	— 0 —	— 0 —	455,377	455,377	— 0 —
Total	1,379,144	1,211,085	1,397,476	(168,059)	(186,391)	3,334,097	3,256,607	77,490

FIGURE 10.6

Cost schedule status report.

337

events occur to alter project performance. The ECAC and ETAC values are estimated monthly to help the project management team to determine the impact of any slippage. The ECAC represents the best estimate of the total cost of the project at completion and is calculated using the formula

$$ECAC = \frac{ACWP}{BCWP} \times BAC \qquad (10.3)$$

Thus the ECAC exceeds the BAC if we are overspending, that is, if the ACWP is greater than the BCWP. Immediate action to change the performance of the project by the management team is required.

Similarly, we find the ETAC by the formula

$$ETAC = \frac{BCWS}{BCWP} \times TAC \qquad (10.4)$$

Here we see how long we may expect the project to be delayed—that is, how behind schedule it will be when the BCWP is less than the BCWS. Because the delay may affect the project's time-to-market estimates negatively, the project team must immediately recommend a new course of action to correct the trend to senior management, who may be willing to spend even more to shorten the delay and meet market objectives (see Chapter 7).

10.5 MANAGING TECHNICAL SPECIFICATIONS AND ASSURING QUALITY

In previous sections of this chapter, we explored techniques that can be used to monitor and control the cost performance and schedule of a project. Assuring the project's *quality* or that its technical performance complies with agreed-on requirements and standards is another of the project management team's responsibilities. As we defined in Chapter 5, one definition of quality is conformance to specifications. The assessment of how well the product in production adheres to its original design depends, in part, on how adeptly the project management team has implemented a number of different measurement systems: (1) frequent design review meetings, (2) change control and exception configuration management, and (3) performance testing and quality assurance procedures.

10.5.1   Technical Reviews

In general, project experts who are not members of the design team conduct *technical reviews*. These outside experts examine the technical integrity and assess the quality of the product being built. Many different types of technical reviews are conducted over the different phases of the project development and production life cycle. We conveniently classify technical reviews to illustrate the basic purpose for which they are sponsored. For example, reviews scheduled during the design phase are referred to as *design reviews* and are held primarily to aid the design team in soliciting recommendations from other technical experts to improve the product's design. Formal technical reviews that are used to validate project objectives and assess compliance with design requirements are known as *compliance reviews*.

Most projects hold compliance reviews at key project decision points when authorized approval is needed from the organization to proceed to the next phase of the project. Often, this may be concurrent with the end-of-phase reviews mandated by the new product development processes described in Chapter 4. Whereas design reviews are held to answer the question of whether project deliverables and their associated milestones are valid, timely, and auditable, a compliance review is conducted to determine whether the product complies with the design requirements specified. Essentially, the purpose of the compliance review is to analyze and recommend alternative solutions to the problems raised by the reviewers and the inspectors of working prototypes.

In a software development project, many different types of technical reviews are conducted during the system design phase of the project just prior to coding. Then, once programming has begun, additional compliance reviews are correlated to coincide with subsequent inspections of the software system. The first of these reviews, the *systems requirements review*, is conducted to determine initial progress defining the design concept and the design team's convergence on an optimum and complete software system configuration. The systems requirements review is an in-process review of the results of preliminary requirements' allocation and integrated test planning.

Later, when the concept definition has proceeded to the point that the requirements and design approach are fairly precise, the design team conducts a *system design review* to evaluate the completeness, correlation, traceability, and risks associated with all of the functional requirements allocated. A system design review for a commercial software development would also verify that the design has thoroughly specified the system's user interface and performance, reliability, serviceability, and installability requirements. In general, this design review is conducted to verify that each of the following items is included in the program configuration: (1) data processing requirements, (2) development specifications, and (3) preliminary test plans.

Just prior to the start of coding, once the detail design is complete, the programmers hold additional technical reviews. These include *critical design reviews* to establish the integrity of the system design at the level of flowcharts and logical diagrams. Then during programming this team conducts further *functional configuration audits and code walkthroughs* to verify that the program actually performs in accordance with the requirements that are outlined in the development specifications and to ensure that the various functions produced in the definition and design phases are included. These inspections, or physical audits of the system, identify potentially costly mistakes, such as errors or typos in the code structure that the software developers can correct immediately to reduce overall project rework.

Finally, to help assure the quality of a software system, the project team may choose to conduct a *formal qualification review* with the end users internal to the firm who have commissioned the work. Internal auditors can also help to determine whether the system, as constructed, has addressed at least minimum specifications by trying out prototypes. Prototypes of commercial systems can also be demonstrated to key customers who have signed nondisclosure agreements with the organization. This kind of audit emphasizes hands-on display of the working system rather than passive study of written documents. By properly conducting this form of user inspection, programmers can provide themselves with closer and continuous user contact over the life of the project, resulting in greater understanding of the permanent system.

10.5.2 Change Control Techniques

How does the project management team control changes in a project's technical design that affect the scheduled completion or delivery date of the product in development? How is the loss or gain of project personnel and other resources tracked and recorded? How are requests for new features and functions handled once project work has begun? Who is responsible for accepting formal requests for engineering design changes due to project rework?

In the best-managed projects, the project management team coordinates the activities of a formal *change control board* vested with the responsibility for standardizing change procedures of all projects. *Change* is defined as any alternation in baseline data that begins with a request and ends with a formal signing off on the approved documentation. *Change management* is defined as the task and activities associated with gathering, analyzing, prioritizing, and categorizing the assignments of modification or *change requests* (CRs). As we described in Chapter 5, the project management team documents how the project manages design changes in the *change management plan* subsection of the project summary plan.

If the design and testing of a specification increases the cost of a commissioned hardware development project, the project management team should immediately notify the customer and draft an *engineering change proposal* (ECP). The ECP documents the proposed change in scope and allocates time and money for the modification. Hardware engineering or design modification requests are usually classified as either *major* or *minor* (*class A* or *class B*) by the change control board to clarify the imposed impact that accepting or rejecting the change will have on the overall project.

In a commercial software development project, CRs that result from executing unit and system tests are also assigned severity and priority levels for resolution by the developers or testers who discovered them and are published for review by the change control review board. The board either accepts or rejects the proposed corrections with input from marketing representatives on the new product development core team who will have final approval of any modifications of the original project specifications.

It is important for the project management team to keep senior management appraised of CRs that constrain project resources, cause a delay in the scheduled end date of the project, or affect the timing of the product launch. We recommend that project management log all baseline data changes into a computerized database that serves the project as a formal change control system. The system should be formatted to produce change control reports and maintain historical data files concerning the different design modification requests and executed engineering changes. It is also important that the change control system in a software project be linked to the project's *configuration management system*, which is used to maintain software source code configurations (see Section 10.5.3).

Systematic tracking of all proposed and executed changes is a required project management activity. Historical data should be traceable to the work package level of the project's WBS so that the designers and systems engineers initially responsible for the project's technical specifications can be kept informed of product enhancements or end-user requests for improvements. Because change management systems help ensure that the established baselines for the project—its schedule and budget—are reviewed whenever engineering design changes are approved, the discipline integrates the administrative activities of the project management process with the technical development work. Change is inevitable and should not be regarded as a problem, in that it is controllable.

10.5.3 Configuration Management

Closely related to the concept of engineering change control is the process of *configuration management* (CM). The U.S. Department of Defense (DOD) initiated the practice years ago to uniquely identify and manage replacement parts for specifically engineered equipment. As software became more prevalent in the equipment the Defense Department built, the practice became a mainstay requirement for contractors producing software. Similarly, recently experienced problems associated with repeated failures of in-house software programs and off-the-shelf shrink-wrapped applications has stimulated the commercial marketplace to force the software development industry to ensure that their product configurations contain all the *correct* (accepted and documented) changes maintained under their projects' change control systems. During the early 1970s and 1980s, the tools used for configuration management were basically a library where programmers stored their software files. The tools that the software industry employs today not only must track the information about the changes to the source files but simultaneously, maintain multiple versions of the code. Configuration management systems track whether a software change to a specific version of the product is configured in the version released to the marketplace. These systems also are used to (1) maintain the integrity of the software components shared between release configurations; (2) avoid what is known as *change regression*, which refers to changes in the software being lost or overwritten as a result of multiple copies being modified simultaneously; and (3) control the software build process by ensuring that the dependent components of the program are linked properly before being compiled and that all the *correct* source versions are utilized for creating their executables.

10.5.4 Project Status Reports and Reviews Meetings

The best report form is one that management will read and understand. Status reports should present control-oriented information in summary form and should be in a format requested by the sponsor. Among the important items to discuss are:

- Risk or problem areas of concern to management and the probable cause(s) of any exceptions to baseline schedules, budgets, or future profits
- Qualitative and quantitative impact of development statistics to the baselines
- Action taken or recommended to alleviate the unfavorable impact
- Indication of what management can do to help or to convey the exception to top management or notify the customer and other pertinent clientele

The report might conclude with a set of exhibits that present:

- Budget or resource usage curves
- Updated project schedule versus planned schedule
- Summary master schedule
- Technical development statistics, test results, and working features versus expected action items, open issues, and so on

An example of a status report is shown in Figure 10.7.

Another widely successful technique used by project managers to measure and compare progress with baseline data is the periodic project *review meeting*. The purpose of a project review meeting is to:

- Assess common goals and objectives
- Review progress and determine current status
- Identify problems
- Solicit recommendations for improvement
- Follow up on previous action assignments
- Communicate with fellow team members
- Educate and/or motivate project personnel

To be effective, review meetings should be scheduled regularly at specific times when project team members can attend. An appropriate agenda and advanced distribution of relevant data are mandatory. The project management team must clearly define the meeting's purpose and encourage participants' contributions to this regard. The team must work together to document any changes in the schedule or budget, report all technical findings, and publish a list of action assignments. The action item list must specify (1) what the recommended action is, (2) how problems will be

WEEKLY STATUS REPORT

Project Manager:

Task/Activity Name/ ID (from Project Schedule):

Reporting Period:

General Status and Summary of Task/ Activity:

Accomplishments (Deliverables, Milestones, etc.):

Potential Problems and Areas of Concern:

Personnel Involved in Task/Activities:

FIGURE 10.7

Report form for recording project status.

solved, (3) who is responsible for their resolution, and (4) what the scheduled action will be and when it will be completed. The project's technical success depends ultimately on the project team working together cooperatively to alleviate unavoidable customer delays, technical failures, and budget or schedule overruns due to nonnegotiated changes in the scope of work.

10.5.5 Project Documentation Control

The timely and valid project control process includes provisions for the project management team to manage and control all the different documents associated with the project work. Project communications must be controlled because the verbal transfer of information between developers, marketing, and functional management is prone to the human failings of forgetfulness and misunderstanding. Some reasons why standardized control must be exerted over written and verbal communication in the project environment are:

1. Checkpoints are required to measure the cost and technical performance of project work over time. The status information passed upward at regular intervals provides project leaders with a documented picture of how the team is progressing.

2. As project requirements change, the documents recording the agreed-on modifications must also be filed and maintained.

3. Initial and current product descriptions that detail every function and probable enhancement must be accessible throughout the development phases to every member of the design team.

4. Project tracking and reporting procedures along with corporate policy statements concerning the project review process need to be communicated to all those contributing to project work.

5. Economic studies, technical specifications, and test plans, as well as the administrative and installation guides, provide project planners and designers with a common road map against which to plan, integrate, and measure their work.

6. Technical documentation that accompanies the product's transfer from development into manufacture must be accurate and up to date as well as traceable to the engineers or software developers responsible for the design work.

Project documentation can be divided into two classifications or categories. One category is related to the actual technical work: specifications, quality performance management standards, test plans, and so on. The other documents project management activities associated with the project endeavor—among which are work authorizations, reports on project status, signed agreements, jeopardy or breakdown reports, and review meetings minutes.

Control of the documentation is achieved by using a project library where documents are filed either electronically or physically in a project office. The library contains all documentation relative to the project along with previous schedules and other useful literature pertaining to how projects are implemented in the organization. These documents are often bound in a separate reference manual called a *Project Handbook*, which describes how the project is conducted. The manual is also a useful source for specifying how to control and file the various technical documents that the developers generate.

It is somewhat difficult to distinguish between these two documentation categories in that the contents of some technical documents also pertain to how the project is managed. Certainly, we can classify meeting minutes as either technical or project-related documentation. The same is true of project reports and other signed agreements. If the technical and project documentation must be available for review by internal and external auditors, it is best to have the project librarian establish a filing system that cross-references all the documents. In general, the project management team should not worry about how to classify the documents, but should, instead, concentrate on establishing an electronic database system to control and track access to what is filed.

10.5.6 Quality Control and Performance Testing

In Section 10.2.2 we emphasized the importance of establishing (1) a performance measurement system to monitor whether the individual components of the hardware or software system under development comply with relative quality standards and (2) a plan to eliminate the causes of unsatisfactory performance. We accomplish this by conducting formal functional tests of the hardware or software system over the development phase of the project and during a specifically dedicated testing phase that is designed to detect errors and defects in the system once development is completed. Before the system leaves manufacturing, additional testing is conducted to assure that the assembled system will perform according to specifications.

In a software development project, where design errors must be detected to prevent defects from being configured into the integrated system, each programmer performs a *unit test* of the individual software modules to determine whether the piece parts do what they should. Once all the unit testing is completed, either the developers or an integration group execute a separate *integration test* designed to assess the feature functionality and intermodule interfaces. High-severity problems that were found by the development team must be resolved before the integrated system is turned over to a separate system verification team of dedicated testing professionals who *system test* the software. Any unresolved problems must be assigned a change request (CR) and referenced in the project's change control system.

To ensure the integrity of the testing process, we recommend that the integration and testing teams be autonomous from the developers and have the responsibility of creating their own test scripts and testing procedures. Also, with the demand today to reduce the overall time to market, software and hardware deliveries to the test teams should be planned in testable increments.

The system test is a series of baseline test scripts or cases designed to assess whether the system meets functional and performance requirements. Test goals include measuring:

· If the flow of the program or application is correct
· Whether program interfaces are properly designed and implemented
· Whether the program performance measures (execution time and program size) are within acceptable bounds
· If the software is sufficiently error-tolerant
· If the software is adequately error-resistant

The test cases also verify:

· The accuracy of the documented installation procedures
· The system's ability to achieve capacity and reliability targets
· Its behavior under error and stress conditions
· Conformance with user documentation, including on-line and separately supplied media

Project management will need to estimate the total number of incremental passes through the software system verification procedures, set up by the test team, that the software system must complete. For instance, the total number of *system test intervals* needed to complete a testing iteration is based on (1) the number of defects in the product at the time it is turned over to testing, (2) the percent of defects that will be detected by the system test team, and (3) the lag time between fixes by the developers. The system verification team will also conduct *regression tests* with the system to assess the effect of changes or additions to a previous version of the software.

When a software system is integrated into the hardware supplied to a customer, the team must perform *environmental tests* to verify that the system meets physical and electrical standards. Qualification tests, performed jointly by systems testers, end users who commissioned the development, or customer representatives in the customer's environment, are called *beta tests*. The testing must cover the same tests that the customer ultimately uses to accept delivery of the system. When a software system must function with other currently operational systems, the verification team may conduct *intersystem* or *interoperability tests* to verify how the system works with these field systems. In many professions, such as telecommunications network management, *operational pilots* are conducted at a customer site to test how the system operates with live traffic.

10.5.7 Quality Assurance

Throughout this book we have employed several now classical definitions of *quality*. One such definition is conformance to specifications. The achievement of a project's technical objectives is assessed by using a variety of performance measures, of which *quality* is foremost to success in the marketplace. As we introduced in Chapter 4 and describe in more detail in Chapter 12, classical quality theory defines *quality* in terms of *fitness for use*. However, as the leading consultants to industry, who are recognized for their contributions to both the fields of quality process improvement (Crosby, Demming, Garvin, Juran) and software process assessment (Boehm, Jones) point out, *fitness for use* is determined by what the customers say it is — not senior management, marketing, or engineering.

Customer expectations ultimately shape an organization's *quality policy*. The initial steps taken during the definition and design phases of the project to understand the quality characteristics desired by a customer also serve as reference points against which the technical performance objectives are specified. The project's commitment begins by planning quality designs and thoroughly testing their execution throughout development according to the control procedures outlined in Section 10.5.6.

It is the responsibility of the project management team to verify that the project is following corporate quality directives and to ensure that all the contributing units and functional organizations implement policies and processes that can be audited for compliance with international and

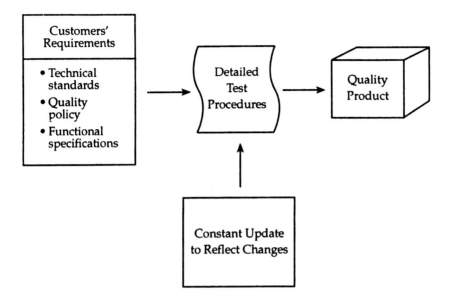

FIGURE *10.8*

Quality assurance process.

national quality standards. Project management must share these standards with the project team to provide the framework — the required tools and processes that engineering will use to establish functional specifications that fall within the boundaries of customer expectations. Since the customer's acceptance of product quality varies, the organization must make every effort to update its standards and policies regarding defects, reliability, performance, installability, and maintainability (serviceability).

As shown in Figure 10.8, recognition of the importance of early and consistent quality assurance procedures is the first step toward carrying out a well-conceived quality improvement process on a project. The project management team must be willing to improve the project team's performance by taking on the responsibility for preventing the perpetuation of design errors into development, prototype fabrication, production, and manufacture.

The following quality appraisal measures help to assure the team's use of feedback on error and defect detection for design improvements of future product releases:

· Technical reviews and inspection
· Regression and functional tests

- Fault insertion tests
- Performance and stress tests
- Reliability and maintainability tests
- Customer and end-user tests

We have also found projects that implement corporatewide performance goals for their products and respond in a meaningful way to formal audits of their project management and development processes experience statistically valid improvements in defect measurements. How to engage a separate quality assurance group, government-sponsored auditor, or outside evaluator to conduct such process audits is explained in Chapter 12. Nevertheless, if project management must ultimately assure that the product it delivers to the marketplace conforms and meets its quality objectives, we cannot emphasize enough how important it is for the entire project team to be involved. One way to accomplish this is to establish the policy that only when specifications are met is an activity complete. Following this policy will not only shorten the formal verification schedule, but limit the system test activities to the discovery of only minor bugs that can be traced and corrected. When the organization does not adhere to the standard policy of "doing it right the first time," the *debugging* activities take an inordinately long time.

10.5.8 Measuring Customer Satisfaction

The ultimate assessment of user satisfaction with the product or service the project team has introduced into the marketplace is to measure the customer's perception of relative value against alternative choices. This measurement differs from measuring the number of software of hardware defects in many ways:

- Gathering user satisfaction measures requires resources outside the project.
- The data collected ask customers to rate all of the surrounding components of value to them: warranty support, service policies, financing, availability, and so on.
- The staff that collects the data is autonomous and reports to a separate quality assurance organization.
- The data computation results in a *customer value analysis* (CVA) metric that the organization may use to predict market share for the new product or service introduction.

Initial feedback from potential and actual customers who purchase a newly released product or service may be obtained from the sales force or through user feedback questionnaires mailed in by customers completing the warranty agreement enclosed in the package supplied to the customer. The members of a new product development core team also may survey the customers who participate in a controlled introduction program or qualification site testing with the new product or service. The actual gathering of the formal CVA metric involves annual assessment and analysis by the quality assurance organization. In many commercial companies, the results are reported quarterly to compare performance against current and potential competitor's performance in meeting customer needs and to identify or redirect corporate strategy to create products and services with distinguishing value for customers that win sales away from competitors. Many consulting companies, such as Software Productivity Research (SPR) also can be engaged to conduct user satisfaction surveys.

Project teams will want to use the data gathered from these multiple sources to learn if high levels of user satisfaction correlate with low levels of product or service defects. As illustrated in Figure 10.9, a simple scatter graph of user satisfaction and defect data can be quite informative for the project team.

Most projects should have scores plotted within the lower left quadrant. Occasionally, a project may release a new software application or integrated hardware system that has limited feature functionality, and hence the user satisfaction score may be low but the product has few or no defects. The team's subsequent enhancements in the next releases of the product may correct this position. Few projects should fall in the upper right or left quadrants. These positions indicate that the project has failed. Perhaps the software application or integrated hardware system was rushed to market before a satisfactory level of defects were corrected — which may be the

	High Level of User Satisfaction	Low Level of User Satisfaction
High Level of Defects	Urgent repairs required	Urgent replacement required
Low Level of Defects	Excellent overall value	Functional enhancements required

FIGURE 10.9
Scatter graph of defect and user satisfaction.

case initially when important new functionality is released. Worse yet, the project may have been rushed into development with little understanding of the customer's needs or willingness to purchase. Usually, the market has no tolerance for products that exhibit these characteristics, and projects are rarely forgiven for mistakes of this magnitude.

10.6 CONTROLLING PROJECT RISKS: COST, SCHEDULE, AND TECHNICAL ASPECTS

Risk management is an important and integral element of project management. Project management teams who use sound risk management techniques to identify, analyze, prioritize, and quantify project activities that may compromise the cost and time to complete the project are likely to advert disasters and maintain control when events get out of hand. As we have explained in Chapters 3 to 7, all project activities have an associated element of risk—usually unwanted—whose likelihood of occurrence will affect the profit of a project in terms of lost revenue and schedule delay. To manage and control a project's risk exposure, the management team may choose to assess the impact of activities that have the highest probability of occurrence and largest negative impact on either the project schedule or budget. In this section we pull together the most important techniques introduced so far in the book and show how to formulate a risk-response plan that prepares the team for more informative decision making and structural problem solving in times of stress.

10.6.1 Implementing a Risk Response Plan

Risk management should be built into a project management team's prowess. Risks need to be managed appropriately and not overlooked. At a basic level, the probability of risk occurrence can be assessed simply as low, medium, or high (see Chapter 7). At an intermediate level, the assessed probabilities are given values either in the range 0 to 1 or as a percentage and are used to calculate the expected value of the risk event. At a comprehensive level, the management team must describe the probabilities by specifying a distribution function (see Chapter 6). To measure risk this precisely, simulations to the project schedule using software tools as add-ons to one or more of the popular project management systems are performed (see Chapter 15). The level selected by the project management team to manage project risk will vary depending on the urgency to come to market and the nature of the business and technical risk events that might occur.

Using probability and impact estimates to calculate the project's expected value without implementing a response plan to manage the risks can leave it exposed to events that may put an organization's strategic business plans in jeopardy, endanger the lives of employees and customers, or cause the organization considerable financial harm. The more that is at stake, the tighter and more precise the control measures should be. Specifically, unacceptable risks that have a high probability of occurrence and high impact must be quantified and tracked. Also, because the occurrence of such high-impact events would be catastrophic, the organization must make every effort to mitigate their occurrence or stop them from happening by (1) deflecting the risks to another party, for instance, through the purchase of insurance; or (2) terminating the project before contracts are finalized. On the other hand, as we explained in Chapter 3, high-impact risks that are known and have a low probability of occurrence can be managed using contingency funds. Undocumented, high-impact risks that were not planned to occur can be managed with management reserves.

A project management team must understand that risks also have a number of different causes, all of which may give rise to other risks of equal or greater impact to the project. For example, a schedule risk may result from combining technologies in a pioneering effort to bring a product to market. The project would initially have uncertain estimates concerning how long it will take to integrate the technologies, which, in turn, may constrain the number of hours assigned to project personnel asked to complete system test. The risk associated with the need to run extra tests may cause a schedule delay and even affect project finances if no contingencies were set aside to bring on additional subcontractors to complete the tests. A mitigation strategy that may be employed on the project would be to integrate the two technologies earlier and assume a cash flow penalty to expedite or crash the project (see Chapter 7).

Project managers should consider the interrelationship among project risks when developing a risk-response plan, as a risk due to a schedule delay may also be:

- *Contractual*, because these risks can have an impact on project finances since penalties may have to be paid
- *A cost-savings opportunity*, because benefits may not need to be paid due to a delay completing the work
- *Resource limiting*, since no new subcontracted labor may be required to begin new work
- *Financial*, because additional cash may be needed to crash the project and start work earlier

Typically, the impact values that the project team assigns to each of these risks are quite different, as is the probability that all four events will occur on a particular project. The team may also find it necessary to detect and track the occurrence of predictor events that act as *triggers* of risk events. Fortunately, triggers can easily be mitigated. With plenty of advance warning, the management team can eliminate a trigger activity before its occurrence causes the adverse occurrence of other unwanted events. By interviewing and recruiting staff two months prior to the start of system test, the management team will prevent the bottleneck of too few testers being available when needed by the project.

The need for risk management must be integrated into the project tracking and controlling processes that occur throughout all the stages of the project's planning, contracting, development, production, and manufacturing processes. Since risk management activities include analysis, identification, evaluation, and control of all risk events, calculation of the *risk exposure* helps the management team arrive at the project budget and the estimates needed for contingencies that must be covered in the price a contractor must charge to remain competitive. The information collected in Figure 10.10 should be used to understand how to establish a risk budget for the project for each life-cycle phase.

10.6.2 Taking Corrective Action and Evaluating New Risks

In Section 10.6.1 we enumerated several risk resolution strategies used by project teams to control the unwanted occurrences of events that can cause an adverse outcome. As we stated, the project management team must select a technique that matches the given implications the risks may pose to the business. A project team may take corrective action by initiating any number of different approaches. These may include:

- Group idea generation activities, such as brainstorming, nominal group technique, and affinity diagramming
- Coordinated reviews of all project documentation to identify contradictions or voids in inspection or acceptance criteria, performance specifications, delivery clauses, supplier prices, contract clauses, and so on
- Decision-tree analysis using probability and expected value to assess the overall impact on the project of the occurrence of each risk event
- If the risk events identified occur, quantification of the project's actual or *net present value*

RISK RESPONSE FORM

Risk Item Description: (Describe the event that could occur.)

Risk Assessment: (Give significant impact on project and probability of the event's occurring.)

Planned Risk Resolution Strategy: (Indicate mitigation technique chosen for project: avoidance, retention, control, deflection, and list advantages, drawbacks, and order of preference.)

Risk Owner:

Date Completed:

FIGURE 10.10

Risk response form used to establish the risk budget for the project.

- Prototyping and simulation of technical alternatives to learn probable outcomes of untested laboratory solutions
- Implementation of a strong monitoring program based on established metrics and decision points that indicates to the team when to take a predetermined course of action to cost-effectively avoid negative ramifications to the project

Of the methods cited above, risk monitoring or reporting is one of the more tangible means for a project team to manage risk events that occur over the life cycle of an R&D project. A predetermined set of performance goals for the project helps bound the parameters for taking corrective action. When test results, for instance, on a software project indicate that problems have occurred which may jeopardize specifications to reuse legacy hardware or the program's size inefficiently uses the processor's resources, the development team must immediately inform project management that these events are outside tolerance limits. This permits the project management team to take the time, therefore, to react to the problem and implement the most appropriately identified risk response strategy.

Another effective risk reporting technique for high-priority research and development projects is to report regularly, upward through the management chain, the top-10 risk events associated with the project. The project team must track every item using a report form such as the one given in Figure 10.11. We suggest updating the form weekly for high-risk project endeavors. It is also imperative for the project team to share with senior management the planned resolution strategies to mitigate each risk and any consequences to the project. This lets the senior managers assess the financial ramifications resulting from the project team's plan of action and to determine what the strategic impact will be for the organization in the marketplace.

A chronological recording of the cause and effect of actions taken to mitigate project risks becomes a useful database of historical information for future projects. Postproject reviews should also collect these data systematically. There may be no better source of information in the future than past experience to influence the course of action of the project management team. Organization-wide maintenance of a database of this information can be quite helpful in drawing conclusions as to what types of responses were most successful. We have found that leading-edge companies use such data to learn how to make improvements in their new product and service development processes — risk management approaches that they believe are critical to their profitability and lead to a keener ability to satisfy customer expectations. Project teams may even compare the risk

HIGH-PRIORITY RISK REPORT

Rank This Week	Rank Last Week	Risk Item	Potential Consequences	Planned Risk Resolution Strategy	Actions Completed	Next Actions to Take	By Whom

FIGURE 10.11
Risk report for high-ranking risks.

357

mitigation responses they identify with internationally gathered data. Appendix 10.1 highlights prevalent software-related risk activities tracked in annual SPR software process assessments that may be informative for software development projects.

10.7 CONCLUSIONS

Successful management of a project's ongoing performance is achieved by exercising daily control over the quality, schedule, and cost of the work performed. Throughout this chapter, we have presented techniques for evaluating, measuring, and assuring proper implementation of a project's control system. In Chapter 11 we explore the communication tools and skills that may be of use to disseminate this information quickly among the many geographically disbursed organizations and project teams that are charged with carrying out the work to realize the project's success.

BIBILOGRAPHY

Addis, W., "Design Revolutions in the History of Tension Structures," *Structural Engineering Review*, Vol. 6, No. 1, pp. 1–10.

Guide to @RISK for Project Advanced Risk Analysis for Project Management. Newfield, NY: Palisade Corporation, 1997.

Berreby, D., "The Great Bridge Controversy," *Discovery*, Vol. 13, No. 2, February 1992, p. 26.

Carter, B., T. Hancock, J. M. Morin, and N. Robins, *Introducing Riskman Methodology: The European Project Risk Management Methodology*. Cambridge, MA: Blackwell Publishers, 1994.

Cooper, R., *When Lean Enterprises Collide: Competing Through Confrontation*. Boston: Harvard Business School Press, 1995.

Fairley, R., and P. Rook, "Risk Management of Software Development," in *Software Engineering*. Los Alamitos, CA: IEEE Computer Society Press, 1997.

Fleming, Q. W., and J. M. Hoppelman, *Earned Value Project Management*. Upper Darby, PA: Project Management Institute, 1996.

Larson, A., "Advances in Aeroelastic Analyses of Suspension and Cable-Stayed Bridges," *Journal of Wind Engineering and Industrial Aerodynamics*, Vol. 74–76, No. 73, April–August 1998, pp. 73–90.

Laseter, T. M., C. V. Ramachandran, and K. H. Voigt, "Setting Supplier Cost Targets: Getting Beyond the Basics," *Strategy and Business*, No. 6, First Quarter 1997, pp. 18–24.

Widerman, M., *Project and Program Risk Management: A Guide to Managing Project Risks and Opportunities*. Upper Darby, PA: Project Management Institute, 1992.

Wiegers, K., "Know Your Enemy: Software Risk Management," *Software Development*, Vol. 6, No. 10, October 1998, pp. 38–42.

EXERCISES

10.1. Why is a measurement system necessary in a project control system?

10.2. Refer to Section 10.2.1. Explain how this project's problems may have been avoided by regularly monitoring and reporting progress against a baseline plan?

10.3. Why are test plans so crucial to proper project control?

10.4. What different test plans can you think of?

10.5. What should a test plan include?

10.6. What is meant by the statement, "Infrequent reporting can result in unstable project control"? Illustrate by giving an example.

10.7. Why is each of these statements concerning the concept of target costing important to the project management team?

    (a) Cost targeting sets the price at which a proposed product with specified functionality, performance, and quality requirements must be produced.

    (b) Cost targeting determines the selling price of a product or service in the marketplace based on competitive benchmarking and a thorough understanding of the customer's willingness to purchase.

10.8. A project has the precedence–cost relationship given in Table 10.2.

    (a) Plot the precedence diagram and determine the critical path.

    (b) Plot the BCWS.

TABLE 10.2 DATA FOR EXERCISE 10.8

Task	Precedence Relationship	Duration (weeks)	Planned Cost to Complete
A		4	$10,000
B	A	6	12,000
C	A, B	8	16,000
D	B	4	8,000
E	C, D	6	14,000
F	D	8	16,000
G	E, F	4	8,000

TABLE 10.3   DATA FOR EXERCISES 10.8[a]

Task	Week						
	2	4	6	8	10	12	14
A	$5,000	$5,000					
	50%	100%					
B	0	0	$4,000	$4,000	$5,000	$6,000	
			25%	50%	75%	100%	
C	0	0	$5,000	$5,000	$5,000	$5,000	
			25%	50%	75%	100%	
D	0	0					$8,000
							50%
E	0	0					
F	0	0					
G	0	0					

[a]Entries are the actual expenditures in U.S. dollars for each task and the percent complete as obtained from the periodic reports.

The project is monitored every two weeks (Table 10.3).
(c) Plot the ACWP and the BCWP on the same graph as the BCWS.
(d) How late is the project?
(e) Is the project over cost? By how much?
(f) What is the estimated date of completion?
(g) What is the estimated cost of completion?

10.9. If BCWP > BCWS, is the project early or late? Explain.

10.10. If ACWP > BCWP, is the project over or under cost? Explain.

10.11. If ACWP < BCWS, what conclusion, if any, can you draw? Explain.

10.12. The project's designated change control board performs all of the following functions except:
(a) Proposal of corrections to the original design specifications
(b) Assessment of any deviations in project deliverables that may affect the quality, cost, or scheduled end date of the project
(c) Maintenance of historical data files concerning different design modification requests and engineering change requests executed
(d) Verification that the change control system in a software

project is linked to the configuration management system used to maintain software source code configurations

10.13. In a software system, maintaining 100% integrity of the configuration depends on having the correct version of the components as well as the correct components themselves. Which of the following responsibilities does the software project management team perform?

(a) Verification that the project uses a configuration management tool to support and track multiple versions of the files under development

(b) Verification that the software build process maintains the integrity of the software configurations so that all the necessary changes are incorporated in a designated release

(c) Holding a critical design review and a functional configuration audit before establishing a baseline of the software item

10.14. What is the most appropriate source of funds to cover work that was not identified in the initial project plan?

10.15. What actions might a project team include in its risk response plan to:

(a) Reduce the severity and impact of technical risks that might occur during the life of the project

(b) Reduce the probability of the occurrence of a minor business risk that might cause the project to overrun its budget

(c) Maintain the scope and quality specifications of the project in the event of an untimely delay

(d) Protect the value of the project due to work stoppage, natural disaster, or political strife

10.16. All of the following statements are true except one. Explain why.

(a) Quality control is concerned with monitoring specific project results to determine if they comply with relevant quality standards.

(b) Quality assurance is concerned with evaluating overall project performance on a regular basis to provide confidence that the project will satisfy relevant quality standards.

(c) It is the responsibility of the project management team to verify that the project is following corporate quality policy.

(d) The formal CVA metric, gathered in many commercial companies by the quality assurance organization, assesses performance meeting customer needs.

(e) There is no correlation between the defect levels found in an integrated software system rushed to market and the level of user satisfaction.

10.17. In your opinion, what process has the greatest affect on the quality of a software development project?

(a) Coding

(b) Configuration control

(c) Resource estimation

(d) Defect removal

(e) Requirements analysis

10.18. Explain why on a software development project it is good management practice to perform code walkthroughs.

CASE STUDY: A SYSTEM IN EVERY HOME*

The excitement of entering the field of digital speech technology and what new possibilities it would afford the computing world was the talk of all the engineering community at E&S. The long-cherished dream of "talking computers" was rapidly becoming a reality. Competitive prototypes were emerging from all the current vendors, and E&S had decided that including natural language in the interface would allow greater penetration into the new home appliance market. A small team of scientists had been working since the 1980s, painstakingly teaching the early programs of that decade to recognize specialized vocabularies. The programs that were sold first to brokerages for use in their call centers were now the basis for new customer applications. E&S would be at the forefront of introducing a new hands-free interface that would permit customers to call up information on a voice-controlled display screen strategically located in any room in the house.

E&S had decided to begin experimenting with a series of minimally complex designs that the company would first test in the laboratory. This was necessary to stabilize the interface before the programming of the

*This study was written for class discussion purposes only. It is not intended to reflect management practices or products of any particular company or organization.

feature sets selected could be scheduled. Once the design was settled on, the assigned project team could then optimize the cost of the physical components that make up the desired functionality of value to the new consumer segment. As a result, E&S would offer the new market a value-based innovation with functionality matched to the consumer's willingness to pay.

Using these new techniques would allow E&S the flexibility of allocating the cost of each subsystem's development and fabrication across the entire project. The development team would be asked to keep the projected costs in-line with the set targets. Then since there was relatively little competition in the market, the management team could time the market introduction, making any needed trade-offs with the scheduled release dates for the proposed system. The initial design proposed by marketing included bundling some very basic applications that would draw this large and, up to now, novice group of users to E&S's new computer offer.

At a minimum, the package would include (1) personalized and household expense tracking, (2) family calendar of events, (3) "to do" list maintenance, (4) World Wide Web and Internet access, (5) telephone message retrieval, and (6) security alarming. The development team assigned to the project was most eager to take its research out of the laboratory, form a venture alliance with one of more providers of telephony/Internet services, and bring the computing platform to market.

The Project Unfolds

Marketing had determined nearly a year ago that there was a significant new consumer market segment interested in acquiring an easy-to-use apparatus that could be located strategically in any room of the house and serve as a nerve center for controlling key activities that occur in the home. In fact, within one year from the start of the project, the programming team eagerly anticipated completing the proposed list of application programs that would need to run on the new "talking box." Although certainly not a final development effort for the physical components of the project, the initial software estimates were reasonable. The chassis would resemble the popular set-top boxes used in the home to connect to the Internet and be an all-in-one design with no provision for expansion. The state-of-the-art serial bus would give users the means to connect the box to a flood of other devices that E&S would sell as add-ons to the system. The device would be simple, yet sophisticated. No one, the engineers thought, would relish the early days of home computing, when floppy drives were used to load applications and control was only through a keyboard, joystick, or mouse.

For an all-inclusive price, the software applications would come loaded onto a removal storage device supplied with the hardware.

What E&S had not figured on when the project effort was sized and scheduled were the lengthy negotiation periods with the new venture partners. Each venture partner wanted exclusive rights to resell the much-talked about home appliance, which would increase traffic on their networks. So although the initial hardware architecture was independent of the partner selected, the human factors engineers insisted that the software interface to each network might not be the same. For each day that there was a delay in deciding which service providers would distribute the system, there was a corresponding delay in architecture design, prototype development, and testing.

The rework and new design options generated by the conflicting interests of the service providers were starting to be disruptive. It was beginning to look as if the project would never get off the ground, even though E&S most senior engineers were assigned to address these concerns. Meanwhile one of the market mangers learned that a formidable competitor to E&S was researching the technology. If E&S Computing was to maintain its lead position in the emerging, home appliance market, a product would have to be realized by the third quarter of next year, in time for holiday sales. Because the computing venture was considered so important to E&S's future, the company president requested that the development team get the project back on schedule whatever the cost.

Response to the President

While preparing for their meeting with E&S senior management, the leaders of the software team assigned to the project learned that engineering was about to release a final prototype chip set, which would allow the software developers to complete the specifications for several of the planned applications. Unfortunately, it was at least a six-month effort before the actual coding would be complete. The development leaders did, however, persuade the electrical engineers to assist human factors in the creation of several finally agreed upon prototypes. Earlier, they had insisted that their designs were independent of the endeavor to test the system's user interface. Fortunately, some progress could be reported to E&S senior management, because testing, which hadn't yet started, could begin immediately. The human factors team could gain ground choosing the best user interface design and the engineering team would be able to use the feedback from these market studies to redesign the system to match customer price parameters.

At the meeting, senior management decided to consolidate the project's budget into a single cost account. To move the development along, they asked Sue Johnson to manage the project. As project manager, Sue would tabulate all cost estimates and schedule projections. If necessary, to meet the scheduled market introduction, she would ask for authorization to optimize the project timeline. To gauge progress, a rigorous series of reporting measures would be initiated. Each of the lead engineers, software developers, and physical designers would be asked to report progress weekly to help move the project through the new product development decision gate process.

Organizing the Project

The following week, Sue requested representatives from each of the cross-functional organizations involved in the project to attend a project status meeting. She knew that proper controls were mandatory to manage *Genie*, the code name that she and her new team assigned the long-awaited development project. But the only documentation that was available for her review was an initial Gantt chart (Figure I) and personnel schedule (Figure II) prepared after marketing first discussed the concept with engineering. The development team leaders had used these estimates to work with the electrical engineers to obtain an estimate to complete an initial prototype design of the system. This resulted in the preparation of a high-level

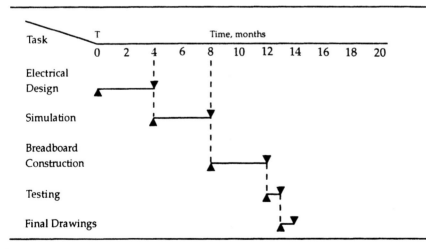

FIGURE I

Preliminary Gantt chart: electrical engineering tasks.

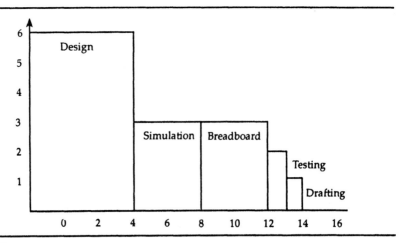

FIGURE II

Staffing profile: electrical engineering tasks.

summary chart (Table I) by the physical designers and human factors engineers.

Sue would have to update these plans to track the project through the testing of the interface design and the overall development and monitoring of the fabrication of the first production samples in the factory. To realize E&S's market objectives, Sue would also need to integrate the efforts of the seemingly independent team members. As project manager, it would be necessary for her to gain their respect through open and honest, two-way communication if she was actually going to estimate how long it would take and what it would cost to complete the project. To accomplish this objective, she called together the new Genie team:

> Ladies and gentlemen, welcome to the team! We're about to embark upon a new endeavor that will bring E&S to the forefront of the computing industry. Before the end of the month, we will need to meet with senior management to report on our design activities and predict whether we can get a product out and on the market within the next 12 months. All our market directives indicate that Genie is what the modern consumer is looking for in household electronics. How exciting to advance the state of the art while offering voice-initiated, true ease-of-use speech technology to our new customers. So let's devise a plan to move our laboratory invention into manufacturing so that Genie can become a household reality!

> What's needed at this point in the project is a master schedule detailing every activity necessary for successful product launch and the eventual release of additional add-on applications for our customers. Therefore, I would like from each and every one of

TABLE I   MILESTONE CHART GENIE PROJECT

Task	Planned Duration (months)	Staff	Uncertainty (standard deviation, months)	Precedence
1. Electrical				
A. Electrical design	4	6	1	Start
B. Simulation	4	3	$\frac{1}{2}$	A
C. Construct breadboard	4	3	1	B
D. Testing	1	2	$\frac{1}{4}$	C
E. Final drawings	1	1	$\frac{1}{4}$	D
2. User interface design				
F. Design	1	2	$\frac{1}{4}$	Start
G. Simulation	2	2	$\frac{1}{2}$	F
H. Testing prototype	1	2	$\frac{1}{4}$	G
I. Final specifications	1	2	$\frac{1}{4}$	H
3. Application software				
J. Design	4	1	1	F
K. Programming	2	2	1	J
L. System integration and testing	1	2	$\frac{1}{4}$	K
M. Documentation	1	1	$\frac{1}{4}$	L
4. Mechanical				
N. Chassis design	4	2	1	A
O. Construction	2	2	1	N
P. Final drawings	1	1	$\frac{1}{4}$	O
5. Assembly				
Q. Complete system	2	2	$\frac{1}{2}$	D, L, P
R. Quality assurance testing	1	2	$\frac{1}{2}$	Q
S. Documentation	1	4	$\frac{1}{4}$	E, I, M, R

you a list of the work activities that you've assigned your teams and a detailed report outlining the status of any outstanding assignments.

In addition, since Genie is expected to bring E&S relatively sizable margins, we need to review the cost targets for all the components going into the system. Senior management will expect to see weekly cost reports that describe the purchasing organization's progress ordering the materials for the system. I also want to learn the results of the preliminary user interface tests. If there are going to be any concerns over the design because of the difficulty signing up only two of the three service providers selected to distribute Genie, I need to know immediately. We must also manage this project's risk exposure. What are the risks that management must be made aware of? Will our small staff of programmers be able to complete all six applications in time for the first release of the system? I'd like you to report these data in your weekly status report so that I can consolidate the information across the entire project.

I strongly recommend that if we're ever going to track the progress of all the different work assignments, we meet with the entire project team and prepare a work breakdown of the project. Not only will we instill among the participants a sense of trust in E&S's commitment to Genie, but everyone involved will come away from the meeting with a clearer understanding of the project's goals, timelines, and organizational structure. Can we schedule this session before our meeting with senior management? Please get back to me by the end of the week.

In response, the lead programmers and engineers went back to their work groups. In an IPP session with the entire project team held the following week, the work breakdown and Genie's new schedule were planned. Table II presents the activities, expected duration, and status of the completed tasks collated at the meeting.

Questions

1. As a member of the Genie project team, review the data shown in Figures I and II and Table I. Prepare an early- and late-start schedule for all activities listed. Determine the staff months needed to complete the work.

2. Taking into account the slippage revealed by your colleagues in Table II, how many months are thought to be needed to complete this project?

3. Determine the probability of completing the project on time.

4. Using the revised schedule, prepare a three-level WBS; compare your results with those presented to E&S senior management given in Figure III.

5. List some of the possible technical and apparent business risks facing the release. What specific management strategies to mitigate these threats and take advantage of possible opportunities in the marketplace would you advise the project manager to use?

Tactics of Managing Change

The new project plans and revised estimates for completing the project work helped Sue to steer the project through the inevitable—a major design change needed to meet required market specifications discovered

TABLE II REVISED GENIE SCHEDULE

Task	Duration After 8 Months	Staff	Uncertainty (months)	Precedence
1. Electrical				
A. Electrical design	—	—	Completed	—
B. Simulation	—	—	Completed	—
C. Prototype	2	3	1	B
D. Evaluation and testing	1	2	$\frac{1}{4}$	C
E. Final design	1	1	1	D
F. 15 boards assembled and tested	1	2	$\frac{1}{2}$	E
G. Drawings	2	1	$\frac{1}{4}$	F
2. User interface design				
H. Design	—	—	Completed	—
I. Simulation	—	—	Completed	—
J. Testing prototype	1	2	$\frac{1}{4}$	I
K. Final specifications	1	2	$\frac{1}{4}$	J
3. Application software				
L. Design	—	—	Completed	—
M. Programming	2	2	1	L
N. Unit test and system integration	1	2	$\frac{1}{4}$	M
O. Documentation	1	1	$\frac{1}{4}$	M
4. Mechanical				
P. Preliminary design	—	—	Completed	—
Q. Model made	—	—	Completed	—
R. Final design	—	—	Completed	—
S. Final model	2	2	1	R
T. Drawings	1	1	$\frac{1}{4}$	S
U. Tool-made sample	2	2	1	T, O, G
V. Test and evaluation	1	2	$\frac{1}{4}$	U
W. Quality assurance	1	2	$\frac{1}{4}$	V
X. Documentation	1	4	$\frac{1}{4}$	W

during the human factors testing of the system's user interface. The testing revealed great customer interest in Genie, with keen customer desire to purchase additional functionality. To satisfy the market's request, however, E&S would need to bundle these features into the initial product offer that it would bring out next year. This meant that the previous strategy of releasing a minimum of two applications or programs with the initial

370

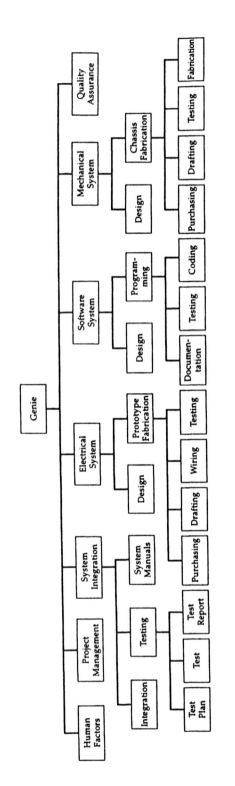

FIGURE III

Genie work breakdown structure.

systems would prove dissatisfying for their customers. Sue would therefore, have to ask for additional staff to optimize the schedule for her team to make the target release date. To prepare for the monthly meeting with senior management, she immediately requested the functional manager of the software development group to assist her in estimating the cost of crashing the project. Once she secured the additional funding to expedite the development, she would ask the system designers to revise the applications' technical specifications to reflect the new tactical plan.

Genie's new designs brought the project team accolades at the monthly meeting with senior management and, just as important, a pledge for the corresponding increase in funds to complete the work over the next 12 months. At the meeting, Sue also agreed to implement an automated tracking system to monitor and report progress. Immediately, she set about the task of installing a project management system to facilitate more timely communication between each of the engineering and fabrication groups and to monitor expenditures for project work through centralized data collection. Each week Sue's new project assistant collected the designers' most recent estimate of time remaining to complete project tasks. The revised estimates proved to be extremely useful for calculating any variances from the planned estimates and reporting progress to senior management and the entire Genie organization.

Questions

6. Using the data given in Figures I and II assign a standard rate of compensation, and plot the budgeted cost of work scheduled (BCWS) by the electrical engineers.

7. Using the data given in Table II for the electrical engineers, plot the actual cost of work performed (ACWP) and the budgeted cost of work performed (BCWP) after eight months have elapsed. What is the cost variance? The schedule variance? The estimated cost at completion? The estimated time at completion?

8. Assume that the final staffing count assigned to program and test the software appications in now four programmers and three unit and system testers. Assign a standard rate of compensation, and plot the BCWS, ACWP, and BCWP after eight months have elapsed. What is the cost variance? The schedule variance? The estimated cost at completion? The estimated time at completion?

9. Now, measure the cost and schedule variances for the mechanical engineers after eight months. Assign a standard rate of compensa-

tion; plot the BCWS, ACWP, and BCWP. What is the cost variance? The schedule variance? The estimated cost at completion? The estimated time at completion?

10. What advise would you share with Sue and her team given the current trends to help E&S manage the project and bring Genie to market?

Importance of Proper Planning and Control

The team successfully overcame a variety of scheduling and planning difficulties imposed by the need to bring the new technology out of the laboratory and meet the market introduction deadline. Key to their success was the firm leadership exhibited by Sue Johnson and her team of lead specialists, chosen because of their years of design experience and excellence in an appropriate business or technical discipline. Her team strove constantly to maintain open communications among all involved parties and went about measuring and forecasting the results of any deviation from established benchmarks. When necessary, corrective action was taken to realize project goals. What was vital to the health of this project was the establishment of appropriate checkpoints against which to measure and report progress to those in a decision-making capacity who were consequently able to act on the information and to help plan an alternative workaround to the problem.

After several false attempts to bring their first speech technology systems out of the laboratory and a near miss in sales projections, E&S had learned the value of proper planning and control of its new product developments. Eventually, Genie went on to be a very successful business venture, earning the company well over $300 million during the holiday season. Enhancements to the system in the form of new application programs followed the introduction, as steady growth and strong demand for the voice-activated home appliance surpassed analysts' early projections. In the following year, E&S translated the interface into five additional languages, making Genie a truly international product.

APPENDIX 10.1: PATTERNS OF RISK ACROSS SIX GENERIC SOFTWARE CLASSES[a] (PERCENTAGE OF PROJECTS AT RISK)

Risk Factor	MIS Projects	System Software Projects	Commercially Marketed Software Projects	Military Software Projects	Contracted or Outsourced Software Projects	End-User Software Projects
Canceled projects	55	35				
Cost overruns	80					
Creeping user requirements				70		
Error-prone modules		50				
Excessive paperwork		60		90		
Excessive schedule pressure	65					
Excessive time to market			50			
Harmful competitive actions			45			
Hidden errors						60
High maintenance costs					60	
Inadequate configuration control	50					
Inadequate cost estimating		65				
Inadequate user documentation			70			
Legal ownership of software and deliverables					20	20
Litigation			30			
Long schedules		70		75		
Low productivity				85		
Low quality	60					
Low user satisfaction			55			
Nontransferable applications						80
Redundant applications						50
Unanticipated acceptance criteria					30	
Unmaintainable software						60
Unused or unusable software				45		

[a]SPR classifies software projects into six generic classes: (1) management information systems (MIS), (2), system software projects (i.e., operating systems), (3) commercially marketed software projects, (4) military software projects, (5) contracted or outsourced software projects, and (6) end-user software projects.

Source: Capers Jones, *Assessment and Control of Software Risks* (Upper Saddle River, NJ: Yourdon Press, 1994).

APPENDIX 10.2: TECHNIQUES FOR PREVENTING AND CONTROLLING COMMON SOFTWARE RISKS*

· *Creeping user requirements.* Base contracts on the number of function points; use reusable designs; build requirements around common features shared among applications.
· *Schedule pressure, long schedules,* and *excessive time to market.* Use reusable code libraries; conduct reviews and inspections, and apply other quality control measures.
· *Cost overruns.* Use function points to estimate the cost of the program; outsource internationally to development contractors.
· *Low-quality and error-prone modules.* Apply defect prevention, detection, and removal technology: QFD, software quality assurance methodology; object-oriented languages; prototypes; structured analysis design techniques.
· *High maintenance costs.* Use technologies to lower risks to tolerance levels for repairs such as structural complexity analysis; reverse engineering; error-prone module analysis and removal.

APPENDIX 10.3: MOST SERIOUS SOFTWARE RISKS*

1. Inaccurate metrics such as using lines of code (LOC) to aggregate the productivity and quality data when multiple languages are involved.
   · LOC lack:
     · Standardization for any single language
     · Conversion rules of cross-language comparisons for the most commonly used languages today
   · LOC behave paradoxically in the presence of a high-level language or object-oriented language.
   · LOC doesn't consider the sizing of such noncoding tasks as writing specifications, plans, and user documentation.
   · LOC is unable to measure the time required to fix software bugs.
   · LOC has difficulty in determining source code size during requirements, at the time when this information is needed for estimating purposes.

---

*Source: Capers Jones, *Software Quality Analysis and Guidelines for Success* (London: International Thomson Computer Press, 1997).

2. Inadequate measurement of the tracking and cost collection systems for software projects is the second most serious problem that affects the accuracy of the data collected and fails to prevent future development efforts from using the flawed historical data.

3. Irrational schedules and excessive schedule pressure occurred on more than 65% of all large projects assessed by the SPR, which became the key contributor to *poor quality, canceled projects, low morale, fatigue, burnout, and high attrition rates among software personnel.*

4. Management malpractice contributes to the mismanagement of all software projects. There is little academic training in the basic tasks of software project management: sizing; estimating; planning; tracking; measurement and assessment.

5. Inaccurate cost estimating is more of a problem for commercial software producers than for military and defense contractors, who are the world's leaders in the use of these tools.

6. Creeping user requirements occurs at about 1% per month (on a three-year project, one-third of the delivered functionality may not have been specified in the original requirements). This is worse for commercial program developers, who work even shorter schedules and have the disadvantage of having to say "no" to their own executives, customers, and marketing representatives.

7. Low quality is currently measured in the United States as a defect potential of about five defects per function point with a cumulative removal efficiency of less than 85%. Military software is more complex than many civilian projects and tends to have a higher defect rate of six to seven defects per function point. However, U.S. military project ranks among the best in the world for defect removal—99%.

8. U.S. military projects typical rank among the best assessed projects for test planning, test methods, number of defect removal operations carried out, defect recording, and use of formal quality methods.

9. Low productivity is highest among U.S. military projects, which average three function points per staff month; for systems software the rate is about four function points; and for information systems, eight function points.

10. *Canceled projects:* The cancellation rate for software projects is directly proportional to the overall size of the system. Fifty percent of all large systems in excess of 10,000 function points or 1,000,000

source statements, such as operating systems, telecommunication systems, and major defense systems, are canceled. An average canceled project is about one year late and approaching or exceeding twice its planned budget at the time of cancellation. Military and civilian projects of the same size are roughly equivalent in the frequency with which they may cancel.

ORGANIZATIONAL AND INTERPERSONAL PROJECT COMMUNICATION: SPANNING ACROSS BOUNDARIES

I know that you believe you understand what you think I said, but I am not quite sure you realize that what you heard is not what I meant....

ANONYMOUS

11.1 INTRODUCTION

Effective communication is the cornerstone of any organization, any project, and, certainly, any relationship. In fact, communication is the single largest factor determining the quality, efficiency, satisfaction, and productivity of a project team.[1,2] Project work presents its own communication challenges simply because of its unique characteristics and organizational requirements. Overlapping responsibilities, decentralized decision making, multiple reporting and nonreporting relationships, and distance-spanning "virtual" teams place a strain on more traditional communication processes and ideals. Successful project management requires a reexamination of our communication processes from a more systemic project perspective and calls for the creation of strategies and skills for today's global projects' information highways.

---

[1] D. S. Kezsbom, "Communications on Communications (C2)", *Communications Society Magazine,* Vol. 20, No. 3, May 1982, pp. 47–48.

[2] D. S. Kezsbom, "Bringing Order to Chaos: Pinpointing Sources of Conflict in the Nineties," *Cost Engineering,* Vol. 34, No. 11, November 1992, pp. 9–16.

Distance-spanning communication technologies have created new territory for "working together apart."[3] Today we find that work is diffusing rather than concentrating, as we move from industrial to informational products and services. Although the use of project teams may indeed be on the rise, the face-to-face aspect of "normal" working relationships is changing dramatically. Global teamwork has become an everyday reality to employees in both big companies and in small. Videoconferencing, e-mail, the Internet, corporate intranets and sophisticated groupware all make it possible for people to work together no matter where they are based. Although technology creates business opportunities and enables us to communicate with partners in faraway places, we cannot rely on technology alone to create a sense of commitment or ensure effective communication. Human relations and interactions remain paramount.

Effective communication is difficult in the best of times and circumstances. Teamwork depends, in part, on members' abilities to communicate effectively and to trust one another. Technology cannot substitute for the actions and relationships that foster trust. One critical role of a project manager that has remained over the decades and may indeed be timeless is that of communicator. The extent to which project managers can effectively obtain, update, and disseminate accurate information continues to affect team coordination, integration, and performance.[4,5] Figure 11.1 illustrates the project manager's role as communicator within the project structure.

It is essential to project success that the project manager serve as a liaison and as a communicator to senior management, team specialists, functional managers, customers, and people outside the project who have an interest or stake in project results. Project managers must therefore expend considerable effort in assuring that they and their teams are communicating effectively and appropriately. In this chapter we present some of the communication strategies and tactics for enhancing communication in today's information-age projects.

---

[3] R. Grenier and George Meters, *Enterprise Networking: Working Together Apart* (Bedford, MA: Digital Press, 1992).

[4] D. S. Kezsbom and R. G. Donnelly, "Managing the Project Organization of the Nineties: A Survey of Practical Qualities of Effective Project Leadership," *1992 Project Management Institute Proceedings*.

[5] J. Lipnack and J. Stamps, *The TeamNet Factor: Bringing the Power of Boundary Crossing into the Heart of Your Organization* (New York: Wiley, 1993).

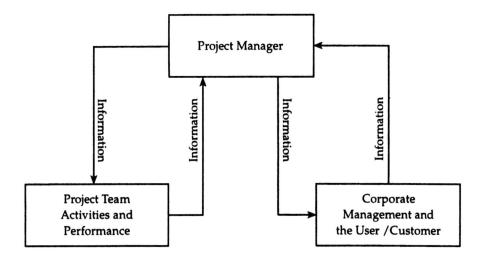

FIGURE 11.1
Project manager's communicator role.

11.2 DOCUMENTATION VERSUS COMMUNICATION: A CONCEPT DEFINED

Communication has been defined in several ways. Many of our readers are probably aware that there are several sciences of communication: kinesics (body language), proxemics (distance or spatial language), linguistics, theater, graphic arts, and others, most of which are beyond the scope of this chapter. Our objective is first to pinpoint a process that assists in creating more effective project communication. Some definitions of the communication process include:

- The exchange of facts, thoughts, opinions, or emotions
- The ability to transmit ideas, thoughts, or commands to the degree to which the other person understands precisely what is being transmitted
- Getting the right information to the right place at the right time and in the correct manner, using various methods of feedback
- A two-way transfer of ideas and information through an acceptable medium to establish a mutual understanding

One of the most inclusive of definitions is: *"Communication is any behavior that results in an exchange of meaning."*[6] This encompasses what is said and

---

[6]C. E. Leslie, "Negotiating in Matrix Systems," in D. Cleland (ed.), *Matrix Management Systems Handbook* (New York: Van Nostrand Reinhold, 1984).

what is not said, what is done and what is not done, how it is said and how it is done. Through this definition it is apparent that everyone communicates, although not everyone communicates effectively.

Communicating with specialists who share similar job responsibilities poses few problems, especially if one works in the physical sciences, such as engineering. Because engineering is a science based on precise symbols and equations, technical jargon helps to convey the difficulties involved with the system. Unfortunately, spoken words used in interpersonal communication are not as precise in meaning as the symbols used in engineering or in other technical project communication. In many cases, technical judgments and decisions must be defined and presented through a complicated verbal communication process. If the parameters of the project system cannot be conveyed clearly to colleagues, contractors, customers, and end users, costly errors and unproductive circumstances are likely to result.

Verbal communication, the ability to talk, is based on a common system of spoken symbols that are used to relay both ideas and emotions. It is a vital and powerful tool for the passage of information, the delivery of explanations, and the imparting of instructions. It is the vehicle through which a negotiator gleans information and develops an understanding of the problems or motives of the other negotiator. Moreover, face-to-face verbal communication helps to forge trust. This is, as we discuss later, one reason why global teams have that much more difficulty establishing trusting relationships.

There are many advantages to verbal communication. The choice of words, for example, and the clarity of expression can be combined carefully to strengthen meaning and impact. Field studies[7,8] indicate that technical specialists, such as software developers, physicists, or engineers, spend between 50 and 75% of their time communicating verbally with others. Verbal communication can be an efficient medium for the transfer of information. It permits timely exchange, rapid feedback, and the immediate synthesis of information.

Verbal communication has disadvantages as well. As the complexity of the project increases, for example, the ability to transmit precise verbal descriptions diminishes rapidly. Jargon develops, which affords greater verbal accuracy among technical people but limits the range of understanding beyond the technical community. It is important for project managers, especially those with more technical backgrounds, to be able to communi-

---

[7]R. W. Rosenbloom and F. W. Walek, *Information and Organization: Information Transfer in Industrial R&D* (Boston: Graduate School of Business Administration, Harvard University, 1967).

[8]L. Sayles, *Managerial Behavior* (New York: McGraw-Hill, 1964).

cate effectively with all aspects of the project, including marketing, management, sales, the customer, and/or the end user. Effective communication skills are not something that is achieved in a classroom setting alone. It requires a mix of concepts, practices, and skills, supplemented by determined effort, conscious practice, and firm resolve.

11.3 UNDERSTANDING THE BASIC COMMUNICATION PROCESS

As Figure 11.2 indicates, effective communication is not as simple as moving from point A to point B. Project communication features a series of distinct, yet interactive, and at times overlapping events. These events, in turn, are subject to a variety of influences that may either hinder or expedite communication. The basic components of the communication model include the following: a source, an encoder, a message, a channel, and a decoder, a receiver, feedback, and noise.

Communication naturally originates at the source. The source may be some member of the project team, the customer, upper levels of management, or just about anyone that has information to transmit to someone else. An idea, a need, or a concept that is, perhaps, unique and has yet to be communicated may trigger the process. Because of the newness of the idea, it may indeed lack clarity, or be somewhat ambiguous or incomplete.

- *Source:* some member of the organization who has information to communicate or convey to someone else in the organization
- *Encoder:* translation of the original ideas into a set of symbols that express what the source wishes to communicate

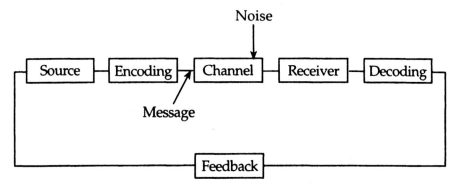

FIGURE *11.2*

Communications model. (Adapted from James L. Gibson, John M. Ivancevich, and James H. Donnelly, Jr., *Organizations: Structure, Processes, Behavior*, Business Publications, Dallas, TX, 1973, p. 166.)

- *Message:* actual physical product of the encoding process; the message is, basically, what the source hopes to communicate
- *Channel:* medium through which the message is carried from the source to the receiver; in an organization this may be face-to-face communication, a telephone call, word of mouth, group meetings, and so on
- *Decoder–receiver:* provides for the process of communication to be completed; the message (sent by the source) must be decoded by the receiver–recipient in terms of a schema or meaning; the recipient will interpret the message in light of personal life experiences, value system, frame of reference, and so on
- *Feedback:* insight into the degree of fidelity that the message has; message fidelity is rarely perfect; feedback gives the source the opportunity to evaluate whether the message produced the intended purpose
- *Noise:* anything that causes a breakdown, interference, or distraction within the communication process; as in electronic communication, noise in interpersonal communication is anything that reduces the fidelity and clarity of the message, such as misunderstanding or misinterpretation by both receivers and senders, preconceived value judgments on the part of both sender and receiver, and so on

Despite these obstacles, the idea we have needs to be *encoded*, or translated into some language or set of symbols that hopefully expresses what the source wishes to convey. Consider, for example, that the source is a commercial customer who needs changes in the performance of a particular information system or product. Background, education, profession, age, and other demographic characteristics influence the choices the sender makes to encode their messages. The sender must use his or her own realm of life experiences from which to draw the appropriate language or code.

The purpose or intent of the source is expressed in the form of a *message*. This description must be relayed to the receiver through a medium or a channel. This may be face-to-face communication, as in videoconferenced meetings, a telephone or conference call, or written communication in the form of memos, e-mail, or groupware. *Noise* on the channel is any occurrence that causes a breakdown, interference, or distraction within the process. As in electronic communication, noise is anything that reduces the clarity or fidelity of the message. This may take the form of misunderstandings or misinterpretations based on preconceived value judgments of both the sender and the receiver.

Finally, it is up to the *receiver* to pick up the message and *decode* or interpret it in light of his or her personal life experiences, values, or frame of reference. The *feedback* that the receiver provides the source presents insight into the degree of message clarity and fidelity. Message fidelity is rarely perfect. Feedback therefore gives the source the opportunity to evaluate whether or not the message has achieved its intended purpose.

In the feedback process, the receiver and the sender switch roles and the entire communication process begins again. Feedback must not only be complete, but objective and useful to the receiver. Certain active listening responses, discussed later in the chapter, provide the sender with helpful techniques that assure all concerned that the message has achieved its maximum fidelity and that the communication process is complete. Communication is complete when the sender compares the initial concept with the final feedback message. Effective communication implies that a receiver's understanding of the meaning is equivalent to the sender's intent. Achieving effective communication is not an easy task, nor is it solely the responsibility of the sender if the sender and the receiver have agreed to work effectively with one another.

11.4 TODAY'S CHALLENGES TO EFFECTIVE COMMUNICATION

Just a decade ago, when you said that you worked with someone, one could assume that you meant in the same office or the same department within the same organization. Today, however, technology permits people who no longer work in the same location, or who fail to be "collocated," to be able to *work together* on a project team. Now, many people work in virtual teams that transcend distance, time zones, and even organizational boundaries.

The characteristics of the hybrid cross-functional matrix organization, discussed in Chapter 2, heighten the possibility of communication difficulties. Today's project teams, moreover, may not only be cross-functional, but global and transcultural in nature. These "virtual" teams, in fact, create entirely different images from the one of people working together in the same organization in the same place.

In the best of circumstances, cross-functional teams consist of people from distinct occupational specialties. The specialized language or jargon that is characteristic of each creates barriers when messages from one specialty must be translated by the next. Training and loyalties of each profession further result in a set of acquired values that creates an occupational frame of reference. This frame of reference affects the way we view subjects such as cost, deadline, priorities, and technical performance.

Unlike the conventional face-to-face teams, virtual teams routinely cross boundaries through an array of interactive electronic communication technologies. Working in these multinational, transglobal teams poses certain challenges not usually encountered when a group of people work together in the same building or even in the same city. Some of these challenges may be quite obvious, as when a group of people are working in different time zones across the world. Team members who are in Shanghai or Singapore are 12 hours ahead of those in New York or Toronto and have no real opportunity to call one another during normal business hours.

Today, many companies use time differences to their advantage, by transacting business virtually around the clock. But for professionals other than bond traders, for example, time differences can be frustrating. Certain types of projects, for example, require overseas affiliates (or team members) to participate in meetings that may be scheduled in the headquarters' time zone. Thus, a 2:00 P.M. conference call with New York means that team members located in Australia and Singapore are sitting around in their pajamas in the early morning hours waiting to take part.

Other problems encountered by teams whose work virtually spans the globe can be more subtle, yet equally important. Nonverbal communication can account for as much as 60% of the message a person conveys. This can entail the furtive glance, the reddening neck, or the twitching face — clues that often convey a plethora of important emotions. Even in the best videoconferencing, facial expressions can be difficult to pick up if the transmission is poor, if one is off camera, or when the mute button is pressed.

Often, compressed schedules and conditions of uncertainly lead many project participants to believe that little or no time exists for effective communication. Misunderstandings naturally arise not only when individuals perceive a lack of time, but also when there is a lack of specificity in the communication sent. A memo that ends with "ASAP" can mean "as soon as possible" in a number of different ways: "by Thursday noon," "within a day" or even "within a month from now" especially when we are dealing across continents and cultures. The point is that considerable leeway exists for misunderstandings in communication and that a very high probability of miscommunication exists. Project planning tools, such as a participatively created work breakdown structure, a highlighted critical path, and collective updates serve as the best means by which to verify each specialist's perception of what is to be done and when it is expected. Such tools as the intranet or Internet, when used to share project information, are not a substitute for, but a vehicle through which, communication takes place.

Completing projects through electronic groupware means that team members are isolated from one another, which increases the chances for misinterpretation. Groupware allows information about a project to be fed into a huge structured database that can be accessed by all team members. When databases fail to contain the newest information, one can assume that the virtual team isn't communicating effectively. This may lead team members to hoard what they know, or share information within their function or discipline, rather than share with all team members. Project managers must assure that the data available to team members are both accurate and current.

Another barrier to effective communication, even within similar cultures, is the many meanings we associate with certain words. It appears that with many words, the meanings are not so much those that lie within the dictionary, but within the values and perceptions of the receiver. Connotations imply that words possess different values—positive, negative, or neutral—and that the real value of the word depends on how the listener perceives and decodes it. Words and phrases that a person may choose to use may or may not evoke the same images in the next person's mind. A word such as *management,* for example, may hold one meaning to a CEO and still another to the head of a labor union. "Framing" memos and presentations, or selecting words carefully to depict a particular mood or feeling, is helpful to any member of a team, especially when communicating with senior management, the customer, or across cultural boundaries.

11.4.1 Trust and Accuracy

In an earlier chapter we discussed the importance of trust in developing influence and establishing power within a project team. Several studies[9,10] have revealed the degree of trust that exists between project specialists and their various levels of management to be strongly and directly associated with the degree of open, accurate, complete, or undistorted communication. In addition to trust, the research has explored several other variables, including level of security, autonomy, and overall organizational climate.

More specifically, the literature supported the following assumptions:[11,12]

- The amount or degree of authority delegated in project work is *inversely* related to the distortion of upward communication. That is, in organizations where authority is distributed widely across project teams, *less* communication distortion is likely to occur.

[9]R. Might. "An Evaluation of the Effectiveness of Project Control Systems," *IEEE Transactions on Engineering Management*, Vol. EM-31, No. 3, August 1984.

[10]S. Brandt, "The Matrix Manager and Effective Communication," in D. Cleland (ed.), *Matrix Management Systems Handbook* (New York: Van Nostrand Reinhold, 1984).

[11]P. Muchinsky, "The Interrelationships of Organizational Communication Climate," as stated in Brandt, *The Matrix Manager*, p. 584.

[12]J. Athanassades, "The Distortion of Upward Communication in Hierarchical Organizations," *Academy of Management Journal*, Vol. 16, pp. 207–226.

· The widely held belief that a supportive communication climate helps to build feelings of trust is supported empirically. Research indicates that accuracy of communication is positively related to organizational climate and trust.[13,14]

Trust has always been an important element of group life. The virtues of having trust in organizations appear to be quite evident. Teams with higher levels of trust blend more readily, organize more quickly, and manage themselves better. Low levels of trust, on the other hand, make it more difficult to generate and sustain cross-functional, cross-boundaries, and cross-organizational teams.[15]

11.5 COMMUNICATION TOOLS IN AN INFORMATION ERA

What makes today's teams new and unique is the array of interactive technologies at their disposal. Teams now use a myriad of electronic technologies to deal with the opportunities and challenges of transglobal work.[16] These include:

· *Pagers or beepers:* useful to teams in conjunction with the use of telephones, e-mail, and other forms of two-way communication
· *Facsimiles:* a standard means by which to provide documentation to the team
· *E-mail:* a fundamental building block for virtual teams
· *Voice-mail:* an immediate and timely means of communicating
· *Telephone:* may allow for more emotional transmission than e-mail or voice mail
· *Audioconferencing:* a form of group telephone calls, useful for disseminating information to many at once
· *Videoconferencing:* a multisense communication medium, integrating telephone communications with visual access
· *Groupware/document-sharing systems:* provides a database of information unique to the team and the project

---

[13] K. L. McMahon, "Effective Communication and Information Sharing in Virtual Teams," *http://www.bizresources.com/learning/evt.html*, August 1998.
[14] S. Jarvenpaa and D. Leidner, "Communication and Trust in Global Virtual Teams," *http://www.ascusc.org*, June 1998.
[15] J. Lipnack and J. Stamps, *Virtual Teams: Reading Across Space, Time, and Organizations with Technology* (New York: Wiley, 1997).
[16] Ibid.

- *Web Conferencing:* a means to share team work spaces over the World Wide Web
- *Intranets:* internal corporate information systems.
- *Internets:* worldwide information communication system

These communication tools have an assortment of qualities and characteristics that affect a team in a variety of ways. Speed, immediacy, and documentability are just some of the characteristics or properties of communication tools that project managers and the teams they lead must consider when choosing which tools to use at what time. A telephone call, for example, is immediate, personal, common in use, transitory, and rapid. It has, in fact, a radically different effect than a fax on the team or the individual. Certain tools chosen, moreover, may engage a team in more team blending activities than others. The interaction desired from the team must be taken into consideration when choosing a communication medium. Table 11.1 illustrates how a project manager may match the interaction desired with the tool on the appropriate level.

TABLE 11.1 INTERACTION MODES AND COMMUNICATION TOOLS

Time/Location	Interaction Mode	Vehicle(s)
Same time/ same place	Traditional face-to-face environment	Flipcharts, overhead projector, handouts, meetings
Same time/ same place	Using technology, participants in face-to-face meeting, using linked computers to input data or apply a variety of software applications	Decision-making support software, computer graphics, electronic white board
Same time/ different place	Participants join scheduled synchronous meetings from any location equipped with necessary access devices	Video/audio teleconferencing, phone bridge, online chat(s)
Different time/ different place	Participants communicate asynchronously at any time, from any place, often in a password-protected environment accessible to their team	Web conferencing, bulletin boards, newsgroups, voice mail, e-mail

Source: Adapted from Lisa Kimball, "Boundaryless Facilitation: Leveraging the Strengths of Face-to-Face and GroupWare Tools to Maximize Group Process," *http://www.tmn.com/ ~ Lisa/bnd.htm,* April 1999.

11.5.1 Leveraging Communication Tools for Productivity

Having a variety of information technologies for communication does not guarantee productivity. Tools must be used in ways that build the team and encourage continued interaction. Teams must search for the most appropriate media for their communication needs, with consideration of the limits of cost and availability. Multiple media should be used to offer many pathways for interactions and the development of relationships. Tools need to be chosen carefully and used for its intended purpose. Some suggested tools and means to leverage their team use are discussed next.

Voice. Since the telephone has become such a basic tool in work and social settings, we rarely reflect on the importance of this technology to the team and to the productivity of the project. The telephone allows the transmission of a somewhat more emotional tone and may provide more feedback than, for example, e-mail or voice mail. It is a good medium for building trust or for any interaction where spontaneity and informality are needed.

Voice mail is one of the fundamental building blocks of virtual teams. It is immediate and timely, and because it is so common can be used readily to establish expectations and trust among team members. Voice mail systems may be enhanced to allow people to send a single message to any subgroup of the team or, if appropriate, to the entire team.[17] In individual voice mail systems, remember to change the voice mail greeting each day to establish the day's date and agenda. This sets reasonable expectations for the other team members about your availability and when you can be reached. Ground rules should be established for the use of voice mail. Shouting at another team member through voice mail gives them no chance to defend themselves or diffuse the situation. What is more, leaving an angry voice mail message "documents" the emotion and can backfire on the sender all too easily.

Audioconferencing. Audioconferencing has the qualities of the telephone but reaches far more team members. The format is good for building trust, brainstorming and other group decision-making processes, and for reaching consensus. It has, in fact, become a very common and therefore a user-friendly medium. The qualities of familiarity and immediate feedback can be leveraged if regularly scheduled audio conferences are established to check, among other things, the emotional "tone" of the team. This is a great

---

[17]J. Lipnack and J. Stamps, *Virtual Teams.*

and relatively inexpensive tool but does not replace face-to-face communication.

Videoconferencing. One primary characteristic of the cross-functional team is lots of face-to-face meetings. Videoconferencing is a multisense communication medium that permits you to see, hear, and talk to other people. Traditional videoconferences link project partners through satellite and telecommunications networks. In the past, videoconferencing had been relatively expensive to use, tightly scheduled, and with few adequate facilities. Recently, however, desktop videoconferencing has been developed that allows a small number of people to hear and see one another through small windows on their personal computer screens. The camera image transmitted by a camera mounted on top of a person's computer monitor creates a more informal atmosphere. Scheduling is less of a problem than with full group videoconferencing, allowing the medium to be used for more ad hoc team meetings.[18]

Facsimiles and "Snail Mail". Today, facsimiles have become a standard means by which to disseminate information and documentation throughout a team. It is useful in sending nonelectronic, hard-copy documents to a relatively limited group. Distribution lists must be arranged as part of a team building effort, to assure that all who must receive the printed information are indeed included. Postal mail, naturally, is slower than sending facsimiles but may still bring to the team that certain impact that an overnight letter or parcel carries.

Computer. In addition to the numerous audio and video links, virtual teams may use a variety of secured computer-based communication connections. Basic e-mail and that with the capability of embedding information or attaching the equivalent of multiple packages of any size to a letter may be available to everyone on the team. Intranet discussion groups also keep participants updated on the status of the project. Groupware is a software application that operates over the company's internal Web-based intranets. Groupware documents can store pictures, graphic presentations, CAD files, sound files, and a variety of other programs. Groupware allows virtual teams to create a "group space" where they may work on the same documents, store team information, and create a sense of collaboration. Web conferencing further provides secured team work spaces over the

---

[18] Ibid., pp. 83–84.

World Wide Web. Web conferencing brings the advantages of newsgroups and computer bulletin boards to the more manageable confines of the virtual team. Web conferencing includes such features as electronic white-board, video, and audio transmission; chat groups; file transfer– shared applications; and voice mail and e-mail capabilities.

As exciting and effective as these new forms of technology and means of shared information prove to be, they still do not replace the need to get together periodically, preferably as a whole team, in a particular site. Physical connections between team players is much more than a nicety. It is a necessity for building credibility and trust, which, in turn, may affect the team's decision-making processes positively or negatively. Face-to-face meetings may also include more of an element of fun and celebration, reinforcing the importance of the relationships and the intangibility of human bonds.

11.6 NEGOTIATING FOR PROJECT SUCCESS

In almost all project organizations, especially those that are of a hybrid nature, there is shared responsibility, authority, and account-ability. Project participants drawn from different operating, administrative, and technical sectors possess differences in perceptions. These differences may be in regard to the technical processes employed, or over the resources, time, or effort expended. Project managers must often negotiate these differences not only among team specialists and managers contribu-ting to the project, but with other project managers in setting disputes associated with, for example, conflicting project priorities or overlapping responsibilities.

Negotiation is a pervasive and important process not only in business transactions, but also in everyday life. Negotiating in a project situation may involve such issues as budget, resources, schedules, and/or technical methodology. It may also include such intangibles as shared project ownership, recognition, and continued effective and productive relation-ships. One criterion of the successful negotiator is the ability to adapt negotiating style to fit the needs of the particular situation and the people involved.[19] This requires a firm grasp of the personalities and group dynamic that come into play in each negotiation. Certainly, knowledge of the negotiating process will prove vital to project success.

---

[19]C. E. Leslie, "Negotiating in Matrix Management Systems."

Effective negotiators recognize that "winning" with regard to the immediate issue in not the totality of the negotiating process. Short- and long-range implications, and associated tasks, must be weighed carefully against the success of an instant win. An astute negotiator may assess that a present loss may be a future gain. Project managers and team leaders who operate in a matrix environment must remember that winning in the short term may not be as advantageous as maintaining harmonious relationships for the duration of the project.

11.6.1 The Negotiating Process

Negotiation is a two-part process. The first part deals with all the activities, beginning with the initiation of a request, demand, or offer and ending when agreement has been reached. The latter part of the process concerns the postnegotiation period and is, in many instances, considered to be the most important aspect.[20] This is the period when a solution is implemented, when any misunderstandings or hard feelings that still remain may escalate, and when relationships can improve or worsen.

Naturally, most negotiators strive to "win" in the first phase of the negotiating process. In fact, negotiating strategies are generally classified around the act of winning. These strategies are labeled according to the results achieved by the negotiating parties. Thus the three basic strategies are win–lose, lose–lose, and win–win. Table 11.2 provides an overview of the characteristics of each of these strategies.

Effective negotiators realize that merely winning in the immediate phase of the negotiation process is not what the art of negotiation is all about. Short- and long-range considerations must be taken into account. Risks must be calculated and assessed before actions are taken. Although losses should be minimized for all parties involved, it is possible that an astute negotiator may want to consider taking an immediate loss to support a long-term win.

Many of us have been indoctrinated with the concept of winning through our sports and leisure activities. Golf, tennis, chess, bridge, and most other leisure activities concern a winner and a loser. But negotiators for a software system or product development effort, for example, must realize that a loss may be as advantageous as a win, if it maintains harmonious and productive relations in the long run.

As indicated by Table 11.2, the win–win strategy naturally has several advantages. Both parties achieve a resolution that makes each feel like a

---

[20] Ibid., p. 631.

TABLE 11.2 CHARACTERISTICS OF NEGOTIATING STYLES

Win–Lose, Lose–Lose	Win–Win
Controlling orientation (we vs. they)	Problem orientation (we vs. the problem)
One party's gain is seen as the other party's loss	Mutual gain viewed as attainable
Arguments over positions leads to polarization and entrenchment	Seeking various approaches to an end, increasing chances for agreement
Each side sees issue only from its own point of view	Parties understand one another's point(s) of view
Short-term approach, focused only on immediate problem	Long-term approach, seeking good ongoing relationship issues
Usually considers only task issues	Considers both task and relationship issues

winner. That is, each obtains what they perceive they need and want from the negotiating situation. This approach is, of course, contingent upon developing cooperative attitudes between all negotiating parties.

Win–lose and lose–lose strategies are used frequently in business negotiations. They evolve from our competitive nature, past behaviors, a resistance to change, and a failure to understand how all parties can win.[21]

In project team efforts, these attitudes and behaviors may result from a lack of understanding of the common goal and a failure to view the effects that our actions have on the project organization.

11.6.2 Examining Negotiation Strategies

By examining the effects of negotiation strategies on attaining project goals, a project manager will be better able to choose the strategy that is most appropriate to the situation. The key issue in each of these strategies is not how much each negotiator obtains, but the degree of satisfaction felt as a result of the process.

Win–Lose. The typical win–lose strategy regards negotiation as a form of competition. In many win–lose instances, the balance of power between or among negotiating parties determines the outcome of the negotiation.

---

[21]W. S. Kirchoff and J. R. Adams, *Conflict Management for Project Managers* (Drexel Hill, PA: Project Management Institute, 1986).

Coercion, ignoring the issue at hand, or using some from of majority rule are all examples of win–lose strategies. Voting, although a frequently used strategy, is actually a win–lose tactic and may prevent team members from becoming personally involved in the solution. Often, in the win–lose approach, one party's gain is seen as the other party's loss. Frequently, win–lose discussions revolve around positions rather than seeking strategies for mutual gain. Win–lose is a short-term approach that tends to focus on the immediate problem, and may, at times, prove to be harmful to long-term team-building efforts.

Lose–Lose. In lose–lose strategies the negotiating parties believe that to obtain consensus and achieve a common goal, each must forfeit something they want. Lose–lose strategies frequently are conciliatory in nature. They stem, many times, from a basic desire to avoid conflict and smooth over issues of dissent. In this way, compromise and arbitration may indeed be regarded as lose–lose strategies; that is, everyone gives up something in the bargaining for the good of the common goal. Once again, however, it is not how much one wins or loses that makes a negotiation a success. More important, success must be measured by the degree of satisfaction that each party derives from the negotiated settlement. Naturally, compromise may bring some degree of satisfaction and is, at times, quite necessary in negotiation. However, if project specialists or any negotiating party feel as if their loss was not worth the gain, commitment to the settlement or negotiated agreement will suffer.

Win–Win. Win–win strategies are characterized by a "we versus the problem" orientation. Mutual gain is viewed as attainable, as negotiators strive to generate alternative approaches that may lead to the common goal. Parties attempt to understand one another's viewpoints and seek a long-term approach, which no one really opposes. The objective of the win–win strategy is to find a solution that each participant can live with. Win–win strategies are those that seek consensus and begin with an attitude of mutual gain. The win–win strategy is a long-term approach that naturally builds good relations.

The steps to be taken in win–win negotiation are as follows:

Step 1: Identify the needs of conflicting parties. Avoid taking positions (i.e., arguing over means) and instead, try to identify the ends or needs of both parties. Put yourself in the person's position and ask "why?" as well as "why not?" Recognize the need for respect.

Step 2: Brainstorm a list of possible alternative solutions. Instead of working against one another ("How can I defeat you?") work together against the problem.

Step 3: Evaluate alternative solutions. Work for an answer to meet the important needs of all the parties.

Step 4: Implement the solution. Make sure that everyone agrees to and understands it.

Step 5: Follow up on the solution. Revise, if necessary.

11.6.3   The Postnegotiation Period

Satisfaction with a win or a loss is only a part of the total negotiation process. The real success or failure comes *after* the negotiation, when negotiated plans (sometimes referred to as *settlements*) must be implemented. Agreements reached must be within the acceptable range for all parties involved if the negotiation is to be considered successful. When contemplating negotiation strategies, balancing the importance of the instant gain to the long-term win is paramount to success.

The way in which the negotiation process occurs, the posture taken, the words chosen, and the emotions displayed are all important pieces affecting this process and influence our perception of the sweetness of our victory or the bitterness of defeat. Implementation of settlements can be enhanced or discouraged by what has transpired during the actual negotiation process. Harmonious relationships need to be maintained and credibility of agreements established. In truly successful negotiations, all parties involved achieve feelings of satisfaction with the end product or negotiated result.

11.6.4   Tactics for More Effective Negotiation

Many tactics or strategies have been developed to assist in the process of negotiation. These tactics must be chosen to fit the needs of the particular circumstances. In all instances, however, maintaining effective communications and working relationships is essential to effective negotiation. Some tactics for negotiation are:

- Whenever possible, negotiate on home ground or a neutral site
- Prepare an agenda. A well-planned agenda adds structure and purpose to the negotiation and is a benefit to the negotiating parties. Provide the agenda well in advance of the meeting. If supplied with an agenda by another negotiator, submit input or acceptance promptly

- Consider the seating arrangement. Arrangement of seating can have a direct effect on the negotiation process. Sit in view of others as well as in view of your team members
- Prepare and ask questions to probe feelings; discover attitudes and determine negotiating positions
- End on an upbeat note! Summarize points of agreement and convey a positive tone

Some steps to further consider when preparing for an effective negotiation are:

Step 1. Determine what you truly need.

Step 2. Determine what you would like.

Step 3. Establish trade-offs between what you would like and what you really need.

Step 4. Know your negotiating partner. Bone up and research (if possible) problems the other participants may have.

Step 5. Record what occurred during the negotiation.

Step 6. Maintain a warm and friendly negotiating climate.

Step 7. Practice listening and patience.

Step 8. Focus on the issues.

Step 9. Set out to obtain the entire package, but be prepared to make concessions and trade-offs.

Step 10. Maintain good relations.

Since negotiation is almost a daily occurrence in a matrix or hybrid project environment, the dynamic project manager must continually maintain organizational cooperation and harmony. In this way, instances that call for negotiation may begin on a more positive note, and resolution may actually improve project relations and performance.

11.7 LISTENING FOR BETTER COMMUNICATION

Listening is one of the most important and powerful managerial and personal skills. It is a powerful way to "stroke" co-workers, customers, or management. Effective listening creates a better environment for negotiation and problem solving. People who are listened to feel valued, respected, and important. Effective listening is as important to the communication process as effective "sending." Most of us "hear" a lot of things, but few of us really listen.

Effective listening is an active give-and-take process. To listen effectively, the listener must interact with the speaker in an attempt to understand both the content and the meaning of the message. This requires a mental and a verbal paraphrasing of the message, with careful attention given to nonverbal cues. Failure to listen effectively not only prevents adequate responses or counterdemands, but also affects the business relationships that are established in the process. Let's examine the example below.

Example 11.1: Listening Builds Good Relationships and Productivity. One day, Ted Burros, a systems engineer working on Teline Incorporated's latest computer telephony integration project, invited Joan Brown from the development group to his area. Ted wanted to explain his department's plans for the new system. He described how he thought the system architecture that was typically used should not be changed. Joan's only response was silence and a frown.

Ted realized that something was wrong and sensed that Joan might have something to say. "Joan," he began, "you've been in this business just about as long as I have. What's your reaction to my suggestion? I'm listening."

Joan paused, eyed Ted mistrustfully, and then began to speak. Ted had opened the door to communication and she felt comfortable offering her suggestions from years of experience. She was also in a more relaxed and trusting mood to be open to Ted's countersuggestions. As the two exchanged ideas and comments, a mutual respect and trust developed along with solutions to many of the technical problems.

Several deterrents stand in the way of effective listening. All people have difficulty listening when dealing with a conflict situation, dealing with an emotional colleague or customer, criticism is being directed at them, being disciplined, or feeling anxious, angry, or fearful. Other deterrents to effective listening include assuming in advance that the subject is uninteresting, mentally criticizing the speaker, or overreacting to certain words or phrases.

11.7.1 Active Listening for Better Professional Relations

Active listening is a strategy that involves reviewing the central themes or content of the message, reflecting on what has been said, and assuring one's understanding through paraphrasing and testing one's assumptions. Active listeners are effective listeners because they become involved in the communication process. In their own words, listeners restate or paraphrase the message for the sender. In doing so, the sender receives feedback concerning the clarity or fidelity of the message and if the communication process is, indeed, complete. Active listeners develop a deeper appreciation of what the other person is thinking and feeling. Paraphrasing and asking questions, a strategy of active listening, is a good way to clarify meanings whenever they are in doubt.

Let's examine the following "Yes, but..." situation.

Gary: I just don't know what I'm going to do with my boss. He's always picking on me for the little things I do wrong.

Kent: You should talk to him about why you are so upset.

Gary: Yeah, but I couldn't do that. He'd make life miserable for me.

Kent: Well, you ought to ignore him and not let him bother you.

Gary: Yeah, but then I'd be letting him get away with his lousy behavior and he would never change!

Kent: Well...maybe you should quit and get another job.

Gary: Yeah, but I need the money and it would take months before I found another job.

Kent: (by this time completely exasperated) Well, why don't you just take a gun and shoot him?

Notice that each time Kent generated a solution, Gary was spending his listening time thinking of rebuttals, and the two co-workers established a circular and frustrating listening pattern.

To be a more effective listener, especially when confronted with a problem or an emotionally charged situation, one may benefit by taking an active responsibility in understanding both the *content* and the *feeling* behind the message:

$$\frac{\text{content} + \text{feeling (empathy)}}{\text{message}} \rightarrow \text{active listening}$$

Now, let's see what happens when Kent uses a more empathetic active listening approach.

Gary: I just don't know what I'm going to do about my boss. He's always picking on me for the little things I do wrong.

Kent: Sounds like you have trouble handling your boss, especially when he points out things you do that he doesn't like.

Gary: Yeah, and he does it a lot. I don't want to tell him about it because it might make him mad. Then he would probably make life miserable for me.

Kent: Hmmm, boy. That's a double bind. On one hand you want to tell you're boss what you don't like, and on the other hand, you don't want him to get upset with you. That's frustrating!

Gary: Yeah! That's exactly how I feel.

Kent: It's a rough spot to be in. What kind of choices do you have? Let's talk about them.

Notice that Kent's summarizing of Gary's response got him out of the "Yes, but . . ." trap and gave Gary an opportunity to really discuss what was bothering him. Kent responded to Gary's feelings by restating the problem in his own words. As he continued to do so, he moved the interaction toward a more problem-solving mode.

Empathetic active listening is an excellent technique to defuse a conflict situation, understand a negotiating opponent, or improve customer relations. The active listening process accomplishes several things, both for them and for you.

They:

- Get the problem off their chest.
- Feel better understood.
- May gain more insight into their situation.
- May solve their own problem.
- Are in a better frame of mind to accept any advice.

You:

- Know them better.
- Can be better informed about their problem to offer more appropriate advice.

For active listening to succeed, however, it requires genuine concern and respect for the person and his or her situation. It also means that you as the listener must remain neutral on all subjects and assume an objective stance. It requires more time to listen and to relate and means that you never make the decision for them but help the person verbalize his or her own decisions.

Some benefits of active listening are that it:

- Encourages honesty
- Develops better understanding
- Creates a nonjudgmental atmosphere
- Encourages greater information and insight into the problem
- Diffuses tension-provoking situations
- Creates a sense of security
- Encourages feelings of self-confidence
- Reduces stress
- Develops mutual respect and trust

Some active listening techniques that help forge better professional relations are as follows:

- *Nonverbal listening "remarks":* eye contact, facial expressions, gestures, body position.
- *Casual remarks:* "I see," "Uh huh," "Go on," "Tell me more."
- *Echo:* repeating verbatim the last few words of what the speaker said: "Okay, from what I understand . . . ."
- *Questioning:* encouraging further contributions from the speaker: "What was your reaction to that?" "When did this begin to occur?"
- *Paraphrasing:* stating in you own words what the message means to you. "So, from what I understand, you believe that the order will be delayed due to these changes."
- *Reflecting:* paraphrasing both the content and the perceived feeling of the message to exhibit empathy: "You seemed concerned about the schedule and the changes to be made."

11.8   CONCLUSIONS: IMPROVING PROJECT COMMUNICATIONS

There may be formidable barriers to overcome in making project communication click. Project managers help to create an environment within the project structure that has a tremendous effect on the success of communication and interpersonal relationships. Clearly defined channels, well-established reporting mechanisms, and good working relationships naturally enhance organizational and project communication. Communication breakdowns can be avoided if people know with whom they should communicate and what information needs to be transmitted.

Communication becomes tenuous under a variety of circumstances. These include:

· Confused lines of authority and reporting relationships
· Lack of proper planning
· Restructuring and/or reorganizing the organization on a too-frequent basis
· Centralized control with little project authority
· Cumbersome bureaucracies
· Need for interpersonal communication skills among employees
· Physical distance, and cultural and language barriers

To encourage more effective organizational and interpersonal communication, project managers and team leaders must model two-way communication strategies, demonstrate a sincere interest in what project specialists have to say, and reward project teams in their efforts to communicate. Requesting reports and information and providing feedback in a regularly scheduled and timely manner further improves the communication process and leads to feelings of trust, goodwill, and an eventual sense of independence for the team. A trusting environment that demonstrates respect, where people feel comfortable with one another, may just be the key to more effective communication.

EXERCISES

11.1. *Nonverbal communication.* Take a monument to reflect on the following questions. Write your initial responses. Reflect on your responses prior to attending your next meeting.

(a) How do people let you know nonverbally that they are listening?

(b) How do people let you know nonverbally that they are not listening?

(c) How do they let you know that they have a problem?

(d) What do they do to let you know that they want to terminate the conversation?

(e) How do they let you know that they aren't interested in what is being said?

(f) How do they let you know that they have responded to "red flag" words?

(g) How do you know that they have been embarrassed or insulted?

(h) How do they let you know that they are daydreaming?

11.2. *Communication tools in an information era.* Having a variety of information technologies for communication does not guarantee effective communication or productivity. Tools must be used in ways that build the team and encourage interaction. Describe briefly the advantages of each of the following communication tools in building the virtual team.

(a) Voice

(b) Facsimiles and Postal Mail

(c) Video Conferencing

(d) Audioconferencing

(e) Computer technology

11.3. *Listening practice.* The following situations will give you a chance to practice identifying empathetic active listening responses. Circle the responses in each situation that you think are empathetic. Check your answers at the end of the exercise.

1. "It happens every time the manager appears in my department. He just takes over as if I weren't there. When he sees something he doesn't like, he tells the employee what to do and how to do it. The employees get confused and I get upset."

(a) "You should discuss your problems with your boss."

(b) "When did this start to happen?"

(c) "The boss must be the boss, I suppose, and we all have to learn to live with it."

(d) "It can be upsetting when one manager overrides another's authority. It's difficult to confront it, too."

2. "It's happened again! I was describing an office problem to my boss when she started staring out the window. She doesn't seem to be really listening to me because she has to ask me to repeat things. I feel she's superficially giving me the time to state my problems, but she ends up sidestepping the issue."

(a) "You should stop talking when you feel she's not listening to you. That way, she'll start paying attention to you."

(b) "You can't expect her to listen to every problem you have. Anyway, you should learn to solve your own problems."

(c) "It's frustrating to have your boss behave this way, especially when you talk about problems that are important to you."

(d) What kind of problems do you talk to her about?"

3. "I think I'm doing all right, but I don't know where I stand. I'm not sure what my manager expects of me, and he doesn't tell me how I'm doing. I'm trying my best, but I sure wish I knew where I stood."

(a) "Has your boss ever given you any indication of what he thinks of your work?"

(b) "Not knowing if you are satisfying your boss can really make you feel insecure on the job."

(c) "Perhaps others are also in the same position. Don't let it bother you."

(d) "If I were you, I'd discuss it with him."

4. "He used to be one of the guys until he was promoted. Now he's not my friend anymore. I don't mind being told about my mistakes, but he doesn't have to do it in front of my co-workers. Whenever I get a chance, he's going to get his!"

(a) "To be ridiculed in front of others is always embarassing."

(b) "If you don't make so many mistakes, your boss wouldn't have to tell you about them."

(c) "Why don't you talk it over with a few people who knew him before and then go talk to him about this situation?"

(d) "How often does he do this?"

Answers: **1.** (d), **2.** (c), **3.** (b), **4.** (a).

11.4. *Empathetic listening.* Using the empathetic listening approach, design listening statements in response to the following situations. Compare your response with those at the end of the exercise.

(a) John, a fellow supervisor who is conscientious and strong-willed, has lost the ambition to be productive. He responds irritably to routine problems and as a result has become difficult to work with. You are talking to him during lunch when he says to you: "Nobody cares. There are no efforts being made to improve conditions. We get no information or leadership from management and we seem to have the same problems over and over."

(b) You are a supervisor in the corporate quality and manufacturing department. One of your engineers, Derek, is outspoken, fault finding, and demanding. His co-workers respond to him by being offended and avoiding him as much as possible. Derek says to you: "The people who work with me are lazy and unfriendly. I tell them what they ought to do, but they don't listen to me."

(c) A fellow supervisor, who is conscientious and responsible, says to you in an angry voice: "I don't know how to deal with this program manager. He doesn't seem to listen to me when I tell him I'm not able to meet the date for his requested action. I have to wait for the vendor to deliver the parts, and yet he won't accept my reasons why the action cannot be accomplished."

Discussion: Here are some possible empathetic responses for the situations given above.

(a) "Seems you feel discouraged about the way things are being handled, and that you're fed up with the same old problems."

(b) "If I understand you correctly, Derek, your co-workers don't do things the way you feel they should. How do you let them know what you think they ought to do?"

(c) "It's annoying to think you've explained your problem in a way the other person should understand, but he won't seem to listen to how you are restricted."

QUALITY IN THE PROJECT ENVIRONMENT

*Quality is never an accident;
it is always the result of
high intention, sincere effort,
intelligent direction and
skillful execution; it
represents the wise choice
of many alternatives.*

WILLA A. FOSTER

12.1 INTRODUCTION

Quality is the foundation of competitiveness for the twenty-first century. We need only to look at the ebb and flow of the history of quality management throughout civilization to understand how this rich heritage with its origins in antiquity has become business's best investment. Quality management is the most basic of business concepts; however, it is misleading to think that the perfection of these concepts is a recent invention.

· In China, from the Xia to Qing Dynastys (twenty-first century B.C. to 1911), an autocratic bureaucracy influenced the quality of control of a feudal handicraft industry. Independent departments, each with its respective special functions, extended central control over the entire progression of processes through which end products were produced. Decrees were promulgated to ban the sale of inferior products; articles that did not meet the standards were confiscated and the cottage craftsmen who created them were punished.

· Ancient quality planning reached its height in the Periclean era, during which the Parthenon, Erechtheion, and Mnesicles at the Acropolis in

Athens were built. The quality of Greek architecture was elevated by an aesthetic sensibility, which involved considerable trial and error by the architects to advance technique and improve construction practices.

- The Romans' execution of large construction projects throughout their empire introduced to civilization standardized building materials and sophisticated quality management techniques. They developed standards as well as measurement methods to produce the construction materials and tools to survey and organize layout of the rural and urban lands that they conquered. Supplier relationships were regulated by strict contractual terms, with planning and supervisory responsibilities performed by the architect. A disciplined army corps of engineers built the infrastructure of strategic importance — the paved brick roadways and public water reservoirs.

- During the Middle Ages the quality concept of inspection and measurement were enforced by Germanic craft guilds through the hidden beginnings of the trained craftsman's mark, which appeared on cloth, tools, weapons, and building materials. The Statue to Protect Trade Marks in the nineteenth century regulated the age-old tradition that essentially allowed consumers at the time of purchase to compare merchandise produced by the various guilds. The Germans are also credited with passing legislation to standardize interchangeable parts as early as 1926.

- The roots of modern quality management were forged in the Industrial Revolution which began in Great Britain and was imported to the United States in the nineteenth century. The need to produce low-cost products to satisfy a large market of indigent people was fueled by the intense process improvement efforts of a number of great industrial pioneers.

- The modern history of the quality movement in the United States began when the United States broke with the European factory system that subdivided former trades into multiple specialized tasks and adopted Frederick Taylor's system of scientific management. The system permitted industry to succeed in increasing production without increasing the number of craftsmen. Because Taylor separated planning from execution and assigned this task to engineers, fewer supervisors and workers were used to execute it. Management of quality also became the responsibility of a separate inspection department that began to use the concepts of *statistical quality control* (SQC) when introduced in the 1920s. Unfortunately, the shortage of goods that followed World War II created a massive decline in the quality of products, as manufacturers gave priority to production volumes. Even though the shortages attracted competition, the persistence to meet scheduled delivery dates remained long after the shortages declined.

· Japan also had a long association and recent history involving quality. As recently as the nineteenth century, craftsman produced high-quality swords, silks, paper, and lacquerware under both feudal governments and during periods of great isolationism. Even the quality of Japanese weaponry during World War II was quite competitive. However, when all of its industry suddenly had to covert to civilian products following the country's defect in 1945, the transition did not go well. To overcome its reputation for poor quality, Japan embraced the methods of U.S. statisticians and quality experts Walter Shewhart, W. Edwards Deming, and Joseph Juran. The turnaround that resulted was spectacular and was accomplished largely through rigorous implementation of modern cost, statistical quality, and technical control techniques.

Today, most organizations recognize quality as strategic, but what will the quality management programs of the twenty-first century be like? Will the goals of these programs echo the teachings of the great civilizations before us and instill in the entire organization an understanding that the customer comes first? Organizations will not succeed in the twenty-first century unless their products and services meet customer expectations for intrinsic function and extrinsic support within the established cost parameters set by the market. Quality cannot be traded off against concerns for total project costs or excessive speed to market — goals that foster failure-driven policies of rushing to market with products and services that must be retrofitted or replaced in the field. In the modern world, when buyers in consumer markets are satisfied, they tell eight other potential buyers. When they are dissatisfied, they share their discontent with nearly three times as many potential buyers.[1]

Because of the far-reaching implications and importance of this message, we have included a separate chapter to cover the most important aspects of the total quality management field as its relates to project management. Topics include (1) tools to help project managers improve the quality of their projects, (2) fundamental assumptions concerning how to conduct process improvement and process reengineering projects, and (3) internationally and nationally recognized quality assurance practices that help elevate standards of performance.

---

[1]A. V. Feigenbaum, "Total Quality: An International Imperative," in *Maintaining the Total Quality Advantage* (New York: The Conference Board, 1991), pp. 38–39.

12.2 INTEGRATING QUALITY AND PROJECT MANAGEMENT

12.2.1 What is TQPM?

Total quality project management (TQPM) uses the various project participants from cross-functional disciplines working together as a team to ensure that a project will satisfy the needs for which it was undertaken. The team carries out the directives of the organization's *quality policy* (see Section 12.2.2) to achieve a set of established project objectives given specific resources. Managing the scope of the quality initiative is well within the project team's purview and is addressed during a *quality planning phase*, where the team decides how to satisfy relevant quality standards and documents the output in the *project summary plan* (see Chapter 5). As the project advances through its life cycle, the project team regularly evaluates its overall performance to assure that it will meet established standards. This means that on an engineering project, the project team must control not only the deviations from specifications that occur during the fabrication of prototype designs, but also unanticipated process variations that occur at the beginning of the production phase.

TQPM embraces the concept of *continuous process improvement*. This familiar concept, known in Japan as *gemba kaizen* (see Chapter 4), embodies the theory of systems management, which likens the dynamics of the business and engineering environments to that of an open-ended cell. *System theory* embraces the premise that a living cell is an open system that maintains a constant state while matter and energy that enter it keep changing within.[2] Continually observing and measuring variations that occur in the business organization and within its related parts and processes may be used to lead management to consider systemwide improvements that affect how quality might be controlled for an entire organization, project, product line, or service.

12.2.2 Interfaces and Basic Definitions

Quality planning occurs at three levels within an organization: (1) the process level, (2) the project level, and (3) the individual level. TQPM is concerned primarily with quality at the project level, even though project management may encompass complex organizational environments where single projects have multiple interdependencies and share common pro-

---

[2] R. J. Cottman, "Process and Systems," in *Total Engineering Quality Management* (New York: ASQC Quality Press, 1993), p. 23.

cesses with other projects. At the project level, the project team must address both product and process quality issues, which must be planned for, monitored, controlled, and improved constantly using specific tools and techniques.

We define *process quality* as improving the productivity, efficiency, or performance of the organization — activities that help satisfy customer needs and deliver a profit for the commercial business. *Product quality* involves translating the customer's perceived requirements for a product's *fitness for use* (see Section 12.3.1) into an acceptable context and tangible parameters that engineering can measure. Such quality characteristics may include product safety, product reliability, product maintainability, beauty, strength, warranty, friendly service, and so on. Establishment of acceptable standards against which to measure actual quality performance and act on the difference is expressed in the organization's *quality policy*. The quality policy states management's overall intentions and direction and defines how to carry out quality functions, track the flow of important information, initiate and effect corrective action, and manage continuous quality improvement.

As we stated in Chapter 5, most projects document issues pertaining to product quality in a *project summary plan*. However, as Figure 12.1 illustrates, procedures pertaining to the quality development of project deliverables must be referenced in a separate *quality management plan* that describes the *quality management system*, the process-oriented procedures that the organization will use to test and control the product quality. Because the quality management system defines the particular development methodology the team will employ to manage all the processes pertaining to the *quality control, quality assurance,* and *quality improvement*, the project's compliance with these procedures is often investigated during companywide quality assurance audits. It is usually appropriate to combine these two documents into one plan, especially when unique combinations of organizations work together on a one-time basis. For such endeavors there may be no need to distinguish between functional and organizational quality procedures and those that are project specific.[3]

Quality control is the process used by the project team to meet the standards established by the organization's *quality policy*. The process consists of observing performance, comparing it to the standard or metric identified, and then taking the necessary action to correct the deviation observed. The cost of failing to prevent, detect, and subsequently correct

---

[3] B. C. Kohn, (ed.), "The Three Levels of Quality Planning," *R&D Quality Planning Guidelines* (Holmdel, NJ: AT&T Bell Laboratories Quality Assurance Center, 1988), p.15.

Project Summary Plan for Project A

Quality Management Plan for Project A

Quality Management System Plan for Project A

- Objectives pertaining to quality improvement
- In-process control limits for metrics
- Responsibilities for quality activities
- Schedules for delivery of reports, feedback, etc.
- Project-specific metrics for product quality
- Provisions of other parameters required by the Quality Management System and other process-related systems

QUALITY MANAGEMENT SYSTEM

Includes:

- Mechanism for obtaining, communicating, and acting upon customers needs
- Guidelines for setting objectives
- Enumeration of the quality functions and procedures for carrying them out
- Assignments of the responsibility for performing the quality functions and procedures
- Specification of important information flows, reports, and feedback through the process
- Procedures for managing control mechanisms leading to corrective actions and quality improvement

FIGURE 12.1

Interrelationship of the quality system to the required quality plans.

deviations from the standard is known as the *cost of quality*. The cost of quality is an important concept for project team specialists to understand, which we explain in Section 12.3.1. Keeping errors out of the process is far less of an expense for an organization than inspecting and rejecting entire production lots to prevent defects from reaching customers. The basic tools used to control quality include: (1) inspection, (2) control charts, (3) pareto diagrams, (4) statistical sampling, (5) flowcharting, and (6) trend analysis. We discuss each of these techniques in Section 12.3.2 along with additional activities that may be used to drive products and processes to meet quality objectives. The output of using these tools to control product and process quality results in, among others, improved tolerances for variations in production processes, determination of a product's *fitness for use*, conformance of the product to specifications, and product acceptance by the customer or end user.

Whereas *quality control* is concerned with monitoring specific project results and eliminating the cause of unsatisfactory performance, *quality assurance* is concerned with evaluating overall project performance to provide confidence that the project will satisfy the relevant quality standards. Therefore, to guarantee performance, the quality assurance process must address all the interfaces in the upstream operations or processes that are both internal and external to the organization. This includes managing internal suppliers within the business family and outside vendors or supply houses. The suppliers may be either contracted sources, for equipment, parts, and raw materials needed by the project to produce the product, or indirect distributors who will warehouse and supply the final product to customers. The project team should document in the quality management plan the requirements against which each supplier will be evaluated. The team should also make suppliers aware of the process operations they'll be involved with over the life of the project. This includes informing the internal supplier or external vendor when contracted supplies are needed, the expected quantity and quality levels, and the corrective action process that will be enforced when obligations are not met. As the team's relationship with the suppliers becomes long term, there may be many instances to improve the process itself. For this reason, the team should also identify a procedure for the supplier to document opportunities for improvements.

Figure 12.2 illustrates the interrelationship of the inputs and outputs associated with TQPM. Because the quality planning, quality control, and quality assurance processes are actually ongoing activities, the project team is bound to experience many initially unresolved issues. As such, team members must share a sense of commitment to the project and be willing to assume full responsibility for surfacing and communicating any issues

FIGURE _12.2_

Input/output view of total project quality management.

affecting the quality of project deliverables to other responsible team members or partners who can resolve the problem. (See Chapter 10 to review the procedures that we recommend for logging quality problems and tracking their resolution.)

We cannot emphasize enough the importance of creating an atmosphere where quality-affecting data can be shared in a timely manner. An official inquiry into the tragic launching of the _U.S. Challenger_ space shuttle in 1986, which included the lost of seven lives, prestige, and billions of U.S. dollars, revealed that there were many problems associated with the quality launch of the craft other than its technical design. The managers in the space shuttle office had worked at the NASA site for a very long time. They employed a "command and control" decision-making style that precluded them from accepting advice or yielding to any intervention or opposition that may have persuaded them to abort the mission. Before an organization can focus on improving the areas of greatest concern to its performance and productivity, its managers must commit to establishing an atmosphere of trust where weaknesses within the project can be revealed without reproach.

12.3 QUALITY IN THE TWENTY-FIRST CENTURY

12.3.1 Enduring Advice from the Gurus

The international quality revolution that began in Japan following the end of World War II and gained momentum throughout the rest of the world two decades later will reach a new pinnacle in the twenty-first century. What projects can glean from the teachings of the time-honored quality gurus who initiated and advanced this philosophy becomes another foundation for renewing the team's commitment to total quality project management. Since there isn't a single approach to total project quality management, to be competitive, organizations today have embraced many of the time-proven principles and techniques laid down by these experts. The challenge, however, is for projects to adapt the teachings that best fit the culture and climate of the larger organization in which they operate.

The Deming Philosophy. W. Edwards Deming was an internationally renowned quality consult who abhorred corporate management's lack of commitment to sustain long-term investment at the expense of short-term profit taking. His work after World War II, first with the American Telephone and Telegraph Company and later with General MacArthur in rebuilding Japanese industry is a prelude for the twenty-first century's emphasis on long-term investment in quality as a competitive advantage. The Union of Japanese Science and Engineering (JUSE) honored his contributions to Japanese society by instituting the annual Deming Prizes for quality improvement and dependability of product. In 1980, the American Statistical Association established an annual Deming Prize in the United States for improvements in quality and productivity.

Deming set out to transform management style, particularly U.S. management, when he advised in a 1982 publication, *Out of Crisis*,[4] that "performance of management should be measured by potential to stay in

---

[4]W. E. Deming, *Out of Crisis* (Cambridge, MA: Center for Advanced Engineering Study, Massachusetts Institute of Technology, 1982, 1986).

business, to protect investment, to ensure future dividends and jobs through improvement of product and service for the future, not by the quarterly dividend." He encouraged companies to make a long-term commitment to transforming business by taking the necessary corrective actions to reduce costs and improve productivity. His famous 14 key points to achieve this transformation are summarized as follows:

1. Create constancy of purpose toward improvement of product and service, with the aim to become competitive, to stay in business, and to provide jobs.

2. Adopt the new philosophy. In a new economic age, Western management must awaken to the challenge, learn their responsibilities, and take on leadership for change.

3. Do not depend on mass inspection as a means of achieving quality; rather, build quality into the product in the first place. Demand statistical evidence of quality being built into manufacturing and procurement functions.

4. End the practice of awarding business on the basis of price alone. Instead, move toward establishing a long-term relationship with a single, qualified supplier that is based on loyalty and trust.

5. Improve constantly and forever the system of production and service. Begin by building quality in at the design phase. Regard every product as one of a kind. Teamwork in design is fundamental. There must be continual improvement in test methods, in methods to understand better each new customer's needs, and in the ways the product is used and misused. It is management's job to improve the system continually, to make every job better than the one before.

6. Institute on-the-job training. The greatest waste is failure to use the abilities of people. Management needs training to learn about the company, all the way from incoming materials to the customer.

7. Adopt and institute leadership. The job of management is leadership, not supervision. Leaders must know the work their supervise, must act on corrections proposed, and show that they are not just concerned about the numbers. Because they have the confidence of the people they supervise, leaders help production workers improve their work.

8. Drive out fear, so that everyone can put in his or her best performance in a secure environment where communication is open.

9. Break down barriers between organizations — research, design, procurement, sales, and production. Institute teams where people in design, engineering, production, and sales accomplish important improvements in product, service, and quality.

10. Eliminate slogans, exhortations, and production targets for the workforce. Most problems that occur have to do with processes and systems that are beyond workers' control, and such exhortations only aggravate employees.

11. Eliminate work standards that prescribe numerical quotas. In their place, use statistical methods to study and prescribe change and knowledgeable and intelligent leadership.

12. Remove barriers that rob people of pride of workmanship. Abolish merit reviews, performance appraisals, and management by objectives. Allow the workers to take pride in their work.

13. Encourage education and self-improvement for everyone. Advances in competitive position have their roots in knowledge.

14. Take action to accomplish the transformation. Management must empower the workforce to carry out the new philosophy to achieve the 13 points above. Enough people in the company must understand that continual improvement of quality is accomplished through iterative cycles of accumulating knowledge; planning a course of action, acting on the change, and measuring the results. Otherwise, management is helpless and the transformation will not take place.

Deming's quality philosophy was influenced by his Bell Telephone Laboratories' mentor, Walter Shewhart, whose work with *statistical process control* (SPC) taught Deming to look at and clarify quality variation according to two important causes: (1) uncontrolled variation due to assignable or special causes such as changes in procedures or operations, raw materials, breakages, and so on; and (2) controlled variation due to the unassignable or common causes in a process that results randomly, from chance. Because in both these conditions there is too much deviation from normal operation, quality improvement can only come about by management reworking or reengineering the system and by involving employees to bring the variation into statistical control.

Over two decades ago, Deming attributed 94% of all quality problems to management and advanced today's movement toward participatory decision making to reduce variation, so necessary for team-based project management. Deming believed that setting targets which are beyond the capacity of the system will knock the system out of control. For this reason, he named the annual performance appraisal as one of several "deadly diseases" that hamper the personal contribution of team members and become unfair when there is a great deal of common cause variation. Improving the system is a managerial activity!

The Juran Trilogy. Juran is famous for what is known as the Quality Trilogy (a trademark of Juran Institute, Inc.): (1) quality planning, (2) quality control, and (3) quality improvement. His work, first with Bell Telephone Laboratories, then AT&T's Western Electric Division, and later with Japanese multinational corporations, advanced the quality revolution around the world. He emphasized continuous improvement of interrelated processes as the framework for an organization to establish quality functions and accomplish quality goals.

Juran believed that the word *quality* has multiple meanings. He argued for organizations to adapt as a primary quality goal the ability to satisfy customers in the marketplace by producing products that are equal or superior to the quality of competing products. We can readily see how use of such a goal is critical not only to an organization's quality planning but also to its strategic business planning. He defined quality simply as *fitness for use* and entertained the possibility that a product could be free of defects and not be salable when completing with a comparable, better-performing product.

As the basis for setting quality goals, he recommended using the market. By establishing early-warning mechanisms, an organization can take re-medial action to attain the *market level of quality* necessary to remain competitive. A project team may discover the market level of quality through (1) inspecting and testing competitor products, (2) putting com-peting products into service to acquire field performance data, (3) research-ing customers' perceptions concerning quality preferences, biases, and values. Juran's recommended practice of establishing a *quality goal* is easily applicable to goods, services, processes, and organizational functions and can be adapted as a banner by all participating units. He coined the long-range goal — *zero defects* — to imply that the need for quality improve-ment is never-ending.

Establishing quality goals, according to Juran, requires joint planning and the participation of all involved decision makers to discover the optimum processes to meet customer needs. He defined *process* as a systematic series of actions directed toward the achievement of a goal. His generic definition is applicable to functions in manufacturing as well as nonmanufacturing organizations and includes all the activities a project team may undertake to: (1) launch new products, (2) recruit team members, (3) fill customer orders, (4) produce goods, (5) service a customer, and (6) control plant operations.

Building on the work of Shewhart and Deming, he advanced the concept of *statistical process control* as it applies to product quality. In a 1988 publication, *Juran on Planning for Quality*,[5] he insightfully explains how processes have an inherent capability for performance that can be evaluated through data collection and statistical analyses. He encouraged organizations to evaluate *process capability* through the analysis of process variability, giving meaning to the need for teams to plan and to develop processes that are guaranteed to perform under all variables and conditions—inherent not just in the laboratory, but in the operating environment as well. For a manufacturing process, a project team may need to measure variables such as temperature, cycle time, or acidity. For a sales process such variables as product type, price, or credit and delivery terms may need to be assessed.

As Juran explained, variation in the product's performance is indirectly indicative of the process that produced it. So data collected on the product variability become a measure of process performance which, for certain quality features, can be expressed as its standard deviation from established norms each time the product is produced. Process capability can be readily set using the measured standard deviations and expressed as a capability index that can then be established as a project team goal. A process capability equal to six standard deviations (see Section 12.3.2) has come to be universally accepted by many manufacturing organizations. Juran admonished U.S. industry to adopt methods equal to those employed by its competitors. It is preferable to quantify product defects in parts per million to drive improvements while evaluating process capability if competitors are using the same standard.

Process design, according to Juran, is the activity of defining the specific means to be used by operating forces to meet product goals. In a manufacturing operation, this requires that the quality planners define the physical equipment and articulate information on how to operate, control, and maintain it. Juran also stressed the need for processes to be clearly understood by internal customers (internal partners who share the responsibility of meeting the same goals) and external customers (outside suppliers who engage in the processes to meet their own goals). Process design should enable an entire spectrum of users to apply the process successfully.

---

[5] J. M. Juran, *Juran on Planning for Quality* (New York: Free Press, 1988), pp. 177–195.

Project teams may apply Juran's concept of *process control* to understand quality control methodology. In contrast to process design, process control is the activity of keeping the operating process in a state that continues to be able to meet product goals. Necessary steps toward achieving control include: (1) evaluating the actual performance of the process, (2) comparing actual performance with goals, and (3) taking action on the difference. Control takes place in a systematic sequence called a *feedback loop* which teams may carry out by evaluating product features, verifying that process components which affect the product features directly are in check, and assessing the impact of troublesome side effects of the product's production environment. Conducting these operations must be in accordance with an established *quality plan.*

Among Juran's legacy to the twenty-first century is his insistence on a companywide quality management policy to drive quality improvement throughout the organization. *Policy,* as Juran used the term, refers to a published statement or guide to managerial action concerning how the organization intends to meet customer needs. Enforcement of organizationwide quality policies not only is the responsibility of the quality assurance department but of every team member. Their deployment requires communication throughout the organization down to the project team level.

Crosby's Making Quality Certain. As vice president of quality of a U.S.-based corporation in the 1970s, Philip Crosby took a pragmatic approach to quality. He focused his program of *zero defects* (after the Western Electric slogan coined by Juran) on *prevention* rather on than *acceptable quality levels* (AQLs) which inadvertently let products ship with known defects. He considered the basic areas of performance in any organization to be cost, schedule, and quality. All are vital for success, and each requires the establishment of a performance standard that cannot be misunderstood.[6] His zero defects concept is based on the fact that mistakes are caused by lack of knowledge and lack of attention — both of which, Crosby states, can be overcome by setting a standard that does not tolerate defects and personally directing top management to improve operations.

His total quality management program defined the ingredients of the *cost of quality* (COQ). According to Crosby, the composite expense for rework, scrap, warranty, service, purchase order changes, software bug fixes, engineering changes, and so on, should total no more than 2.5% of every

---

[6]P. B. Crosby, *Quality Is Free* (New York: McGraw-Hill, 1979), p. 171.

sales dollar. This means that every team member has the opportunity to increase the return on sales by the exact amount by which he or she is able to reduce expenditures for these items. Cosby also advocated conducting a *zero-defect program* that includes (1) quality awareness training for suppliers as well as team members, and (2) mutual pledging of acceptance by management to accomplish quality improvement tasks. The pledges become commitments that can be turned into actions by encouraging individual teams to set goals for their projects. As a result of the activity, team members become personally concerned over the conformance of their product or service to the organizationwide performance standard and the quality reputation of the company.

Crosby also encouraged organizations to appreciate the contributions of those who participate in the quality improvement efforts by recognizing their value. He also realized that by bringing together the appropriate professionals to share quality-improvement information, the organization benefits from their counsel. After all the quality goals are reached, he advocates that new individuals join the quality councils to continue the quality program and communicate to the organization new requirements and standards for improvement.

Garvin's Managing Quality for Strategic and Competitive Edge. Harvard University professor and consultant David Garvin,[7] views quality as a strategic weapon for corporations. Because U.S. industry faces fierce competition from abroad, he defines quality based on eight dimensions that underlie the concept of strategic management: (1) performance, (2) features, (3) reliability, (4) conformance, (5) durability, (6) serviceability, (7) aesthetics, and (8) perceived quality.

A basic tenet of his approach to strategic quality management is that quality must be defined from the customer's point of view. While trade-offs between the dimensions are inevitable, an organization does not have to be first on all eight dimensions to be perceived by customers to be a leader in the marketplace. To be acceptable, a product may need only to be perceived as high in quality by customers, a concept that project managers can use to their organizations' advantage. For instance, the Japanese automobile industry was able to win over many U.S. car owners with this strategy, as the vehicles they introduced into the U.S. market during the 1980s were actually low in *performance* and *durability*, but outstanding in *reliability*, *conformance*, and *aesthetics*.

---

[7]D. A. Garvin, *Managing Quality* (New York: Free Press, 1988), pp. 49–61.

Performance, according to Garvin, refers to the primary operating characteristics of a product. This would include, for example, how an automobile accelerates or the speed or absence of waiting time for service at an airline ticket counter. A customer will often judge this dimension of quality through personal preferences, needs, or interests. A similar connection exists between *performance* and *features* — the secondary characteristics that supplement a product's basic functioning.

The third dimension, *reliability,* reflects the probability of a product's malfunctioning or failing within a specified period of time. As we have discussed in Chapter 10, project teams will often use as common reliability measures for durable goods — the *mean time to first failure* (MTFF) and the failure rate per unit time. Although the measure is less relevant for products and services that are consumed instantly, this quality dimension garners attention because of Deming's and Juran's success with Japanese manufacturers. Their superiority in this dimension has given their industries a competitive edge over less reliable contenders.

Garvin's fourth dimension, *conformance*, also discussed at length in Chapter 10, refers to the degree to which a product's design and operating characteristics meet preestablished standards. Garvin points out that American project managers typically equate the quality dimension of conformance with meeting specifications, whereas the Japanese adapt a tighter definition of the tolerance limits around which products outside the specified conformance limits are found acceptable. This definition of conformance embraces the work of the prize-winning Japanese statistician Genichi Taguchi. His work applies tighter process controls and measures the variability of a product's performance around its control limits. It also includes the concept of measuring the *functional loss* to society when defective product is shipped. Both approaches require that project teams monitor production with the factory and rely on the indices of defects to track conformance to specifications and improve performance.

Durability is a measure of product life and has both economic and technical dimensions. Durability, technically, can be defined as the amount of use one gets from a product before it deteriorates physically, and is closely related to reliability. Unreliable products are more likely to be scrapped earlier, repair costs will be higher, and when faced with a series of competitive choices, consumers are less likely to choose a brand that has a high failure rate, because of the hidden expense of frequent repairs.

Serviceability, the sixth dimension of quality, considers the speed, courtesy, competence, and ease of repair with which a manufacturer corrects product defects. Projects often measure this quality dimension by surveying customers who use their service organizations. Establishing a complaint-handling process to measure customers' ultimate satisfaction with the products becomes a true test of the acceptability of this quality dimension.

The final two quality dimensions, *aesthetics* and *perceived quality*, are very subjective and are related to how users actually judge quality. Project teams may need to study potential users to understand what specific attributes of their products individual customers prefer and associate with quality. Quality may also be inferred from intangible aspects of a product. This might include where a specific product is manufactured, brand name recognition, similarity with other manufactured product lines, producer's reputation, and so on.

Garvin advises companies and projects to examine the interrelationships among these eight dimensions when competing on quality. Each dimension is a multifaceted variable itself that teams must consider given the constraints imposed by the various markets they enter. An organization need not compete on all eight dimensions to be successful, nor is it advised unless exorbitant prices are charged. Optimizing one dimension may force a trade-off against another. For this reason, Garvin recommends that projects single out a few quality dimensions that are achievable and strive for excellence in these categories. Across-the-broad superiority seldom is a competitive necessity.

12.3.2 Traditional Tools of Quality Management

Our intent in this section is not to replicate exhaustively the excellent references covering the most prevalently used quality management tools listed at the end of the chapter. Instead, we will review what are known as the seven traditional quality tools: (1) cause-and-effect diagrams, (2) flowcharts or process-flow diagrams, (3) histograms, (4) pareto charts, (5) scatter diagrams, (6) run or trend charts, and (7) control charts. Deploying these tools on a project offers several advantages. Each technique can be used in addressing quality issues of concern to project management and the organization sponsoring the R&D, production, service, or business development project. A team's choice of tool should match the purpose for collecting and displaying data to enhance management's understanding of the information gathered.

Cause-and-Effect Diagrams. *Cause-and-effect diagrams,* also known as *fishbone* or *Ishikawa diagrams* after their shape and Japanese originator, are used to systematically analyze the underlying or root causes of a problem. They are helpful in displaying and grouping multiple causes according to categories, which lets project teams identify the relationship between the effects and potential causes of an existing problem. Project management can use the graphical tool to explore numerous factors related to a problem, as the completed diagrams are effective as a framework for recording the results of a brainstorming exercise with project specialists who have detailed information about the different aspects of a problem.

For each effect identified, team members must think of a cause. Grouping and recording this information on the diagram's branches, in the manner shown in Figure 12.3, will help stimulate specialists from the various functional areas represented to think of potential causes of the known effect — in this case unacceptable defects in the software application. The specialists will continue to populate each subbranch of the high-level branches, adding more specificity to the diagram until all actionable causes of the defects are identified. The team may use the completed diagram to highlight and assign for further investigation the cause(s) it believes contribute(s) to the most likely effects.

Process-Flow Diagrams (Flowcharts). *Flowcharts* are the visual representation of the steps in a process. Project teams will discover that they are particularly useful tools for discovering how processes actually work and the cause of specific difficulties that may be impeding progress. There are several types of flowcharts that teams may choose to map the stages of a process. These include (1) the top-down flow diagram, (2) the detailed process-flow diagram, and (3) the decision-path flow diagram.

As illustrated in Figure 12.4, in a *top-down flow diagram* only the most fundamental steps in a process or project are represented, with the key activities listed below each task. These charts can be prepared rather quickly and are useful diagrams for examining only the significant or value-added activities. When a team needs to understand very specific information about process flow, including feedback loops and decision points, it is best to represent this information in a *detailed process-flow diagram* (see Figure 12.5). Constructing these diagrams requires that the project team identify the essential activities that are necessary to complete the process, the starting and end points of these activities, and all their subtasks. Typically, specific symbols are used to represent operations, data flow, and equipment. Teams may use this type of chart to depict procedural events that must be followed precisely; however, if their project tasks follow multiple paths, they should illustrate this information in a *decision-path flow diagram*. These diagrams are similar to detailed process-flow diagrams but show no subtasks. Instead, across the top of the chart, a team allocates all the responsibilities associated with a particular project process that it wants

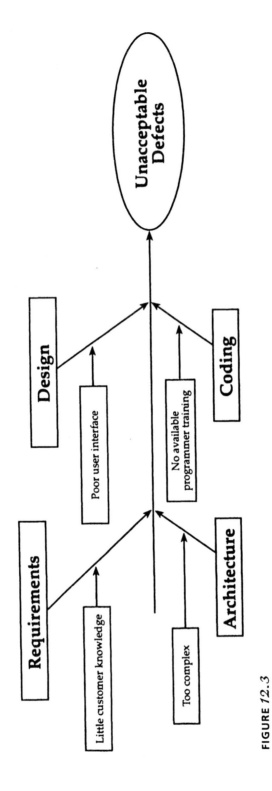

422

FIGURE 12.3

High-level branches of a cause-and-effect (fishbone) diagram used to detect causes of unacceptable defects in a software application.

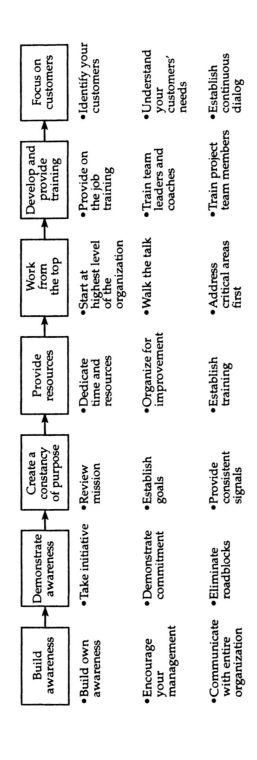

FIGURE 12.4

Example of a top-down flow diagram illustrating key ctivities in a quality improvement project. (Adapted from V. D. Hunt, *Quality in America: How to Implement a Competitive Quality Program*, Irwin Professional Publishing, Chicago, 1996, p. 243.)

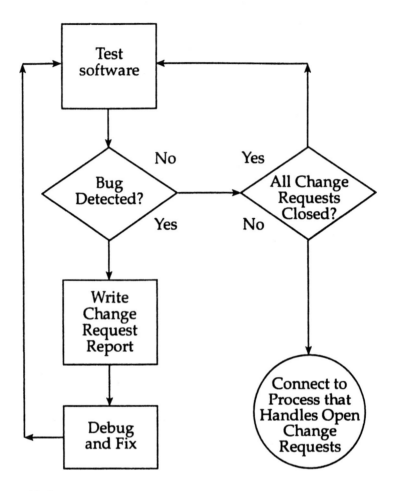

FIGURE 12.5

Detailed process-flow diagram showing the step-by-step activities to test software code. (Adapted from R. J. Cottman, *Total Engineering Quality Management*, ASQC, Milwaukee, WI, 1993, p. 72.)

to improve and lists under the appropriate columns the steps and times that are required to complete them. As shown in Figure 12.6, the chart capably illustrates inefficiencies in a process flow and serves as a very useful tool for planning process flow improvements.

Histograms. *Histograms* visually represent the spread or distribution of variable data—data that a project team may collect for study, such as the frequency or the number of defects discovered during the system test

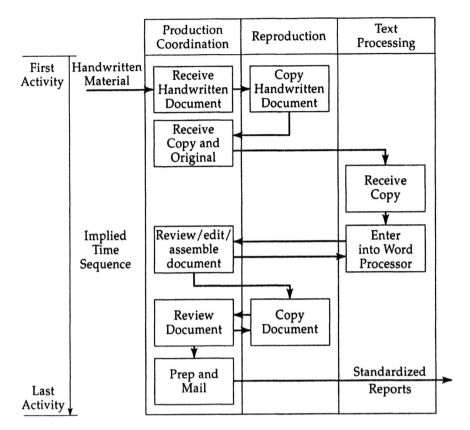

FIGURE 12.6

Decision-path flow diagram. (Adapted from R. B. Ackerman, R. J. Coleman, E. Leger, and J. C. MacDorman, *Process Quality Management and Improvement Guidelines,* Bell Telephone Laboratories, Indianapolis, IN, 1987.)

interval of a product realization (see Figure 12.7). The *histogram* is used to show how variable measurements may be used to establish standards. The simplicity of construction and interpretation of the histogram also makes it an effective tool in the elementary analysis of data. It is used extensively to gain insight into and manage manufacturing processes.

Histograms readily illustrate that data have a natural tendency to fall toward the center of the distribution, with fewer items represented as you move progressively from the center. This permits production managers to measure the central tendency of specified quality characteristics: the mean, median, and mode of randomly sampled data. Cornerstones for understanding process control technology—the centering, width, and shape of

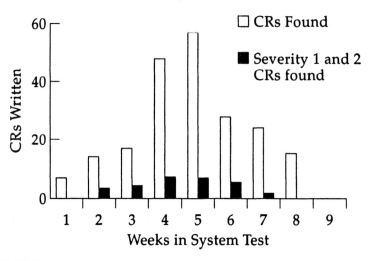

FIGURE *12.7*

Histogram of system test data.

the histogram—provide production teams with enough insight into the manufacturing process to interpret its ability to meet specification limits.

Pareto Charts. Because the causes of quality problems are always unevenly distributed, project teams can conduct a *Pareto analysis* to determine the relative few causes that account for the bulk of their difficulties. This management tool is widely used as a means of understanding the most significant problem areas with the least amount of analytical study. Joseph Juran formulated what he called the *Pareto principle—the vital few among the trivial many*—after a nineteenth-century Italian economist who found that typically 80% of the wealth in a region was concentrated in less than 20% of the population. As Figure 12.8 shows, data under study are plotted in descending order as a bar chart, along with an ascending plot of the cumulative percentage of the height of the bars. This illustrates how the Pareto principle defines priorities for improvement projects and dramatizes that the effort should be deployed unequally over all activities. Because the chart helps teams to isolate their problem areas and narrow these causes down to only a few sets of activities, they can consequently plan appropriate interventions or process improvements to effect a positive change. The results of the intervention can be displayed again in the same manner later and compared.

Scatter Diagrams. The relationship between the data points at one time can be examined by drawing a *scatter diagram*. By plotting points against

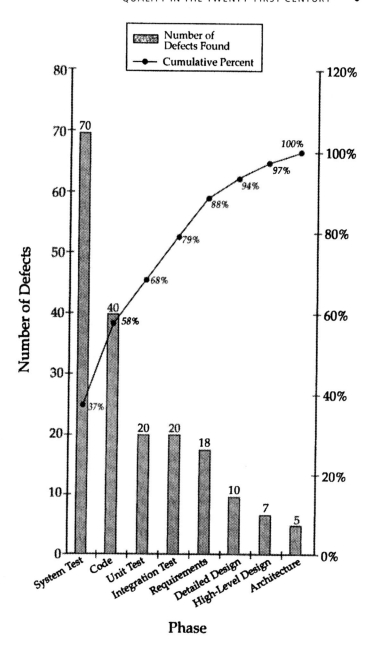

FIGURE 12.8

Example of a Pareto diagram showing that most of the defects during the illustrated product realization occurred late in the cycle.

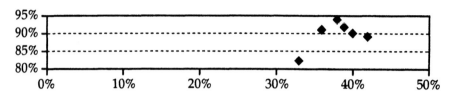

FIGURE *12.9*

Scatter plot of customer satisfaction data (percentage of customers whose expectations were met) indicating strong positive correlation.

two different measures, one on the *x*-axis and the other on the *y*-axis, the *scatter diagram* reveals a visual pattern that is easily discernible, indicating the absence or presence of a positive or negative correlation between the variables. A project team can easily interpret the graph given in Figure 12.9 and conclude that the measures of the percentage of surveyed customers who indicated that they were satisfied with the team's service correlate positively with the measures indicating that these customers' expectation were met. A more quantitative analysis of the correlation would require the assistance of a statistician, however.

Run or Trend Charts. The *run chart*, also known as a *trend chart*, is the simplest way to display data over a specific time period. It is easily adapted to the requirements of a number of different project types and is frequently used to display either a long-range change or the average or mean variation of the process under study. It is produced by sampling the process under investigation for a specified period to ensure that only inherent variation is present in the sample. In the manufacturing environment, this would mean that no operator intervention would occur during the sampling interval.

A run chart that is arranged such that the average value for each sample appears around the middle of the *y*-axis is called an $\bar{X}$ (*X-bar*) *chart*. We use averages rather than individual values because they are more sensitive to change. The average measures the *aim* or centering of the process under investigation. As displayed in Figure 12.10*a*, variation is shown in the $\bar{X}$ chart by connecting a line through the plotted average values. A run chart that displays the range of the measurements in a sample is known as an *R chart*. The range measures variability about the *aim* of the process, as shown in Figure 12.10*b*.

Control Charts. *Control charts* allow the project team to track the variability of sampled data collected over time and determine by observing the patterns of variability whether the processes for which they are

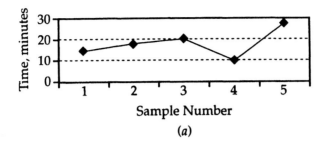

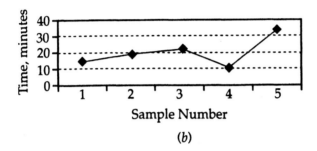

FIGURE 12.10

(a) Run ($\bar{X}$) chart display of the average transaction processing time used to improve customer satisfaction; (b) Run (R) chart display of the range of measures taken to improve transfer processing time.

responsible are in control. Through surveillance of these patterns, teams monitor the variations of manufacturing and a variety of customer-affecting transaction processes in the financial service, banking, and fast-food industries. Problems can quickly be attributed to either common or assignable causes (see Section 12.3.1), and the necessary action can be taken to correct or eliminate them.

There are a variety of types of charts. As illustrated in Figure 12.10a and b, the $\bar{X}$ chart plots the average of a sample and the R chart displays the difference between the highest and lowest values in the sample. Both tools are variable control charts and are used to detect the causes that tend to disturb a process. Project teams working in a manufacturing environment will want to plot the ranges and averages of sampled data together. These charts become very powerful for diagnosing production troubles, such as trends, the interaction of two or more variables, instability, the grouping or bunching of measures, and so on. The patterns that appear in the charts can be associated with particular causes and are used to help isolate specific factors disturbing the production process.

The *percentage* or *attribute chart* is another type of control chart that project teams will encounter in a production environment. Called the *p-chart* (for "proportion"), the chart is often used to display the fraction or proportion of defectives compared to the total. It is less powerful than $\bar{X}$ and R charts because it displays only a single pattern of variability.

There are many other types of *attribute control charts*. These include (1) number of defectives (*np-chart*), (2) defects per fixed unit (*c-chart*), and (3) defects per unit (*u-chart*). Each of these charts is used differently, depending on the variables under observation in the sample. If the samples to be plotted on a *p-chart* are all the same size, the project team may find it simpler to plot the number of defectives found in each sample instead of calculating the percentage. When this is done, the chart is called a *np-chart*. If the number of *defects*, instead of defective units, were plotted, the project team would use a *c-chart*. The *u-chart* is a special variation of the *c-chart* that plots the average number of defects per unit in a sample of *n* units. Appendix 12.1 illustrates how to select the correct attribute chart based on project requirements. Consult the bibliography at the end of the chapter for more detailed explanations.

Although knowledge of statistical measure is required to calculate the control limits associated with these process charts, at a minimum a team should understand how to set the *control* or *specification limits* for $\bar{X}$ and R charts. As shown in Figure 12.11, the standard that should be set must represent three standard deviations, annotated as 3σ (sigma) of sample averages above and below the mean or 6σ total. When a process is operating in statistical control, it is operating with the minimum amount of variation possible and with process capability within these limits. The σ represents how close all the measurements sampled are to the average. The larger the σ value, the more spread out the measurements are from each other and the mean. The values used to calculate σ are given by

$$\sigma = \frac{\bar{R}}{d^2} \tag{12.1}$$

where $\bar{R}$ is the average range and the value of d^2 is obtained from a statistical table (Table 12.1).

In addition to the σ, two other statistical terms are required to construct a control chart. These are the *upper* and *lower control limits* (UCL and LCL) for both the R and $\bar{X}$ charts. These values are determined as follows:

$$\text{UCL } R = D_4 \times \bar{R} \tag{12.2}$$

$$\text{LCL } R = D_3 \times \bar{R} \tag{12.3}$$

$$\text{UCL } \bar{X} = \bar{X} + A_2\bar{R} \tag{12.4}$$

$$\text{LCL } \bar{X} = \bar{X} - A_2\bar{R} \tag{12.5}$$

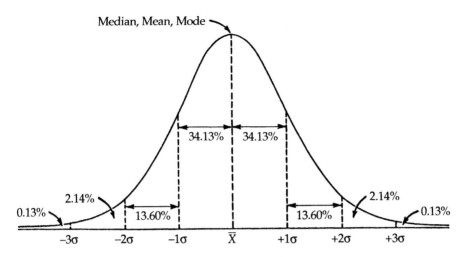

FIGURE *12.11*

Standard deviation layout for a normal distribution. (From *Statistical Quality Control Handbook*, AT&T Technologies, Charlotte, NC, 1984.)

where the value of A_2, D_3, and D_4 are obtained from a statistical table for the size subgroup charted.

Example 12.1. The data for this example are given in Table 12.2. To illustrate how a plot of the values obtained for the UCL and UCL limits are used to indicate whether a process is out of control: Calculate the UCL value for a subgroup sample size of 4, if $\bar{X} = 10$ and $\bar{R} = 3$. Then use the same figures and calculate the LCL value. Construct an $\bar{X}$ control chart. Draw the $\bar{X}$ value as a solid dark line and draw the UCL and LCL values as dashed lines.

TABLE 12.1   VALUES USED TO
CALCULATE σ

Sample	d^2
2	1.128
3	1.693
4	2.059
5	2.326

TABLE 12.2 DATA FOR EXAMPLE 12.1

Subgroup Size	$\bar{X}$-Chart Factor for Control Limits, A_2	R-Chart Factor for Control Limits, D_3	R-Chart Factor for Control Limits, D_4
2	1.88	0	3.27
3	1.02	0	2.57
4	0.73	0	2.28
5	0.58	0	2.11
6	0.48	0	2.00
7	0.42	0.08	1.92
8	0.37	0.14	1.86
9	0.34	0.18	1.82
10	0.31	0.22	1.78

SOLUTION: The set up of the control limits calculated from this sample data are shown in Figure 12.12(a) and a completed chart in Figure 12.12(b). Since the points plotted from this sample fall within the UCL and LCL, the data indicate a normal process. In this case, the UCL represents an area above the mean value in which 49.35%, or 3σ, of the $\bar{X}$ values should fall, and the UCL represents an area below the mean in which the remaining 49.35% should fall. When the project team observes that no values are outside this area, nothing is unusual in the process. However, should the team detect a shift (Figure 12.13a) either above or below the chart's centerline, it can be concluded that something is unusual with the process. Seven or more consecutive plot points going up or down from any one anchor point must be investigated as if it had an assignable cause even though the plotted points may be within the control limits. And a change of more than 4σ between two plotted points indicates that the process has jumped (Figure 12.13b) due to some cause which the project team must identify and correct. A trend (Figure 12.13c) is indicated when six or more consecutive plot points go up or down from any one anchor point. To interpret the causes of these changes, the team may choose to use any combination of the other traditional quality tools.

12.3.3 Beyond Traditional Tools: The 6σ Business Initiative

The statistical control offered by measuring quality within 3σ of the process centerline is rapidly becoming the quality improvement method most used by today's leading and most profitable organizations. The aim of the 6σ

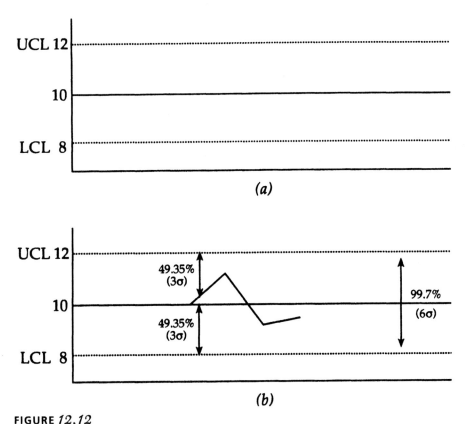

FIGURE 12.12

(a) Set up of the upper and lower control limits for Example 12.1. (b) Control chart with plot points within the upper and lower limits indicating that the process is normal.

initiative is to eliminate nearly all defects from every product, process, and business transaction. As we discussed in Section 12.3.2, when used as a metric, 6σ permits no more than 3.4 defects per million in every product or service that a business produces. But unlike other total quality approaches, projects managed under the 6σ initiative have predetermined cost, quality, and yield metrics set to drive improvement in the organization's overall financial performance. Companywide programs that achieve their targets use the basic quality control tools to chart and mistakeproof standard operating procedures. In addition, project specialists from the R&D and production organizations involved undergo extensive training in effective communication skills to improve how information is shared among assigned team members. Project accountants, engineers, and operations personnel usually are assigned full time and are given various titles, such

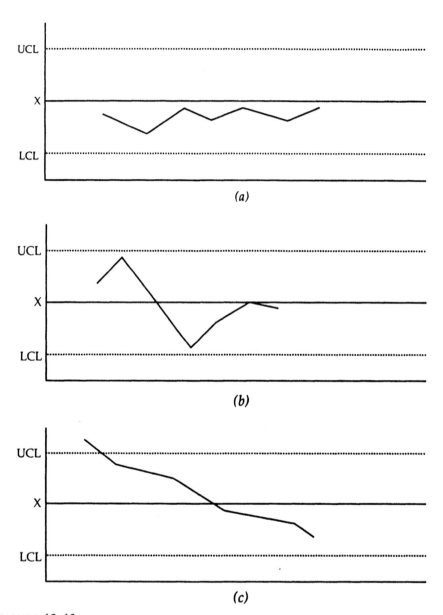

FIGURE *12.13*

(*a*) Control chart with seven plot points below the centerline, indicating an unusual process; (*b*) control chart with more than 4σ between two plotted points, indicating a jump; (*c*) control chart with six plot points going down from one anchor point, indicating a trend.

as "green" and "black belts" at General Electric Corporation and "process improvement masters" at AlliedSignal to designate their leadership role in implementing the 6σ methodology.

One key distinction between the 6σ approach to quality and other quality philosophies is that the quality organization does not own the initiative. Instead, the marketing, sales, or production groups together are responsible for achieving the targeted improvements. Quality becomes the responsibility of the entire project team, and ownership is shared among the integrated disciplines that contribute toward its achievement. Expertise and procedures needed to orchestrate how the entire system works may reside in a separate quality assurance group who play a critical and supportive role in managing the quality system, providing training, interpreting information, monitoring progress, and assisting in gathering competitive quality data analysis. Project personnel are also held accountable for the financial results of the 6σ initiatives. As such, the 6σ business process is more than likely to deliver more tangible results and recognition for projects well into the future than other, less aggressive quality approaches.

12.4 SUSTAINING VALUE THROUGH BUSINESS PROCESS IMPROVEMENT

Business Process Reengineering: Friend or Foe?[8] We would be remiss in not introducing in this chapter the total quality concept of *business process reengineering* which was popularized through the initial work of Michael Hammer and Thomas Davenport in the 1990s at Ford Motor Corporation. Although the process improvement methodology unfortunately became largely synonymous with the rampant and somewhat destructive restructuring of business and industry throughout the last decade, the concept merits discussion.

An ex-computer science professor at Massachusetts Institute of Technology, Hammer originally intended that businesses embrace the principles of *reengineering* to break away from outmoded processes and the design principles underlying them. Cross-functional teams, whose members represented the functional units involved in the processes being reengineered, were engaged to design new processes to replace the old. The output was

---

[8]D. Kezsbom, "Business Process Reengineering: Project Management Friend or Foe?" *Proceedings of the Project Management Institute,* October 1996.

not intended to be haphazard, as the fundamental principle of reengineering was to discover new business processes and jumpstart others. As a means of obliterating outmoded processes, Hammer recommended putting in these decision-making positions personnel who actually performed the work under investigation. He believed that such people were closest to the customer and brought to the team an understanding of customer needs. To control the new processes, he also advocated introducing information technology—not to automate but to enable them.[9]

As defined, this popular management trend embraces many of the fundamental principles of the *total project quality management* (TPQM) philosophy discussed in Section 12.2. However, implementation of the practice to reorganize, delayer, or shrink hierarchical structures as a means of flattening organizations is a misuse of the original concept. As many organizations have discovered painfully in the aftermath, the cost of doing business may be lowered, but crucial lack of corporate "bench strength" may result.[10] For reengineering to be successful, a business must first determine which management personnel and which processes add value in response to market needs. As Hammer has recently acknowledged, customers call the tune to which everyone in the company must dance; and when customers come first in the environment, everything else in the culture has to adjust. Customers care nothing for the management structure, strategic plan, or financial structure. They are interested in only one thing—the value delivered. Hence a customer focus forces an organization to emphasize results and the processes that produce those results.[11]

Is *business process reengineering* worth considering? And what relationship does it have to project management? The answer to both these questions is definitely "yes" if an organization uses the method as designed by the originators to create new processes in tandem with the continuous improvement of others.

It is helpful to understand that the aim of process redesign is to produce breakthrough or dramatic jumps in performance improvements for a few core business processes. On the other hand, continuous quality improvement is an ongoing activity which everyone must address. The team involved in the redefinition and redesign of business processes must

---

[9]M. Hammer, "Reengineering Work: Don't Automate, Obliterate," *Harvard Business Review,* July–August 1990, pp. 104–111.
[10]M. D. Zinn, "Viewpoints: Reengineering Affects More Than the Bottom Line," *Manufacturing Engineering,* Vol. 118, No. 4, April 1997, p. 112.
[11]M. Hammer, "Beyond Reengineering," *Executive Excellence,* Vol. 13, No. 8, August 1996, pp. 13–14.

radically rearchitect how the processes operate without regard for the way things were done previously. In addition, the project effort is often tied to cost-reduction goals using value analysis technology (see Chapter 4) while established quality levels are maintained or improved. The radical recommendations made by these projects often necessitate changes in job design and new management systems that go beyond development of new process flows. In contrast, quality improvement projects tend to happen over a longer period of time without the disruption of jobs, systems, and organizational structure.

Teams frequently find that each project has a different breath of focus. Quality improvement projects trend to focus on specific, often narrowly defined processes or subprocesses. Business process reengineering projects, on the other hand, have much broader goals: to improve customer satisfaction, produce a larger return on investment, or grow market share. Because these efforts have a larger scope, they tend to involve all the critical units within the organization whose performance the project is charged with improving.

The intensity of team involvement may also be different. With quality improvement, team members may take part in the activity as needed. With process redesign, the teams tend to be dedicated to the work full time. Often, the team members decide how often to meet and how much effort each team member is required to contribute.

Both of these initiatives involve employees at all levels of the organization, from the COE to front-line associates. Methodology may come down from top management, but the initiative for specific process improvements may come from anywhere in the business. However, as process improvement ignites the organization, senior management often becomes less directly involved with the daily activities of the team's efforts and resorts to a more reinforcement role. With process reengineering, senior executives must drive the effort from start to finish to accomplish the radical change sought for the organization in the relatively short time frame that they establish for the team.

Neither approach to quality management is best to implement alone. Since TPQM allows for the flexibility of using new tools to implement the philosophy, business process reengineering becomes a valuable tool that organizations may consider within this framework. The combination of the two bears the fruits of success.[12] Consider, for example, that a project team

---

[12] H. D. Allender, "Is Reengineering Compatible with Total Quality Management?" *Industrial Engineering*, Vol. 26, No. 9, September 1994, p. 41.

discovers the need to change radically how products are distributed to customers to satisfy their needs for expediency and quality. The formulation of a new delivery process within the course of only a few months' time that alters all the organization's other distribution operations and systems would take the form of a reengineering project. But once the new delivery process is conceived and put in place, it will no doubt require major tweaking through an ongoing or continuous process improvement. For this reason, the information systems department asked to redesign the support systems and associated operational procedures may choose to release them incrementally, which permits the organization to test the new functionality in an environment where as little disruption as possible to customer service is permitted. It is erroneous for project teams to think that process improvement and business process reengineering need to be done in isolation.

12.5   DRIVING IMPROVEMENTS THROUGH QUALITY ASSURANCE

12.5.1   National and International Standards

So far the focus of this chapter has been on explaining how, through the concept of total project quality management, a project organization establishes a guiding vision for building its approach to quality management. We also examined the tools available to the empowered teams of project specialists who are charged with execution of their organization's total quality approach. In this section we review two important quality endeavors that an organization may pursue in tandem with TPQM to enhance its ability to achieve total quality: the European ISO 9000 registration process and the U.S. Malcolm Baldrige National Quality Award. Pursuit of both of these achievements is compatible with TPQM. ISO 9000 registration is often pursued first as an initial step toward total quality, with the process of achieving registration seen as a means rather than an end to continuous improvement. Attainment of the Baldrige Award often culminates after many years of repeated effort and denotes organizational excellence in all aspects of total quality management and control.

ISO 9000 and Total Quality Project Management. Project teams engaged in the continuous obsession with quality are able to document with little difficulty the procedures they use to subscribe to the *International Standards Organization 9000* registration guidelines. Because ISO 9000 and TQPM are compatible, complying with ISO 9000 gives a project that has not yet implemented TQPM a head start. Standard selection is based on the type of business pursued and customer demand to follow particular standards. It therefore makes sense for a company with multiple operating units, locations, and unique systems, processes, or markets to register on a unit-by-unit basis. To achieve ISO 9000 registration, projects must (1) recognize that project work is accomplished through many different processes, (2) document all quality-affecting procedures, and (3) establish controls to comply with established codes, standards, quality plans, and procedures. The controls that are implemented are left to project management's discretion. Statistical process control as advocated by Juran and Deming is not required.

The objective of ISO 9000 is to assure the fitness of purpose of a supplier's goods or services. A registrar or auditor is engaged to certify that the organization's published, interrelated process steps produce acceptable products. The documentation required must also cover four distinct levels: (1) organizational policy, (2) project procedures, (3) project practices, and (4) proof of execution. Documents may be cross-referenced, referred to in other documents, and kept in either hard-copy or electronic form. Internal reviews and external audits are scheduled semiannually to verify that operating procedures are complied with and that what is done is actually documented. Once certification is awarded, surveillance audits are not normally scheduled in advance. When a company proclaims compliance to the ISO 9000 standard, it must be certified as meeting the requirements. In most European nations, the national government makes this a requirement. In the United States, certification is traceable to the American National Standards Institute.

Hence corrective action is mandated to rectify differences in procedures not documented in the organization's quality system and is also required to obtain and renew a registration certificate. Consequently, projects need to maintain a documentation configuration management system as part of their quality system structure where team members can record the status of self-audits or action taken to correct noncompliance found by either an external auditor, a third-party registrar, or an internal (intrinsic) auditor. The ISO 9000 also requires periodic review of the quality management system in place within the organization for its suitability to the constantly changing business needs of the global marketplace. The up-to-date system would embrace new technologies, strategies, compliance with environmental regulations, and possible social changes in the market.

ISO 9000 also assures the quality of work performed for others, such as product installations or service under warranty and maintenance contracts. Services categories in which an organization may apply for registration include hospitality, health care, communications, finances, purchasing, consulting, trading, professional, and scientific. If a service is delivered too late to correct an imperfection, either the process controlling the delivery was not complied with or employee training was inadequate. To comply with ISO 9000 standards, companies seeking to deliver and execute quality services that meet or exceed customer requirements and expectations must demonstrate control over their service delivery processes.

The Malcolm Baldrige Quality Award. In 1987 the U.S. Congress inaugurated the first national quality award in the United States to raise awareness about quality management and to recognize companies that have successful quality management systems. The award is named after a U.S. Commerce Department secretary, Malcolm Baldrige, who was instrumental in opening global trade and reforming national antitrust laws in the 1980s. The award criteria establish standards for quality excellence in business performance and are designed to help organizations deliver ever-improving value to their customers as a result of overall improvement in business capability and performance. The U.S. Commerce Department's National Institute of Standards and Technology (NIST) manages the award with the assistance of the American Society for Quality Control, in close cooperation with professional volunteers from the private sector. Awards are given annually in such categories as manufacturing, service, small business, and health care. The award criteria are updated annually to reflect the latest achievements in such predetermined categories as quality leadership, strategic planning, process management, business results, customer focus and satisfaction, and human resources. When properly integrated, they define a total set of requirements for an excellent organization.

As part of the appraisal process, companies must provide written proof that they have achieved world-class quality in all the criteria described. Each autumn applicants are given a feedback report, advised of their scores achieved in the examination categories, and are informed if members of the board of independent examiners will visit them in the following year. The purpose of the in-person examinations is to review high-scoring companies' strengths and to evaluate the areas in quality management that need improvement.

As we stated previously, the purpose, focus, and content of the Baldrige Award differs significantly from ISO 9000. Because ISO 9000 registration merely determines whether a company compiles with its own quality system, it covers only 10% of the Baldrige Award criteria. The Baldrige criteria faithfully and fully used can be one of the most important tools at a company's disposal. Implementing the self-examination activity alone can help projects achieve greatness. In the spirit of total quality project management, periodic assessment on an ongoing basis will provide fresh input concerning organizational strengths and what areas need improvement. Instituting an annual self-assessment can spawn process improvement projects and issue resolution that ultimately will affect customer satisfaction. Companies are also encouraged to learn from national award winners. Each year a *Quest for Excellence Conference* provides a forum for business leaders to hear and question award recipients from each of the criteria categories.

12.5.2 Software Quality Management

We encourage leading-edge organizations engaged in software development to measure the quality of their outputs and compare ongoing performance to competitive benchmarks by readily conducting project audits, reviews, and process assessments. In today's highly competitive and cost-conscious business environment, these primary tools can be used to guarantee quality, improve time-to-market intervals, and assess the risk factors associated with success and failure. Reviews and audits conducted during the development of software applications can assure that a project effectively and efficiently meets customer requirements and exceeds market expectations. Postmortems conducted when a project concludes can record valuable input for continuous and ongoing improvement in succeeding projects. Holding annual *software process assessments* (SPAs) allows a software team to benchmark its performance against comparative norms for industry type and project size and to assess the need to adjust performance to remain competitive.

Project Audits and Reviews. A *project management audit* is conducted to help a project team evaluate the effectiveness of its project and quality management plans and the application of its quality management system (see Section 12.2.2). A *technical review* (see Chapter 10) checks for correctness of an in-process product relative to internal benchmarks. A *postmortem review*, although similar to a project audit, is less formal and is done at the end of the project.[13]

Typically, a project management audit is conducted by an expert external to the project who has experience in similar development activities and credibility with senior management. If a project is having difficulty

[13]M. H. Fallah, J. P. Holtman, J. F. Maranzano, D. P. Smith, and G. T. Tucker, "Development Process Audit and Reviews," *AT&T Technical Journal,* Vol. 70, No. 2, March–April 1991, pp. 99–108.

managing interpersonal relationships, an organizational development specialist may conduct an assessment of the human relation issues. Many organizations have independent quality assurance departments dedicated to improving companywide performance with trained professionals on staff who can lead the audits and read out the results. It is important to guarantee confidentiality, so the audit team needs to maintain its independence from the project specialists when conducting interviews and reporting its findings.

Audits may be performed after the planning phase is complete or midway through the project when 20 to 40% of the development is under way. An initial or *planning audit* is performed before development has started to assess the adequacy of the current summary, quality management, and quality management system plans to meet the customer requirements. The in-process or *compliance audit* is conducted to measure a project's actual compliance with these plans. If timed effectively, a *compliance audit* can highlight for the project management team the necessary corrective action that it may take to put the project back on track. An audit held at the request of senior management is known as a *special audit*. It is held typically when a project is in jeopardy and often results in actions to replan, reorganize, or even cancel a poorly performing project that is late to market, grossly over budget, or perceived as unable to satisfy customers' demands for quality.

Although an audit is a valuable exercise for revealing difficulties and potential problems in the project plan, it is inadequate for revealing the detailed technical flaws in the development process or the actual software application. To discover potential issues and control the associated problems, the development team must earnestly conduct the focused *technical reviews* we discussed in Chapter 10. The cost of conducting these reviews and the time spent preparing for them is negligible if customer-affecting issues that could have disastrous effects later when the application is released to the market are detected early in the development cycle. The time spent retracting programs and reissuing corrected code is not only costly, it may also damage an organization's reputation for delivering quality products.

Once a project has finished, generally a *postmortem review* is held to capture valuable knowledge about the project and the effectiveness of the processes that it employed. Project specialists are interviewed and a debriefing report is shared with the entire project team and often with senior management. Then the team may choose to conduct further analysis to discover the root cause of the sources of project difficulties that led to the errors and known faults identified in the review. Subsequently, the team may address the underlying cause using the traditional quality tools that were described in Section 12.3.2 as part of its ongoing TQPM activities.

In the postmortem it is important to capture information on what went well as well as information on processes, practices, tools, or approaches that should have been done differently. The aim is to discover problems and make recommendations to future projects on how to avoid them. If the planning and scheduling did not go well, attempt to learn why. If important technical designs were flawed, why were no contingency plans in place to address and mitigate this risk? The project management team will often go on to manage another release of the software application, and many of the specialists will be retained to perform the new work. Therefore, the debriefing must be conducted in an atmosphere where interactive, two-way communication can proceed so that the results of the review encourage process improvement.

Software Process Assessments. A *software process assessment* (SPA) involves the collecting, analysis, and reporting of information necessary to define and manage or baseline a project's software development process. As we illustrated in Chapter 10, software development organizations across many worldwide industries are using SPAs to identify opportunities for improvement and to obtain insight concerning how their development processes compare with the general software industry. Many organizations have as a quality policy goal the achievement of best-in-class status on two widely applied assessments: the *software productivity research* (SPR) profile and the *software engineering institute* (SEI) *capability maturity model* (CMM). Since their inception, a significant body of data has been amassed to support the fact that widespread implementation of these models leads to improved business results as measured by internal productivity and quality metrics.

The SPR assessment utilizes a knowledge-based tool originally developed by Capers Jones and subsequently enhanced as a result of its continuous application over many years. (Information on how to order SPR tools is given in Appendix 15.1). It assesses eight different dimensions of strengths and weakness and displays the results on a Kiviat chart much like the project data shown in Figure 12.14. The eight categories are (1) customer focus, (2) project management, (3) project team variables, (4) methodologies, (5) quality focus, (6) tools, (7) metrics, and (8) physical environment.

The SPR model represents characteristics that successful software development organizations have implemented to bring applications to market or to satisfy contractual obligations for their customers. Each strength and weakness is assessed and assigned a number between 1.0 and 5.0. A rating score less than or equal to 2.5 indicates that the project is at the leading

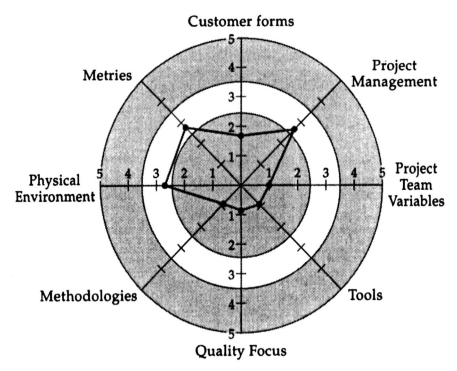

FIGURE *12.14*

Kiviat chart of SPR top 10 software process assessments. (Courtesy of Capers Jones.)

edge for that attribute compared to the software industry. Scores between 2.5 and 3.5 indicate that the project is deficient for that attribute and is trailing the industry. A score above 3.5 is referred to as a high-risk category for the project organization. Since most organizations have interval reduction, defect detection, and fault improvement goals, one of the five answers to the SPR questions may be chosen as a goal that will make determining what improvements and weaknesses to address easier. As a follow-up to the assessment, the project team should write and undertake action plans to prioritize and close the identified gaps between desired states and current practices. However, the project may not choose to make changes to improve some SPR weaknesses. Some gaps, such as the number of physically separated locations involved in the work, can be interpreted as a risk that the project needs to manage.

As shown in Figure 12.5, the SEI capability maturity model has a five-level maturity model that projects can use as a road map for planning improvements. The CMM views process maturity as measurable and as an

Maturity level: 2

x Level 5: Optimizing
  – Process change management
  – Technology change management
  – Defect prevention

x Level 4: Managed
  – Software quality management
  – Quantitative process management

– Level 3: Defined
  ✓ Peer reviews
  – Intergroup coordination
  ✓ Software product engineering
  ✓ Integrated software management
  ✓ Training program
  ✓ Organization process definition
  ✓ Organization process design

✓ Level 2:
  ✓ Software configuration management
  ✓ Software quality assurance
  ✓ Software subcontractor management
  ✓ Software project tracking and oversite
  ✓ Software project planning
  ✓ Requirements management

✓ Level 1:

✓ Fully satisfied
– Partially satisfied
x Not satisfied

FIGURE 12.15

Sample SEI CMM software process assessment showing key process areas that need improving to move upward in the ratings and demonstrate a higher level of maturity (i.e., level 3). ✓, Fully satisfied; –, partially satisfied; ×, not satisfied.

indicator of process quality. Process capability initially described by Shewhart, Deming, and Juran (see Section 12.3.1) is defined as expected performance that can be achieved by following a process established initially at the project or organizational level. Since process capability is seen as a predictor of future project outcomes, the SEI CMM measures the actual results achieved from following project- or organizational-level processes. It also lays the framework for successive and continuous improvement of the software development process across projects.

Most organizations find themselves at level 1, with reported gaps in levels 2, 3, 4, and 5. To assure that activities are performed in compliance with the processes that have been established, a software quality assurance group reviews and/or audits the performance and reports the results, listing activities required to achieve the next levels of the SEI CMM. Project personnel are asked to map key practices to improvement goals that are important to their organizations rather than concentrating on gaps within a specific level. Business objectives and strategic plans should be used to establish work plan priorities for these gaps.

The SEI CMM shares common themes with TQPM: (1) continuous improvement; (2) defined, documented, and used processes; (3) commitment by senior management; (4) stable, measured, and controlled processes; and (5) processes that evolve with maturity. It also shares common concerns with the ISO 9000 standards. Both assessments are driven by the concern that projects say what they do and do what they say. In mature organizations, processes are defined, documented, trained, practiced, measured, maintained, supported, controlled, verified, and validated.

We recommend that organizations conduct SPAs every one to two years to achieve best-in-class software development processes. It is also important that the software development community have a regular forum like that of the industry leaders who achieve the annual Baldrige Award to exchange new practices and share baseline trends. Such an exchange would permit project managers to recommend to other software managers what specific actions to take to affect project performance.

12.6 CONCLUSIONS

The purpose of this chapter has been to explain why quality has become today's best investment in competitiveness. While the total quality project management programs of the twenty-first century must include the rich heritage of the past, they must also meet the demands of a global and competitive marketplace where there is no such thing as a permanent quality level. Unless efforts to improve the performance of product designs and production processes are rigorous and continuous, organizations will lose customers to competitors with superior quality.

The goal of every organization's quality policy must be to instill in the entire organization an understanding that the customer comes first. Projects will not succeed unless their end products and services meet customer expectations for intrinsic function and extrinsic support within the established cost parameters set by the market. It is imperative to understand that customers will not let us trade off quality against internal concerns for total project costs or the need to reduce time to market at their expense. Why follow a receipt for failure by rushing to market products and services that must be retrofitted or replaced in the field? Implementing the concepts and tools introduced in Chapter 12 should arm projects to overcome these huddles and meet the quality challenge of the twenty-first century. The project's quality goals must be pursued relentlessly and the improvement initiatives adjusted continuously as required by market dictates.

BIBLIOGRAPHY

"ANSI ASC Z-1 Committee on Quality Assurance Answers the Most Frequently Asked Questions About the ISO 900 (ANZI/ASQ Q900 Series)," *http://www.asq.org/standcert/iso.html.*

Bank, J., *The Essence of Total Quality.* Hemel Hempsted, Herts, England: Prentice Hall International, 1992.

AT&T Bell Laboratories Software Quality and Productivity Cabinet/Hardware Quality and Productivity Cabinet, *Best Current Practices: Root Cause Analysis.* Indianapolis, IN: AT&T Bell Labortories Technical Publications Center, 1990.

Brocka, B., and S. M. Brocka, *Quality Management Implementing the Best Ideas for the Masters,* Homewood, IL: Richard D. Irwin, 1992.

"Business Reengineering Overhauling Corporate Processes Through Information," *Insights,* Vol. 1, No. 2, Fall 1989, pp. 2–8.

Capezio, P., and D. Morehouse, *Taking the Mystery Out of TQM: A Practical Guide to Total Quality Management.* Hawthorne, NJ: Career Press, 1993.

Cottman, R. J., *Total Engineering Quality Management.* Milwaukee, WI: American Society for Quality Control, 1993.

Goetsch, D. L., and S. B. Davis, *Understanding and Implementing ISO 9000 and ISO Standards.* Upper Saddle River, NJ: Prentice Hall, 1998.

Hoeerl. R. W., "Six Sigma and the Future of the Quality Profession," *Quality Process,* June 1998.

Hunt, V. D., *Quality in America: How to Implement a Competitive Quality Program,* Chicago: Irwin Professional Publishing, 1996.

Juran, J. M. (ed.-in-chief), *A History of Managing Quality.* Milwaukee, WI: American Society for Quality Control, 1995.

Juran, J. M., and F. M. Gryna, Jr., *Quality Planning and Analysis from Product Development Through Use, 2nd ed.,* New York: McGraw-Hill, 1980.

"Malcolm Baldrige National Quality Award," *http://www.quality.nist.gov.*

Ovenden, T. R., "Business Process Re-engineering: Definitely Worth Considering," *TQM Magazine,* Vol. 6, No. 3, 1994, pp. 56–61.

"Software Hell," *Business Week,* December 6, 1999, pp. 104–118.

Sprague, D. A., "Adding Value and Value Analysis to TQM," *Journal for Quality and Participation,* Vol. 19, No. 1, January 1996, pp. 70–72.

Statistical Quality Control Handbook. Charlotte, NC: AT&T Technologies, 1984.

Thomas, M., "What You Need to Know About: Business Process Re-engineering," *Personnel Management,* Vol. 26, No. 1, January 1994, pp. 28–31.

"Who Was Malcolm Baldrige and Why Did They Name an Award After Him?" *http://www.asq.org/abtquality/awards/baldrige/mbguide2/html.*

EXERCISES

12.1. Which of the following is defined as taking action to increase the effectiveness and efficiency of the project so that both the performing organization and the project customer benefit?
(a) Quality planning
(b) Quality control
(c) Quality assurance
(d) Quality improvement

12.2. Which of the following is concerned with monitoring specific project results to determine if they comply with relevant quality standards and with identifying ways to eliminate causes of unsatisfactory performance?
(a) Quality planning
(b) Quality control
(c) Quality assurance
(d) Quality improvement

12.3. Which of the following is concerned with evaluating overall project performance on a regular basis to provide confidence that the project will satisfy the relevant quality standards?
(a) Quality planning
(b) Quality control
(c) Quality assurance
(d) Quality improvement

12.4. Deming and Juran hold to the philosophy that quality improvement can come about only by managing rework, reengineering the system, and involving employees to bring variation into statistical control. Explain how their thinking differs from the more traditional approaches to quality performance practiced prior to World War II.

12.5. *Kaizen* is an approach to quality assurance that emphasizes:
(a) Incremental improvement
(b) Cost control of customer satisfaction
(c) Breakthrough advancements
(d) Participative quality circles

12.6. Which quality tool might a project use to separate the vital few problems from the important many?
(a) Run chart
(b) Pareto chart
(c) Cause-and-effect chart
(d) Histogram

12.7. Statistical control charts are used primarily to assist the production team:
(a) Measure conformance to specifications
(b) Monitor process variation over time
(c) Compare the performance of several sources of suppliers' products
(d) Set quality objectives

12.8. Serviceability is a dimension of quality that considers the:
(a) Speed, courtesy, competence, and ease of repair with which a manufacturer corrects product defects
(b) Economic life and technical dimensions of the product
(c) Degree to which a product's design and operating characteristics meet preestablished standards
(d) Probability of a product's malfunctioning or failing within a specified period of time

12.9. Deming postulated that constancy of purpose is a core concept for continuous improvement. Describe how an organization might illustrate this concept by publishing its quality policy and creating an awareness by all project team members of the vision, goals, objectives, and roles required to achieve this purpose.

12.10. Describe why a quality management plan should cover the following components of the project quality system.

(a) Process-oriented procedures used to test and control product quality

(b) Any operational planning that implements an organization's quality policy

(c) The development methodology a team will employ to manage all the processes pertaining to the quality control, quality assurance, and quality improvement

(d) The requirements against which each supplier will be evaluated

12.11. An audit is an assessment conducted to determine to what extent certain standards or requirements have been meet. Explain why it is important for a project involved in software development to undergo a project audit at the start of the project during the planning phase and again after 20 to 40% of design and coding is complete.

12.12. Why might a project team choose to diagram the symptoms they determine as probable root causes of an unusually high number of defects found during the last weeks of system testing of a new application? Which quality tool best illustrate these relationships?

12.13. Three standard deviations on either side of the aim or centerline of a process that is in control will contain approximately what percent of the total data from the population distribution?

(a) 99.7%

(b) 95%

(c) 68%

(d) 49.35%

12.14. Review the values for the production samples given in Tables 12.1 and 12.2.

(a) Calculate the σ value for a subgroup sample size of 5 if $\bar{R} = 1.59$.

(b) Use the same tables and calculate the control limits.

(c) Construct an R control chart.

12.15. The cost of quality is defined as the sum of the cost of prevention, appraisal, and failure. How can this particular indicator be used as a measure of productivity, variation, and efficiency?

12.16. The ISO 9000 standard establishes the framework for an organization's quality system. Define project management's responsibility in documenting the activities that make up the project quality system and list the levels of documentation needed to comply with the standard.

12.17. Describe how the focus of the Baldrige Award differs from that of ISO 9000. Explain why your project organization may choose to undertake ISO 9000 registration before applying for the Baldrige Award.

12.18. Explain how the popular management trend of business process reengineering embraces many of the fundamental principles of total project quality management.

12.19. All of the following are part of the SPA output report except:

(a) Integrated findings from the assessment tools used

(b) Management and project views of project goals, risks, and constraints

(c) Overall assessment recommendations based on current level

(d) Specific plan of recommended corrective actions based on current assessment

12.20. Review the Kiviat chart compiled by Software Productivity Research in Figure 12.14, showing the top 10 software process assessment scores across such software industries as commercial application development, management information systems software, military systems software, and end-user software. According to the data, 35% of the projects are considered leading edge, and 5%, high risk. How would you compare project scores in your industry with these data?

APPENDIX 12.1: SELECTING THE CORRECT ATTRIBUTE CONTROL CHART*

		Characteristics being examined	
		Defectives	Defects
Sample size	Constant	np-chart	c-chart
	Variable	p-chart	u-chart

---

*Source: R. R. Clements, *Statistical Process Control and Beyond*, Malabar, FL, Warrior Books, Inc., 1988.

where p is the percent defective, calculated as parts found defective/sample size, np the number of defectives, u the count of defects per unit, and c the count of all defects in a sample. To select the proper attribute chart, determine whether the critical characteristic involves defects or if the main concern is whether the product is defective. If consistent sample sizes cannot be obtained, use control charts for variable sample sizes.

BUILDING AND MAINTAINING PROJECT TEAM PERFORMANCE

The truth of the matter is that you always know the right thing to do. The hard part is doing it.

GENERAL H. NORMAN
SCHWARTZKOPF

13.1 INTRODUCTION

Like the economy, the interest in sports has, especially within recent years, increasingly become a global phenomenon. Those of us who are sports fans probably have a favorite team. If we are avid fans, we attend at least an occasional game, watch it on TV, or listen to a radio sportscaster. When the game is over, who better than us can provide a detailed analysis to account for the team's victory or defeat? What irritates us and most sports fans is to watch a team fail to play together. We notice immediately when, for example, one player fails to make a pass, misses a block, or tries to shine at the other team members' expense. We are "experts" in diagnosing the team's weak spots and in providing feedback to the coach or umpire. With all our insight into teamwork, many project and team coaches fail to see the parallel between what is needed to improve, for example, a football team and what is needed to shape up their own project teams. Considering the fervor for team sports, it is surprising to find the low level of carryover into organizational life. Project teams have much to learn from the playing field.

Quality project management is team management. It is a highly integrative process that requires the application of a variety of participative approaches to blend all the technical and organizational components of the project into a cohesive whole. Creating major improvements in project and organizational processes takes more than skill in applying project management technology. What it truly requires is a back-to-basics, hands-on process of integrative planning, coupled with a keen awareness of group dynamics and the value of teamwork.

13.2 FACTORS INFLUENCING TODAY'S TEAM MOVEMENT

Today's organizations are striving to extend employee involvement into that realm which was once clearly management's domain. Several factors have influenced this shift toward more team-based methods of managing and accomplishing work. Among these are:

1. *The focus on quality and customer satisfaction.* Quality management practices emphasize the importance of each employee's contribution to a project and to the final product.

2. *The philosophy that people support that which they create.* This involves the growing awareness that the autocratic, paternalistic organization and management style of the past fails to create productive, committed resources.

3. *Advanced and pervasive technology.* Sophisticated technologies require specialized personnel who must work in a leaner, resource-limited environment.

4. *New ways of designing the job.* Specialization in today's leaner structures demands increasingly opportunities for job sharing and job rotation.

5. *Increased realizations that the project manager alone cannot plan, schedule, or track a project.* In view of today's complicated products and systems, project managers cannot possibly have all the knowledge, time, or resources to bear the burden of project completion alone.

The complexities of the project environment mandate the use of cross-functional project teams, whose efforts span domestic and international boundaries, vendors, partners, governments, and customer organizations. The uncertainties introduced by social, technological, economic, and political factors create an ever-present element of risk, which challenges the formation of any team. New vendor-contractor alliances, moreover, have complicated the team process and have challenged the very nature of how we conduct business. Multifaceted tasks, complex organizations, and the need for synergy encourage an organization to abandon traditional line–staff relationships and discover the benefits of project teams.

13.3   BACK TO BASICS

Before attempting to build an effective team, we must acknowledge a few basic concepts. In our team facilitation workshops we find that the most common response we receive to our simple question "What is a team?" is "A group of people with a common goal," or "A group of people working together to accomplish a common goal." These are all fine responses — except that they are wrong — or just plain *incomplete*. Such a common but erroneous assumption of what a team is prevents us from truly developing a "group" into a productive and cohesive *team*. A team is much more than just a group of people who perceive that they share a common goal. A team is a collection of people who must rely on the cumulative knowledge, skills, and talents of each *interdependent* team player. All teams can be considered groups, but all groups are *not* teams!

A *team* is a special designation awarded to a group of people who are *aware* of the very nature of their interdependent roles and how the skills ands talents each possesses complement their efforts and assure goal attainment. Effective team output is best explained through the concept of *synergy*. Through effective teamwork, a group can generate solutions to problems that are far superior to those developed individually by each team member.

13.3.1   Essentials of a Team

Moving from a "group" to a "team" is a distinction achieved only through awareness and effort. Like coaches, managers must examine team effort with a skilled and watchful eye. They must determine what is blocking maximum output, and with the help of their players, devise a strategy to overcome or remove the obstacles that prevent goal attainment.

Teamwork is not a mystical process. From the research on group dynamics and the process of team building, we know a great deal about how teams operate. In addition, we can pinpoint the essential elements that distinguish teams from groups and assist in leading the team to optimum performance. These essentials basically consist of (1) a charter, a mission, a reason for working together; (2) a sense of interdependency; (3) commitment to the benefits of group problem solving and decision making; and (4) accountability as a functioning unit.

For any group to make the transition to a team, several important factors must be present. First, the group must have a *charter* or a reason for working together. The members of the group must perceive that they need each other's combined experiences and capabilities to accomplish their goal. They must also be committed to the idea that working together as a team leads to more accurate risk analysis and decision making than does working in isolation.

Participation in planning and decision-making helps team members to develop a sense of ownership and commitment to team goals and procedures. Team members must be held in regard for the special talents and abilities they bring to the team and encouraged to contribute actively to team strategy sessions. Team players, with the help of the coach, recognize their respective roles and functions within the framework of the team and realize that through the integration of these roles, project goals may be attained. Candor and trust characterize communication, and feedback regards team-related actions, and not personality traits that cannot be changed readily. Listening occurs for understanding, rather than defense. Finally, the group members are accountable for their collective actions, and as such, demonstrate pride in their accomplishments.

13.3.2 Characteristics of Effective Teams: You'll Know It When You Have It

Why are some teams so successful, whereas others appear to be less coordinated or ineffective? It's sad but true that all of us could easily list a least a dozen or so characteristics of *ineffective* teams as a result of either direct personal experiences or close contact with a losing team. Ineffective teams often waste much of their energy defending themselves against feedback and competing inappropriately with each other. Players tend to demean or diminish one another rather than collaborate toward achieving mutual goals. Instead of supporting or encouraging new ideas, ineffective teams dwell on personality factors that are either irrelevant to the task or just cannot be modified. Energies that should be directed toward winning are funneled into wasteful practices that lead to negative output.

Effective teams, are characterized by clear, well-understood priorities. They also plan for ample opportunities to discuss, clarify, and agree on team objectives. Project team members individually understand their own priorities in relation to the overall team objective and how to accomplish them. Leadership is consistent with the needs of the group and appropriate not only to the project team players' needs, but to the skills required by the team as it proceeds through its own stages of development.

Integrating personal and team goals is a major task of the effective team leader. For the team as a whole to accomplish its goal, team members must first be able to achieve their personal goals within the larger framework of team objectives. Motivation is at its maximum as team members see opportunities for growth and development.

It is true that the quality of the team's output is as good as, if not better than, the diversity and synergy of the ideas it generates. The proper blend of skills and perspectives represented by the collection of team players enables the project team to achieve solutions to difficult and diverse problems by calling on the right person at the right time. Although such diversity usually leads to conflict or disagreements, successful resolution of the disagreements creates stronger bonds and even better solutions!

Productive teams are aware that the climate or quality of team interactions affects not only morale, but also productivity and performance. A climate that produces respect, innovation, and excellence fosters a winning team. Successful teams attend to detail, anticipate problems, and follow through on plans while continually strengthening the bonds of friendship and respect through celebrating their accomplishments.

Creating well-blended, effective teams does not require the elimination of superstars. Top players, however, learn not to dominate the game, but use subtle techniques to accomplish their objectives. A well-blended team is like a well-running piece of machinery; each part of the machine must interact smoothly with the next for the machine to operate at maximum efficiency. If one part is out of synchronization, the machine will cease to function properly, if at all. The real secret of team blending is knowing how much to assist, when to pass to someone in a better position than you, and what to do to improve not only your position, but that of your teammates.

13.4 STAGES OF TEAM DEVELOPMENT: LIKE PROJECT, LIKE TEAM

Just as projects develop through a *life cycle*, teams tend to go through a natural process of development that a skilled or trained eye may readily identify. Team building activities should be planned and sequenced along this natural order.

Stage 1: Establishing identity. This phase of team development, often referred to as *forming* or *inclusion*[1] is characterized by team members' attempts to find their rightful place within the group. In the animal kingdom, one could liken this to the "sniffing" stage of development, where each creature sniffs the other to see if one is among friends or predators. To assist the team through this phase of development, the project manager or team leader should (1) allow ample opportunity for team members to get to socialize and to get to know each other, (2) affirm and legitimize the distinctive abilities and strengths of each participant, (3) clarify work expectation and rules that will govern team performance, and (4) agree on the major mission of the team and determine team objectives and priorities. Failure to perform these tasks may cause team members to "fixate" or get stuck in this phase of development. In that case, they may never progress to maximum team performance.

Stage 2: Questioning authority. This phase is often characterized by the storming, struggling, and infighting that surfaces as like-minded members form coalitions and factions that question team leadership and control. In this "snarling" phase of team development, team leaders may best be served by listening and responding calmly to member challenges while mediating between factions. Leaders should be sure that workloads are allocated properly, according to member skill and task requirements. Mediation is the primary goal of the leader during this challenging but necessary phase. If the split between team members is not properly bound, the team will never reach maturity and eventually will wither.

Stage 3: Productivity. Once team members feel relatively comfortable with each other and mechanisms for task accomplishment and difference resolution are in place, the team focuses on accomplishing its primary objective. Norms or rules governing behavior are well understood as the team sets out to accomplish what it must accomplish. "Sprinting" toward their end result, team members develop feelings of goodwill. Leadership becomes more participative as the leader facilitates the resolution of disagreements and injects fun and variety into the work. Failure to demonstrate appropriate leadership strategies may result in boredom or complaints regarding non-work-related issues.

---

[1] M. Berger, "The Technical Approach to Teamwork," *Training and Development Journal*, Vol. 39, No. 3, March 1985, pp. 53–55.

TABLE 13.1 STAGES OF TEAM DEVELOPMENT

Stage	Characteristics	Leadership Skills and Behaviors
1. Establishing identity	Team members attempt to find their place within the group	• Allowing ample opportunity for team members to get to know one another • Affirming and legitimizing the distinctive abilities and strengths of each participant • Clarifying work expectations and rules that will govern team performance • Agreeing on the major mission and determining the objectives and priorities of the team
2. Questioning authority	Struggles and infighting surface, as like-minded individuals form factions and coalitions; jockeying for control; questioning of team leadership	• Mediating between factions • Listening to and responding calmly and fairly to member challenges • Allocating work according to skill level and task requirements • Focusing the group on the team mission or common goal
3. Productivity	Team focuses on its primary objectives	• Facilitating resolution of disagreements • Injecting fun and variety into the work setting
4. Uniting	Successful integration of team efforts; comradeship	• Providing opportunities for team get-togethers and celebration of accomplishments

Stage 4: Uniting. Feelings of comradeship grow from the successful integration of team efforts. Team members enjoy their association with one another and often come together to celebrate the accomplishment of milestones and other fruits of their labor. Finding opportunities for the team to celebrate their accomplishments facilitates this "snuggling" stage.

Table 13.1 provides an overview of the characteristics associated with and skills required by each of the four stages of team development.

13.5 TEAM DEVELOPMENT PROCESS

Team building is the process of *creating* and then *maintaining* effective team performance. Developing a team takes time and commitment, but so does planning a project. The results, however, are great in terms of performance, productivity, and problem solving, and best of all decreased levels of stress. When done properly, team development addresses both the human and technical aspects of a project. In fact, as discussed in Chapter 5, effective project planning is an integrative approach that blends team players through the process of defining project scope, tasks, responsibilities, and assignments. It is an integrative process that blends the technical aspects of the project with the more intangible people side.

Frequently, there is substantial resistance to building an integrated project plan and, consequently, to building an integrated team. With the global competition and general time pressures that corporations and government agencies experience today, most of us work on multiple projects and don't believe that we have the time for group planning and team building sessions. Oh, they may be good for the organization, but who has the time? In reality, however, logic demands that we don't have the time *not* to do them!

In simpler times, a product could be developed and produced by an individual specialist or tradesperson. One person could possibly have all the skills and knowledge requisite for completing most, if not all, of the tasks, without seeking the assistance of others. If other opinions or talents were necessary, time was available to meet with others on a one-on-one basis, waiting as long as necessary for their expertise and input. But as technology became more complex, competition, time, and cost constraints grew. One specialist could no longer accomplish the total effort alone or in an unspecified period of time.

Today, we recognize the value and depth of expertise of the cross-functional, multidisciplined team as limitless. Yet, building effective, successful teams takes more than an intellectual awareness of the value of teamwork. One fundamental ingredient in team development is the full participation of each team member in accomplishing whatever objective the group sets out to achieve. The most effective means of implementing any plan, process, or strategy is by encouraging the full participation of those who will be responsible for implementation. Participation translates into commitment and creates a psychological bond between the plan and those who generate it. This is the "buy-in" that all project and team managers strive to accomplish. If commitment is the desired outcome, participation in the broadest sense must be achieved. The *integrative planning process* (IPP) described in Chapter 5, not only parallels the steps to team building outlined in the next few pages, but is natural to the work itself and, thus, succeeds in creating committed, well-balanced teams.

13.5.1 Barriers to Team Development

Despite all the attention given to the importance of teams and the process of teamwork, anyone who has ever tried to organize a project knows how hard it can be to get the entire team on board, to ensure that everyone knows where the project is headed, and agrees on what it will take to succeed. Teams seem to spend an extraordinary amount of time and energy on fine tuning the technical details and cost parameters associated with objectives that are often ambiguous to everyone involved. Often, team participants fail to gain clarity about a project's and/or team's mission, not fully understanding the customer or management objectives. Other barriers to developing a team may include:

- Erroneous beliefs that team building is a luxury, rather than a necessity
- Inappropriate ("touchy-feely") and unsuccessful previous team building attempts
- Lack of management support
- Incomplete or erroneous concept of what a team requires
- Lack of commitment to team problem-solving and decision-making processes
- Insufficient time delegated to the team building process
- Logistics, language, and time impediments
- No formal or reinforced project or team processes
- Poor patterns of communication
- Fear of conflict
- Inadequate participation by all who are involved

We have repeatedly found that *an integrative planning process* minimizes many of these barriers by increasing awareness of the project and team's goals, mission, and scope. It results in a common understanding of project scope and hand-offs, and develops stronger relationships among project specialists. In fact, there are multitudes of benefits that result from this duel team planning process. As discussed in Chapter 5, some of the benefits accrued are:[2]

---

[2] Deborah S. Kezsbom, "Integrating People with Technology: A Paradigm for Building Project Teams," *Cost Estimator*, April 1995.

- Clearer roles, responsibilities, and relationships
- A common view of project scope and team mission
- More accurate definition of risk
- More realistic and achievable milestones
- Better integrated, more aggressive plan
- Empowered team members

13.5.2   Six Steps to Building a Dynamic Project Team

Team building is an ongoing process. Simply bringing people together on a specific task is not enough to create a team. Since there is little time available on most projects today for special efforts apart from the actual project work, team building *must be* incorporated into the work. It may start with the project planning session, continue into task scheduling and sign-off, and repeat itself regularly during weekly, monthly, or quarterly reviews, technical reviews, and quality tests. Formal team building consists of activities and steps carried out in the early stages of the project, then integrated throughout ongoing project activities.

The following six steps entail a formal *team building intervention.* Notice how these six classic steps not only parallel but are also realized through the use of the integrative planning process, discussed fully in Chapter 5.

Step 1: Establish a positive environment. Help participants to understand the importance of team building and the benefits of the process. With the help of the team, define the goals of a team building effort, what it will require in terms of time and commitment, and how it will contribute to project objectives. Many team members, especially those from a more technical environment such as software development or engineering, may erroneously believe that team building is a "touchy-feely" process, of no consequential value to the real issues of the project. Care must be taken to explain that the objective of team building is to increase team performance and project productivity. These objectives are accomplished by (1) focusing on the team mission, (2) determining key tasks and responsibilities, and (3) developing team roles.

Step 2: Develop a sense of interdependence. A critical step in the team building process is to develop a sense of interdependence and respect for team players' complementary talents. Provide opportunities to discuss team players' backgrounds, illustrating how their diverse talents contribute to accomplishing the team mission. As early as the project kickoff session, opportunities should be provided for team players to understand and feel confident in the abilities of co-members.

Step 3: Define and clarify team goals. It is critical to project success that team members understand and accept project goals. Jointly develop a mission statement defining what the team expects to accomplish. Specific objectives concerning costs, schedules, and performance criteria are generated from the mission statement. Objectives should be reviewed and clarified at regular team meetings. As objectives are discussed, the team deals with disagreements and strives to collaborate acceptable interpretations of expectations. Goals should be understood by each team player, be as measurable as possible, and be directly applicable to the schedule.

Step 4: Define team roles. Fuzzy, ambiguous roles cause frustration, duplication of effort, and conflict. Each function's responsibilities must be determined, including any informal agreements made among team members. Define responsibilities in terms of objectives and subobjectives, primary and secondary responsibility, and dates expected. Expectations for each other's performance are established and communicated. Difficulties perceived regarding these expectations should be clarified and discussed openly. New roles may, if necessary, be negotiated. Using the work breakdown structure as a guide, a responsibility matrix, discussed in Chapter 2, is helpful for clarifying responsibilities and expectations. Group planning sessions further permit team members to strategize best methodology and practices for the entire project.

Step 5: Develop procedures. A team cannot possibly reach its full potential if procedures are not thought out and agreed upon by the group. Guidelines and policies for recurring as well as special issues are developed. Proper guidelines and procedures help the team to minimize conflict and concentrate on obtaining project objectives. Who should attend meetings, how cost data will be tracked, and what groupware should be used are but some of the plethora of issues that must be determined or included as early during the project team kickoff session. The preliminary project plan developed by the team should outline these and other management procedures.

Step 6: Develop a decision-making process. This final step in the team building process involves determining responsibilities for decision making: that is, determining who should be involved in the decision-making process, and how they will be involved. The scope and importance of the decisions to be rendered will, of course, affect this. As early as when defining the project team charter or mission, team authority boundaries and the degree of *empowerment* the team experiences must be outlined. Decision-making responsibilities should be mapped out for the team's immediate and longer-term future. Misconceptions concerning team decision-making responsibilities and empowerment will only lead to frustration for all parties involved. Clear expectations must start from and be consistently supported by the higher levels within the organization.

The one technique that the authors have used extensively and successfully to accomplish these six important steps with a variety of teams is the *integrative planning process* (IPP). The IPP is directed toward developing a clearer understanding of the team's mission and determining collectively what the team is not only chartered, but *paid* to do. Through this collective process, key managers and specialists involved in implementing project objectives are responsible (with the assistance of an objective facilitator) for determining what is to be done, how it is to be done, and who is going to do it. Mission statements originally generated by the management team define the boundaries of the project and the team, and identify critical elements or tasks that must be done to accomplish project objectives. Through brainstorming sessions, team members explore both internal and external opportunities and threats (risk) in accomplishing their mission. Consensus is reached on the critical factors necessary for the success of the team and the strategies to assure that these factors are achieved. Table 13.2 compares the conventional team-building intervention with the integrative planning process.[3]

13.6 CREATING TEAMWORK IN VIRTUAL TEAMS

As discussed in Chapter 10, new distance-spanning technologies have created new methods for "working together apart."[4] The normal working relationship is changing rapidly, and global teamwork has become an everyday reality to employees in large corporations and small. But effective teamwork is difficult in the best of times and under the best of conditions. Teamwork depends in part on members' ability to trust one another.

---

[3]Deborah S. Kezsbom, "Are You Really Ready to Build a Project Team?" *Industrial Engineering,* Vol. 22, No. 10, October 1990.

[4]Ray Grenier and George Meters, *Enterprise Networking: Working Together Apart* (Bedford MA: Digital Press, 1992).

TABLE 13.2   COMPARISON OF TRADITIONAL TEAM BUILDING INTERVENTIONS WITH THE INTEGRATIVE PLANNING PROCESS

Conventional Team Building		Integrative Planning Process	
Step 1:	Establish a positive environment	*Step 1:*	Create a positive climate
Step 2:	Develop a sense of interdepencence	*Step 2:*	Establish a common vision
Step 3:	Define team goals	*Step 3:*	Determine the "what;" establish the "who"
Step 4:	Define roles and relationships	*Step 4:*	Develop a sense of interdependency; establish precedence relationships
Step 5:	Develop procedures	*Step 5:*	Identify trade-offs and risk
Step 6:	Develop a decision-making process	*Step 6:*	Seek management approval of the team's plan

Source: Adapted from Deborah S. Kezsbom, "Are You Really Ready to Build a Project Team?" *Industrial Engineering*, Vol. 22, No. 10, October 1990.

Technology cannot substitute for the human relationships and actions that foster trust. Successful teams of all types must pay a great deal of attention to building the foundations of sound teamwork. Virtual teams must work even harder to compensate for many of the elements that are inevitably lost when teams work together, yet apart.

13.6.1   Making a Distinction: The Meaning of Virtual Teams

Within the past decade, the word *virtual* has made its way into "virtually" everyone's vocabulary. Although its original meaning stems from the Latin root *virtue* "a personal quality of goodness and power," more recent use has brought newer meanings to the term. These more "cyber" meanings include "not in actual fact, but 'almost like,'" as in *virtual reality, virtual organization,* or *virtual office.*" A *virtual team,* in fact, creates different images from the one of people working together in the same organization in the same place.

When we refer to a virtual team, we do not mean for it to be assumed that the team is "not real, but appears to exist," which is one meaning ascribed to the term *virtual.* Rather, virtual teams attest to fast-moving electronic forces that define the very existence of the team. Virtual teams, in other words, are teams that have "gone digital" in order to function as a team. They are using the Internet and intranets and any other electronic media that are real to the groups that inhabit them. Like every team, a virtual team is a group of people who interact through interdependent tasks and relationships, guided by a common purpose. Unlike conventional teams, however, virtual teams work across time, space, and organizational boundaries with links created by communication technologies.[5]

Virtual teams routinely cross organizational and international boundaries through electronic media. Socially, however, they lag behind everyday reality. There are no chance encounters, no getting together casually for lunch, no passing each other in the hallway or dropping by one another's office. A major reason why many of today's more conventional teams are ineffective is that they tend to overlook the strong implications of the seemingly obvious. Imagine, in the boundaryless virtual team, what occurs when members choose to ignore how really different they are. Virtual teams must adjust to their new realities — or fail!

13.6.2 Identifying Differences

In addition to the many challenges encountered by virtual teams that were discussed and outlined in Chapter 11, cross-cultural, transcontinental team members often do not have the opportunity to know the people with whom they are assigned or expected to work. Often, extracurricular activities do more to cement a team than a cartload of formal team building sessions. When participants are co-located, dinners and outings serve as an invaluable means of breaking the ice. When e-mail or videoconferences replace social contact, team members lose the chance to socialize with their colleagues, form a more realistic opinion of them, and bond. How can one tell *on-line* which team member is crushed by criticism, who is power hungry, or who is in need of some stroking or handholding? Although not all the answers to these types of questions become apparent during the office barbecue, informal gatherings go a long way to developing the

---

[5]Jessica Lipnack and Jeffrey Stamps, *Virtual Teams: Reaching Across Space, Time, and Organizations with Technology* (New York: Wiley, 1997).

understanding and personal trust that team members must develop in each other to weather the conflicts that naturally arise during the course of teamwork.

13.6.3 Same Arena, Different Ball Game

Regardless of their shape, size, composition, or objectives, any team that wishes to perform well must recognize the four essential disciplines or basics that must be covered. As we discussed earlier, those four basics are (1) a sense of interdependency, (2) an appreciation of the benefits of group problem-solving and decision-making, (3) accountability as a functioning unit, and (4) a common goal, a mission, or sense of collective purpose. Generally, the first three can be achieved whether or not team members work in the same place or location. Of course, if the team does work apart, it will need far greater discipline. Several pre-team discussions may be necessary to establish roles and accountability. Frequent tele- or videoconferencing establishes a sense of progress toward goals and helps to get a clearer sense of what must further be accomplished. In fact, virtual teams require more formal communications than do conventional teams. Precisely because there is less informal chatter and social interactions among team members, project team managers may need to change their informal styles of management and adapt a more formal approach. More direct and rigorous project management techniques are needed to assure that people are aware of who does what and by when.

But the fourth team basic, a common vision with a sense of collective purpose, is more difficult to achieve. Although having a purpose is fundamental to *all* small groups, teams are specifically and deliberately results oriented. *Tasks* are the work, and the *common processes* are the means to the results. *Purpose*, however, binds the team. Purpose, in all its forms — vision, mission, goals — lies at the heart of understanding teams. A sense of collective purpose binds team members to the task and to each other. Unlike business objectives, a sense of common purpose harnesses individual pride and seizes team member imaginations as something worth the effort and the sacrifice. It is often truly developed only after team members have struggled with disagreements, debates, and reflection, and work through the inevitable divergent opinions. It creates a sense of connection and deepens trust. It is the energy that delivers a team to its highest levels of performance.

Purpose, however, is notoriously difficult to grasp. It is an intensely personal process and thrives on frequent face-to-face meetings. Therefore, if the performance stakes are high, the cost of holding face-to-face meetings before the work begins is highly justified for virtual teams. Mistakes, mistrust, unexpressed viewpoints, and unresolved conflicts all too easily spring up and become part of operating norms. A belief in spending more time on the front-end and investing in beginnings is widely held by experienced leaders.

The effectiveness of Boeing's huge globally scattered team effort has been widely documented.[6] At the start of its 777 project, Boeing brought members of the design team from dozens of countries to Everett, Washington, and provided them with opportunities to work together. From a practical point of view, for a period of 18 months, they learned how to function within the company's project management system. But the shared experiences also developed a level of trust between team members that later enabled them to overcome the obstacles inevitably raised by their separation. Linked by a network of 1700 workstations spanning more than a dozen countries, the 777 was launched in five years — 30 to 40% faster than comparable paper-based designs. The plan also boasted 33% greater fuel efficiency than that of the 747 and cost 25% less.

13.6.4 Creating Virtual Team Life

In many situations, however, it may be impractical to bring a team together for any meaningful length of time. Travel costs as well as wear and tear on the body rule it out. The question then becomes: Can teams that cannot spend time together physically ever be as effective as teams that do? In theory, the answer is "no." Teams separated by time and space fail to go through the personal interaction of the level and intensity that is required to create and maintain a common purpose. Since less than 5% of teams that do get together reach optimal performance,[7] it is still possible for remote teams to show superior performance *if* they concentrate on attaining the other team basics. In other words, teams that cannot work and play together must compensate in several ways for the loss of physical proximity. Compensating measures may include some of the activities discussed below.

1. *Concentrate on building credibility and trust.* When team members have few opportunities to get to know each other, trust and credibility are, naturally, in limited supply. A lack of trust creates difficulties in decision making, for example, when time delays require team members to miss

---

[6] Richard Benson-Amer and Tsun-yan Hsieh, "Teamwork Across Time and Space," *McKinsey Quarterly*, No. 4, 1997, pp. 18–27.
[7] Ibid.

certain meetings and must rely on their colleagues to best represent their interests. Professional judgments made by team colleagues are accepted on the basis of trust, credibility, and integrity. If one does not have an opportunity to observe performance consistently, one can only judge a person's integrity on the basis of reputation. Team members, especially those in remote or virtual teams, must pay close attention to the many ways that others perceive them. Consistency of actions, fulfilling promises, considering other peoples' schedules, and responding promptly to e-mail and voice messages help to build positive perceptions. Reliability is a virtue, but in the case of the virtual team, it is a necessity! Team members who have been reliable in the past may build strong, positive reputations that can help them combat the inevitable problems they encounter, problems such as poor transmission and delayed responses.

2. *Create·time together.* Team processes are expedited by spending more time on the front end and in reaching consensus in developing procedures. Invest in beginnings! The time spent in the first two phases of the project life cycle will be recouped many times over in the latter phases of the project. A lack of clarity around goals, tasks, and procedures hinders a team's performance in the later, more critical project phases.

3. *Stress cooperative goals.* Cooperation occurs when people have compatible goals or when they perceive "If *you* succeed, *I* succeed." Cooperation generates positive feelings of family, community, and the sense of "good will" that will be critical in the team's future. A wide range of studies across all age groups indicates that cooperation results in higher productivity than competition or even independent work.[8] The old "tooth and claw" Darwinian competition that may have assumed the natural order of life is giving way. Cooperation *at all levels* of biology's kingdom, from our own microscopic cells to the largest of mammals, may be a factor in successful evolution and survival.

4. *Keep communication constant and vary the medium.* A groupware application that offers sophisticated e-mail, conferencing, newsletters, and bulletin board services may encourage more frequent on-line communication. Monthly team reports are helpful and may be shared with stakeholders interested in team progress.

5. *Develop a sense of "shared space."* When people operate in the same place, they never need to think about the space in which they work. They can set up meeting rooms, discuss their ideas over the lunch table, or gather around a model or prototype as someone describes the problem. The shared space is the immediate ground — physical or mental — that people

---

[8] Lipnack and Stampe, *Virtual Teams.*

use when they come together to create new ideas. But when team members are apart, the issue of shared space becomes more critical. Establishing a communication medium, such as e-mail distribution lists or videoconferencing on a regular basis with defined procedures constitutes the team's shared space, if used to discuss ideas and hold informal cyber get-togethers.

6. *Reward performance.* Punctuate the team's progress with milestones, where the team is provided an opportunity to converge and realign its work and purpose.

7. *Reach out and help someone.* Building credibility and trust may mean sharing information or passing ideas on to others who might benefit. Although altruism seldom brings immediate rewards or recognition, it has the long-term benefit of building a positive reputation and accruing trust.

13.7 IDENTIFYING AND OVERCOMING BARRIERS TO TEAM PERFORMANCE

Identifying the barriers that may impede any team's performance is a requirement for building more effective project teams. Common team performance barriers and probable causes include the following:

Barrier	Cause
Differing outlooks, priorities, and interests	Different professional interests or focus
Role conflicts	Ambiguous assignments; lack of planning
Unclear objectives	Lack of mission, understanding of market, or clearly defined requirements
Dynamic project environment	Characteristic of project life cycle; changing project requirements
Competition over team leadership	Lack of team definition and structure; lack of leadership and management skills in project team manager; lack of a "license to lead"
Lack of team definition and structure	Improperly defined tasks, responsibilities, and reporting relationships; multiple contributory organizations

Barrier	Cause
Team personnel selection	Unavailable talent or limited resources; inappropriate assignments
Credibility of project team leader(s)	Lack of experience or track record; poor communication skills or technical judgments; inconsistent or confusing actions and behaviors
Lack of team member commitment	Poor match between professional interests and project objectives; inadequate or inappropriate rewards structure; insecurity associated with project effort; poor working relations between project and functional managers
Communication problems	Multiple causes — difficult to determine without investigation; may be due to logistics, language, culture, differing organizational structures, lack of procedures, etc.
Lack of senior management support	Senior management may fail to understand the requirements of the project environment; lack of timely feedback; shifting priorities; inadequate strategic project plans

13.7.1   Minimizing Barriers to Team Performance

To be successful as a project team manager, one must not only be aware of the potential challenges that lie ahead, but must constantly be on the lookout for symptoms of, or conditions that encourage, what may be deficiencies in the team and/or organization. Team building requires creating an environment conducive to developing teams on a *daily basis*, in conjunction with accomplishing the work itself.

Table 13.3 provides an extensive summary of some of the common causes of poor project team performance, and recommended actions that may minimize, overcome, or prevent them.

13.8   THE ROLE OF PROJECT TEAM MANAGER

Contrary to some popular opinion, managers *are* needed by teams. Teams not only require a manager's experience to guide them through the

TABLE 13.3   BARRIERS TO TEAM PERFORMANCE: CAUSES AND
RECOMMENDED ACTIONS[a]

Barrier	Causes	Recommended Actions
Differing outlooks, priorities, interests of team	Multidisciplined team; different professional interest; multiproject environment	Provide overview of project scope through WBS and project plan; attempt to match project objectives and personal goals of project specialists; sell team concept; define roles and responsibilities early
Role conflicts	Ambiguity over responsibilities; unclear, multidisciplined matrix environment	Provide opportunity for team members to submit input to WBS; assign and negotiate roles early on in project efforts (e.g., kick-off session); reinforce roles through linear responsibility charts and through frequent status review meetings
Unclear project objectives	Lack of mission; unspecified market or customer; poor communication among Marketing, Engineering, and senior management	Jointly establish team mission; define team objectives; develop common sense of purpose perhaps through team name and logo; communicate frequently with top management and other contributing project organizations
Dynamic project environment	Changing requirements due to customer demands or senior management needs; regulatory changes; characteristics of life cycle	Stabilize dynamic influences; forecast environment, perform risk analysis; develop contingency plans; establish a Change Control Board and invite user–customer participation

TABLE 13.3 *(Continued)*

Barrier	Causes	Recommended Actions
Competition regarding team leadership	Lack of team definition and structure; weak leadership skills in Project Manager; problems in project effort and technology	Clear role and responsibility definition; support of Project Manager by senior management; formal training for Project Manager in leadership, influence, etc.
Lack of team structure and definition	Poorly defined roles and responsibilities; unclear reporting relationships; no project focal point	Education in team essentials and team building process; regular team meetings; linear responsibility chart; provide visibility and recognition
Project personnel selection	Limited resources or lack of talent; matrix environment; turf issues	Negotiation of roles early on with Functional Managers and team specialists; present overview of project in kick-off session; replace those who are really not interested; job posting; alignment of project formally within organization
Credibility of project leader	Lack of managerial or technical skills or experience; wavering or inconsistent values; conflict between Project Manager and functional departments; previous track record	Develop good working relationships with key project contributors; train and develop personnel; choose from informal leaders, and provide with appropriate skills and authority
Lack of team member commitment	Poor match between team player abilities and project needs; insecurity regrading project effort; inadequate reward structure	Role model enthusiasm by Project Manager; match skills with needs of project; determine reasons for project insecurity; resolve conflicts that may exist between team members; team rewards

TABLE 13.3 *(Continued)*

Barrier	Causes	Recommended Actions
Communication problems	Logistics of matrix organization; geographic dispersion; failure to determine responsibilities and reporting relationships; insufficient time; poor interpersonal relations and carelessness; personality clashes	Teleconferencing; timely meetings; communication tools, reporting mechanisms; colocation; frequent communication with contributing functions and with customer; social activities; clearly defined and delegated work packages
Lack of senior management support	Lack of understanding of project management needs; shifting priorities; unclear strategic mission; budget limitations	Invite senior management to kick-off sessions; frequent newsletters; keep management informed of milestones and successes

*Adapted from Kerzner, H., and Thamhain, H. *Project Management for Small and Medium Sized Businesses.* New York: Van Nostrand Rheinhold, 1984.

development process, but also require his or her direction, information, and organizational connections critical to successful team functioning. The position of the project team manager is therefore far from obsolete. The role is however changing rapidly.

Today's project managers must place a higher value on the knowledge and experience of all group members. They must, at times, be perceived as yet another group member, while simultaneously encouraging and gently coaxing others to fully participate and perform the functions needed by the group. Project managers must be facilitators, encouraging participation to create a shared vision.

The increased focus on empowered teams requires a new style of leadership. Employee involvement requires project team managers to be more facilitative and less administrative: to direct less and empower more. It means using group processes and team dynamics to maximize productivity, quality, and satisfaction. Instead of managing the project and managing the people, a project team manager must manage the project through other people.

13.9 CONCLUSIONS

High levels of project team performance come from an environment in which team members share a unity of purpose in an atmosphere of interdependency and trust. Team building should be directed at accomplishing the following:

- Establishing a better understanding of the team's role and purpose within the organization
- Developing a sense of interdependency and a better understanding of the complementary roles and talents within the team
- Increasing channels of communication, and creating greater support and sharing among team members
- Discovering more effective ways of working through team problems at both the task and interpersonal levels
- Developing the ability to use conflict as a collaborative rather than a destructive force
- Encouraging greater interaction and collaboration among team members

Technologies have created organizations with overlapping structures. Employees frequently work on multiple teams, with multiple reporting relationships. People shift rapidly from one project team to the next. The project manager who can build a team quickly clearly has the corporate advantage. Teams will continue as long as human beings rely on others to achieve results.

EXERCISES

Getting a handle on team performance begins by assessing where the team perceives it is now and where it believes it should be in the near future. Prior to a formal team building session, it is helpful for all team members to focus on the primary team functions. These include team roles, team goals, team procedures, team relationships, and outside environmental factors. Information may be obtained through completion of the following anonymous surveys. It may be collated and used as the basis for team discussion and action planning.

13.1 *Team factor survey.* This survey consists of a series of team characteristics, processes, and definitions. Each member of the team may rate the critical characteristics of team functioning by indicating the degree to which it is descriptive of the team and its importance to the success of the team. By completing this diagnostic survey, you allow yourself an opportunity to step back and review the status of the critical factors related to team success. These factors include team planning processes, team communication, team relationships, team operations, and external factors that influenced team performance.

Directions: Consider the following critical factors for effective team performance. Using the 7-point scale, assign the number that best indicates the extent to which you agree or disagree with the statement, as it relates to your project team.

1	2	3	4	5	6	7
Strongly disagree			neutral		strongly agree	

Project Team Mission

_____1. Our team mission and its relationship to the general goals of the organization are clear and readily understood.

_____2. Our team has had significant participation in the development of the mission.

_____3. Goal setting is a cross-functional activity.

_____4. The company supports the team in a variety of ways.

_____5. The team is very enthusiastic about the project's chances of success.

Team Planning and Operations

_____6. I have had the opportunity to learn and understand each team member's project goals.

_____7. As a team, we have collectively determined responsibilities and time lines.

_____8. Tasks appear to be very interdependent.

_____9. Roles on the team are well defined.

_____10. There is some overlap in team roles.

_____11. Team members have had input in defining milestones.

_____12. Milestones are measurable.

_____13. Discipline of some type follows the failure to accomplish milestones.

_____14. Business is conducted with a minimum of red tape.

_____15. Progress is reviewed with the team at least quarterly.

Communications

_____16. Feedback is provided to team members on a constant basis.

_____17. Team meetings are well attended.

_____18. Team meetings are participative and no one dominates discussions.

_____19. Team members welcome differing points of view.

_____20. Results of meetings are always published and distributed.

Group Functioning/Relationships

_____21. Team members are comfortable about being open and frank.

_____22. I am satisfied being a member of this team.

_____23. The team members work well together.

_____24. People put forth their best efforts.

_____25. Disagreements are expressed openly and worked through.

_____26. Overall, I would assign a high rating to the effectiveness of this team.

Decision Making/Problem Solving

_____27. I have been very active in the team's decisions.

_____28. There appears to be a no-blame approach to solving problems.

_____29. Everyone on the team has the fullest opportunity to participate in decisions that affect the group.

_____30. The leader on this team makes most of the decisions.

Personnel/Staffing

_____31. There are appropriate skills, abilities, and knowledge on the team.

_____32. The size of the team is effective in accomplishing our mission.

_____33. It appears that members on the team are very accountable for project success or failure.

_____34. Adequate training is provided.

_____35. There is a team reward system.

Leadership

_____36. It is understood how leadership is delegated.

_____37. Leadership seems to be shared among team players.

_____38. The project team leader is effective in getting the team to work together effectively.

_____39. The project manager is always in charge of team meetings.

_____40. Recognition is given visibly to team members.

13.2. *Team building: a group effectiveness questionnaire.* This survey is designed to identify how you feel your team has been functioning: specifically, what areas you feel need improvement and/or maintenance. As you read each of the issues, think about how your group has been working. Respond as openly and as honestly as possible.

    Directions: Where there are multiple choices, check only one answer.

Goals

1. List the goals that require members of the team to work together.

    _____

    _____

    _____

    _____

2. With regard to the goals of this team, the people on the team are:
   _____Very committed   _____Somewhat committed
   _____Somewhat resistant   _____Very resistant
   Team goals that I feel have a low level of commitment are:

    _____

    _____

    _____

   Team goals that I feel have a high level of commitment are:

    _____

    _____

    _____

3. To what extent do you know and understand each team member's goals?

_____Very knowledgeable about team member's goals
_____Fairly knowledgeable about team member's goals
_____Somewhat vague knowledge of team member's goals
_____Very vague knowledge of team member's goals
Goals I would like to know more about are:

_____

_____

_____

4. The goals of members of this team are:
    _____Strongly in conflict   _____Somewhat in conflict
    _____A little in conflict   _____Not at all in conflict
    The key conflicts I see are:

_____

_____

_____

Roles

1. The roles of the members of the team are:
    _____Very clear to me   _____Fairly clear to me
    _____Somewhat unclear to me   _____Very unclear to me
    Areas I would like to have clarified concerning my role are:

_____

_____

_____

Areas in which I am unclear about what others expect of me are:

_____

_____

_____

Areas I would like to have clarified concerning others' roles are:

_____

_____

_____

_____

2. Team members' roles overlap:
    ____Very much ____Quite a bit ____Somewhat
    ____Not at all

Roles that overlap are:

_____

_____

_____

3. Team member's roles that are in conflict are:

_____

_____

_____

_____

Procedures

1. The decision-making process on matters that affect more than one member of the team is:

   _____Very clear   _____Fairly clear

   _____Somewhat unclear   _____Very unclear

   Decisions that need to be clarified are:

   _____

   _____

   _____

2. Thinking of all the communications within this team, I would say they are:

   _____Good with all members of the team

   _____Good with some members of the team

   _____Good with very few members of the team

   _____Not very good with the majority of the team

   Specify what subjects you feel need to be better communicated:

   _____

   _____

   _____

   _____

3. Team meetings

   _____Are very effective

   _____Are fairly effective

   _____Need some improvement

   Specify how you would like team meetings to be improved.

   _____

_____

_____

4. I believe the leadership of this team
   (a) Is helping the team's performance by:

   _____

   _____

   _____

   (b) Could improve the team's performance by:

   _____

   _____

   _____

5. As teams work together, behavior patterns are established that help or hinder the team's performance, such as:

   · Following up or not following up on decisions

   · Raising or not raising sticky issues

   · Facilitating or delaying decisions

   Do you feel that there are patterns that inhibit this team's effectiveness?
   _____Yes _____No
   If yes, what are they?

   _____

   _____

   _____

   _____

Relationships

1. Conflicts within the team are:
   _____Discussed and resolved openly
   _____Discussed somewhat
   _____Very seldom mentioned
   _____Not discussed at all
   A conflict that needs to be resolved to improve the team's performance is:

   _____

   _____

   _____

2. Relationships I would like to discuss are:

_____

_____

_____

_____

3. What is causing stress for you and/or other members of the team?

_____

_____

_____

_____

Environmental Influences

1. What constraints or influences outside the team keep it from working more effectively? Explain.

_____

_____

_____

_____

General

1. What do you believe are the team's key strengths?

_____

_____

_____

_____

2. If you could, what would you do to make this team more effective?

_____

_____

_____

_____

C	H	A	P	T	E	R

14

PROCUREMENT AND CONTRACT MANAGEMENT

An honest man's word is as good as his bond.

MIGUEL DE CERVANTES,

DON QUIXOTE, CHAP. xxxiii

14.1 INTRODUCTION

Today, most high-technology projects are intertwined with cross-national business practices that involve the solicitation, negotiation, and awarding of contracts. These contracts are those that are negotiated by the organization to procure needed resources to produce and deliver products and services to market. The renewed attention on how businesses manage their procurement process is a direct result of the increasingly rapid pace of introducing new commercial products and services by shortening the lead times traditionally experienced by manufacturing organizations when new products move from the laboratory and into production. Another important factor leading to the relentless pressure on project organizations to actively manage the procurement of resources results from the instability of world currencies, which has made the cost of the components that comprise today's products unpredictable. In such times of economic uncertainty, a defensive posture for most businesses is to adapt a global sourcing strategy and expand the role of the project management team to include the logistics of interfacing between marketing, R&D, and the company's centralized purchasing organization as it carries out the strategy.

Over the last decade, commercial enterprises have become less vertically integrated, which means that today, many of these organizations globally outsource needed components and services. This activity subsequently requires closer coordination of marketing activities, R&D, manufacturing, and an in-depth understanding of the new roles played by both the buyer and seller as they enter into a relationship. How project management must manage this new relationship to succeed in the new world markets is the focus of this chapter.

14.2 PROCUREMENT MANAGEMENT

Procurement is defined as the process of obtaining, acquiring, securing, or taking possession of goods and/or services. There are several major process activities associated with project procurement management. At a minimum, these include (1) procurement planning, (2) solicitation planning, (3) solicitation, (4) source selection, (5) contract administration, and (6) contract closeout.

Depending on the nature and scope of the project, the project team leaders will undoubtedly become involved in carrying out some of these activities to customize, purchase, and secure equipment, commercial off-the-shelf components, shrink-wrapped software, or professional services. Because none presently exist in a form that satisfies the project's technical requirements, project teams must often purchase, license, or commission their development for inclusion in certain products. Consequently, to realize these objectives, the team needs to acquire a basic understanding of the organization's formal procurement management process, which establishes the framework and cross-functional foundation for the planning, scheduling, and risk management of the technology. The process applies to all supplier management activities and manufacturing projects, since it also assures delivery of the technical solution to the designated company-owned or alliance partner manufacturing sites anywhere in the world.

On many different occasions, over the development project's life cycle, equipment will be purchased through competitive solicitation. To ensure that the best system is bought at the best price, we recommend that the project team follow the standardized practices that have been established in their companies to select and manage the vendors supplying these commodities. To help understand their organization's procurement strategy, the project team may want to seek the advice of dedicated specialists from outside the project. Most companies have a centralized procurement organization that follows the documented business and commodity strategies that have been approved by a chief procurement officer and makes available to project teams dedicated subject-matter experts who consult on the strategy. These talented specialists can also assist in conducting value analysis or target costing exercises, whose aim is to arrive at the costs, based on competitively benchmarked pricing, that should be paid for the technology required (see Chapter 10).

Although project teams may become heavily involved in formatting the preliminary contract documents provided to suppliers, it is best to rely on the professionals in the procurement organization and its legal division to negotiate the final agreements. However, as we have experienced, project teams often fail to learn the value of this recommendation, unless discovered firsthand. For instance, a project team trying to shorten the interval needed to program a software application that exchanges data with a third party's embedded database may appear to be making great strides by starting work with a specific vendor's product that satisfies all the necessary technical specifications. Unfortunately, when later it is determined that the cost of the high royalties owed to the supplier to license and replicate the database cannot be recovered, a less expensive supplier must be engaged. As a result, any work that the software developers may have completed must not only be redone but must also be assumed as an unbudgeted expense by the project.

A project team may also seek the expertise of the procurement organization when closing out or terminating a contract with a preferred vendor or supplier. No doubt many lessons were learned from the experience that will help other projects, so we advise team leaders to record and share their observations formally.

14.2.1 Procurement Management Process

The simplest way for a project team to understand the procurement management process is to divide it into three stages: (1) *preacquisition* or *planning phase*, (2) *acquisition* or *execution phase*, and (3) *postacquisition phase*, which is also commonly known as the *life-cycle phase*. The preliminary stage, *preacquistion*, is characterized by four sequential but interrelated process steps:

Step 1: determination of need.

Step 2: selection of product or service to fulfill the documented need.

Step 3: selection of vendor or source to provide the product or service.

Step 4: establishment of the contract and/or agreement with the vendor identified to satisfy the need.

It is important for project managers to understand the interrelationships of these activities in the preacquisition stage and the synergy between them. When the steps are performed improperly, a mistranslation of the buyer's needs in the marketplace usually results. Either the vendor's performance and deliverables fail to satisfy the buyer's expectations or the supplier's product is misused due to the buyer's failure to specify adequate functional and performance criteria. To overcome such pitfalls, the project team needs to complete an assessment of the specialized technology that it requests the procurement organization to secure. Then the details, which are documented in the form of functional specifications, can later be attached to the purchasing agreement or contract.

As we have stated, it is best to involve dedicated purchasing professionals in the systematic execution of each step of the preacquisition stage. This helps especially to assure that all relevant issues pertaining to the risks of purchasing at fixed prices or contracting on a cost reimbursement basis will be assessed and covered adequately in the agreements negotiated with vendors.

Each of the three stages of the procurement management process is iterative, meaning that the outputs of the four steps of the preacquisition stage are the inputs to the *acquisition* or *execution phase.* So milestones associated with their completions may be used to measure vendor performance against predetermined standards and included in the contract schedule. When tied to a vendor's successful completion of a task or the delivery of a product, these milestones may be used to trigger an incentive or progress payment. Even though project teams typically are *not* involved in negotiating the final terms of the vendor agreements, they still need to share criteria documented in their project's functional and performance specifications with the procurement professionals who will use them to communicate what is important to the suppliers. We have found that the systematic completion of these planning activities helps to avoid over 90% of the typical problems experienced by teams during a project's acquisition phase.

The primary focus of the acquisition phase is on executing those activities that the team planned during the preacquisiton phase. Although the stages of the acquisition phase may be performed out of sequence, this phase commonly includes three steps:

Step 1: purchasing.
Step 2: inspection.
Step 3: expenditure.

Purchasing is the element of the acquisition phase that involves execution of the planning process steps. Most often a separate division in the procurement organization, which is referred to as *contract administration*, performs this function. The functional responsibilities of this specialized group involve the assurance that the vendors' or sellers' performances meet contractual requirements. This may include ordering prepriced or sourced components or off-the-shelf software, acceptance testing, expediting orders, inspection or quality control, and administrative duties that pertain to accounts payable and accounts receivable.

Although project management spends less of its time completing acquisition phase activities, it is important for the managers to (1) recognize the interrelationship of these activities with the earlier planning or preacquisition phase activities, and (2) understand how the purchasing cycle is affected by suppliers' lead times, the time frames from when an order is placed to when the product is delivered and accepted.

Since each of the vendors' lead times are different, contract administration needs to draw up a time line that can be closely coordinated and communicated to project management to minimize any potential delays executing the project schedule. We also recommend that the project managers become cognizant of the procedures used by contract administration to manage contracts. How to administer contracts needn't be a detrimental battle for the project but a chance for interactive two-way communication. To start, project management may want to have team members briefed, for example, on administration procedures for processing purchase orders and signature authority.

This group also maintains *vendor profiles* of preferred vendors: lists of approved suppliers who can be invited to submit goods or services that the project can consult. Often, when a project prefers to buy from vendors who are not prequalified or included on these *preferred bid lists*, there can be long delays while contract administration proceeds to qualify the vendors. So it is best to allow sufficient time in the project schedule for the purchasing group to verify that a newly selected supplier will:

- Warrant title and issue a proper *bill of lading* (receipt issued by a carrier for delivering merchandise to a party at some destination) to ensure that title/ownership passes according to the established requirements of the Uniform Commercial Code governing the sales of goods and services

- Have appropriate contractual liability and other types of required insurance coverage

- Agree to *implied* versus *expressed warranties* (see Section 14.4.3)

- Recognize and not dispute *back charges* (the cost of corrective action taken by the purchaser and chargeable to the supplier)
- Agree to sufficient remedies to make the buyer whole in the event of a contract breach

Project teams working in large corporations also must understand that the purchase order system is approved annually by the corporate board of directors, and that this process gives validity to the numbers assigned to purchase requisitions. For this reason, team leaders should maintain a system to track the purchase order numbers assigned to project acquisitions to facilitate communication with the contract administration should the integrity of a supplier's order be questioned.

Another activity that is completed during the *acquisition phase* is the inspection process. The objective of the inspection process is to ascertain the quality of the goods and services obtained. While initial inspection of goods is most often performed when the development or production team receives them, the procurement organization is also involved in this process on an ongoing basis. Accepting a product frequently includes continuous testing in the laboratory or factory to determine its capabilities or compliance with stated performance parameters that were delineated in the contract. During the acquisition phase, the project team often will be involved in a review of the supplier's quality processes. A company may request to review a supplier's quality processes as part of their compliance with ISO certification requirements by issuing written notice well in advance of the date of the requested review.

The final activities of the acquisition phase, which are known as the expenditure stage, include the payment for goods and services and an approval and disbursement process. The expenditure stage is very important to the project team because it affects when resources will be released. The risks associated with disbursing payment must be weighed carefully, since the project team may want to leverage progress payments to compel the vendor to rectify problems with an order or a contract.

It is best to think through the expenditure process during the preacquisition phase when the agreement and contacts are drawn up. It may affect negotiation, selection of contract type, and contract terms and conditions. To maintain control over when payment is rendered and to offer incentives to a supplier for desired performance, a project team will often establish progress payment schedules. This might be done, for example, for customized software development by tying a vendor's compensation to the achievement of project specific milestones, such as 100% code completion or delivery to the project's system verification team. We recommend that project management rely on the procurement organization and its legal division to interpret regulations regarding sales terms and statutory law, such as prompt payment acts.

The final stage in the procurement process is known as *postacquisition* or *life-cycle* management. The two main process steps of post acquisition include:

Step 1: product utilization.

Step 2: disposal and termination of the service relationship.

The term *life cycle* is defined as the time frame in which a product or service is (1) identified as a solution for a need, (2) obtained, (3) placed into service, (4) maintained, (5) administered, (6) paid for, (7) utilized, and (8) disposed of according to national and international regulations. Because the life cycle of a product may have a significant impact on the longevity of products internally consumed or commercially offered, it is considered an important element of the procurement contract. It is important that the project team specify life-cycle requirements in the documents used to qualify vendors during the preacquisition phase. Also of concern to project mangers are the ancillary services that might be required to maintain an acquisition properly. It is best to specify that suppliers disclose such requirements when the agreement is negotiated so that the expense is allocated properly.

Often projects need to be concerned with the last stage of the procurement process, *disposal*. This is when the environmental impact of the disposing of precious metals or hazardous by-products must be considered. There may even be salvage value for the discarded materials. Many large companies have environmental specialists on their staffs to assist in this process. In the United States, the interstate commerce and transportation laws and hazardous waste disposal acts assign responsibility and provide for significant civil and criminal penalties for noncompliance with these regulations.

14.2.2 Preparing the Procurement Management Plan

Procurement management plans contain four elements: (1) a written plan, (2) a statement of work (SOW) describing the end product or service, (3) the contract WBS, which translates the scope statement (SOW) into a manageable context, and (4) a budget. Preparation of a procurement management plan should be seen by all project team members as vital to the success of a project. The activity is a normal extension of the planning activities discussed in Chapters 3 to 5 and should be completed once the project team understands the nature and scope of the technical or business problems it is charged with solving. The project's scope is documented in the *statement of work* (SOW), which is attached to the contract and given to the prospective suppliers as part of the *bid invitation and proposal request package* (see Section 14.3).

As soon as the project managers are satisfied with the level of detail of the project tasks outlined by the project specialists in their work packages, they should complete with the purchasing organization a contract work breakdown structure (CWBS) for the project. The contract work breakdown structure illustrates the vendor-related tasks and deliverables that must be supplied or outsourced to deliver the project to market. This sixth level of the WBS is referred to as the *level of effort* because it is often used to track suppliers' work efforts. A CWBS would delineate all the software, hardware, and associated quality documentation that the project expects to procure, along with the time frames for when these deliverables are needed.

The project team must share the project's budget and targeted costs for the goods that the procurement organization is expected to purchase. Procurement will identify sources of supplies that fall within the guidelines (budget, schedule, quality, as well as geography) shared by the project team and determine what is available in the marketplace. It is in the best interest of the project not to share proprietary information, sales forecasts, or expense plans when meeting with a supplier. Providing expense budgets usually sets a vendor's expectations at a certain level and often results in higher prices.

14.2.3 The Role of Project Management

As discussed in Section 14.2.1, centralized procurement management affects the discipline of project management by helping to achieve a balance in the overall process. While project managers typically focus on the preacquisition and postacquisition activities, it is best to leave the purchase of components and software to the procurement organization. However, project managers still need to acquire and perfect many procurement management skills. These include (1) understanding commercial law, (2) understanding contracts, (3) understanding vendors/suppliers, and (4) developing negotiation skills.

Generally, the project manager should follow these five guidelines:

1. Understand the general process and rules of contract/procurement.
2. Rely on the contract/procurement organization's expertise.
3. Ensure compliance and adherence to ethical standards.

4. Have in-depth awareness to guide the project.

5. Assess and minimize risk early to avoid future problems.

14.3 SOLICITATION MANAGEMENT

14.3.1 Understanding the Buyer–Seller Relationship: Pre-award, Award, Post-award

The buyer and seller are in natural conflict that centers on the amount of risk that each party shares and the allocation of resources to carry out the contract. In addition, a buyer must be aware of a seller's agenda. A seller, in competitive situations, is often under pressure to close a sale for a particular product or service. Hence there may be a tendency to overcommit or even to fail to disclose relevant information that, if disclosed, would make the sale less attractive. To minimize or reduce such problems, we recommend working closely with the procurement organization to qualify and rank prospective vendors according to preestablished criteria.

Project teams will find that their needs for vendors vary across projects, as does their relationship with them. Each company with whom the project interfaces also will follow different policies. Project executives often know the suppliers personally, especially if these vendors represent smaller firms in the marketplace. No matter how small the firm, however, we recommend that the relationship be formal and contractual.

Some project organizations may get lower costs, better technology, and even more leverage by maintaining a multiple vendor strategy. Others may choose to issue *sole-source agreements* with reliable vendors who offer uniquely competitive solutions in the marketplace, even though this arrangement does increase the overall risk to the project by creating dependencies on the vendor. Often, the purchasing organization will insist that other, less expensive suppliers be taken on as second sources and that the project's design be reformulated for competitive solicitation. However, the design is often so tightly geared to the primary supplier that it is not possible to make such changes. In this case, procurement will assist in developing a contingency plan in the event that the primary supplier cannot fulfill the terms of the contract. This may include developing alternative suppliers that are able to cover all or partial deliveries.

Overall, the degree of formality in the agreements with vendors will parallel how your project executes its business strategies. As a good practice, software projects should escrow the source code for commercial shrink-wrapped software or off-the-shelf components licensed from a sole-source vendor. In the event of the supplier's failure, the escrow account may be accessed and the product self-maintained.

14.3.2 Preparing the Solicitation

One of the primary activities in solicitation management is for the project team to determine whether it is going to *build* all the components of its technological development or prepare a solicitation to *buy, lease,* or *license* them under contract. When making this decision, the team must remember to calculate both the direct and indirect cost of the prospective procurement as well as the cost of development and production. At first glance the savings that might result from low labor costs elsewhere in the world may appear attractive, but companies mustn't be blind to the possibly unfavorable political, currency, and cultural risks that could affect already tight project schedules. Vendor and source selection must be entered into carefully.

Bid invitations and *requests for proposals* (RFPs) are the most customary forms of procurement documents that large businesses and government agencies use to select the best vendors. As discussed in Section 14.2, selection criteria must be documented prior to issuing a bid invitation or request for proposal if the decision is to be objective. A company will generally use a *bid invitation* to describe a solicitation that is price driven, whereas a *request for proposal* will include additional elements, such as request of descriptions of technical skills, vendor history, plant information, drawings, or even prototypes. These other documents that may be issued include request for information (RFI) and request for quotation (RFQ).

The RFI and RFQ are usually issued when the desired items are readily available in the marketplace. The *request for information* (RFI) is a formal way to request information from a vendor about the system or product the project team wishes to procure. It is used primarily to screen candidate suppliers and to limit the number of respondents who will be invited to provide further details in a formal proposal request. The RFI usually states that any responses received are not binding. The RFI may also be used to gather information from vendors for a product that meets a project's specifications but does not exist in the marketplace. This solicitation documentation may be used to test the waters, so to speak, and to gather information about similar product offers in the market that may be adapted or customized.

The *request for quotation* (RFQ) is a request issued by the procurement organization for a *firm fixed price* (see Section 14.4.4) from a vendor. This type of procurement is used primarily for off-the-shelf products that the team will use with little or no modifications. At other times, this type of procurement is used when an organization is purchasing both equipment and the labor to install it. The vendor relies on his or her experience working in similar project environments and quotes a price to deliver a complete turnkey solution. Because RFQs are usually simple, nonbinding requests by the buyer for price quotes from the vendor, this type of procurement is rarely used to purchase sophisticated technology that will be customized to meet requirements specified specifically. Rather, the RFP, a more complex and elaborate document, is the style of solicitation that is used to procure creative solutions that are significantly risky or cost a great deal. Most technical development projects source hardware components, software, or professional services not currently available in the market through RFPs.

The RFP is actually a package of documents prepared by the project team and the procurement organization to manage the selection and evaluation of prospective vendors. First, the project team prepares a letter of transmittal for each vendor receiving a RFP. Then it translates theoretical concepts into functional requirements that accurately and completely describe the needs that it captures in the SOW. Finally, it breaks the specifications down into the level of effort of CWBS work tasks (see Section 14.2.2) and includes both documents in the RFP, along with the time frame for all deliverables requested. Representatives from the procurement organization see to it that the RFP contains: (1) the contract terms and conditions; (2) an indication of the format and time frame for the formal response; and (3) the rules, regulations, or standards that must be adhered to in the vendor response.

Often, the project team will expect to receive one proposal from the vendor, consisting of separate technical, managerial, and cost proposals, as is frequently required in responses to RFPs issued by the U.S. federal government and its agencies. Each of the proposals complements the other and creates a complete picture of how, when, and at what cost the vendor will comply with the stated business and technical requests.

As the buyer, the project team must insist that the *technical proposal* contain the supplier's understanding of the posed business problem and several technical alternatives or possible approaches to solving it. The technical proposal must also include the program plan and schedule along with the biographical sketches and the hours of the personnel the supplier will assign to the project. The costs and any lead times for purchasing materials needed to complete the SOW should also be explained. The supplier typically does not include wages or salaries of technical specialists or subcontractors who will be assigned to do the work. Rather, the supplier discusses both the *candidate approaches* that are technically feasible and the *approach proposed* to let the prospective buyer know that all the various ways to solve the problem or issues queried have been fully considered. There also should be a response section in the *technical proposal*, entitled *feasibility of approach*, that outlines the risks associated with each element of the design requested. Another extremely important section is the vendor's reply to the CWBS and Gantt chart showing the duration of the activities proposed and start and stop times expected for all major contract deliverables described in the RFP.

The *cost and price proposal* requires the vendor to justify the contract price. The cost information that is supplied should include the monthly or quarterly cost of each deliverable and personnel who will be assigned to design, develop, or manufacture the desired product or service. The intent of the cost and price proposal is to provide the price for the contract and to show the associated cost allocations. Cost categories may include: (1) direct and indirect expenses, (2) labor categories by month and by task, (3) overhead, (4) plant costs, (5) tasks by months, (6) materials, (7) preparation of reports, and (8) travel and subsistence.

Finally, the *management proposal* describes the management techniques and controls, as well as the organizational plan for completing the work. A quick review of the management proposal will indicate the supplier's experience using sound project management techniques to manage similar work. Suppliers often include a table demonstrating to the buyer that they have won related contracts and have met the specifications on time and at cost. When preparing solicitations, the buyer should state that:

· The solicitation can be withdrawn anytime.

· All costs of preparing the bid/proposal and attending a bidder's conference will be borne by the seller and not reimbursed.

· The seller must specifically note any deviations in the proposal.

· The buyer must acknowledge these amendments in a contract addendum.

· All information received is confidential to the buyer and may not be disclosed.

· This is not an offer; any purchase is subject to the negotiation of a written contract.

The RFP should include instructions to the vendor on how to submit the proposal so that the responses received are consistently priced and easily compared. On occasion, the buyer will convene an evaluation team of experts, along with the procurement organization, to review and rate the prospective suppliers' responses. Often, each of the respondents is invited to a face-to-face meeting to review their proposals. Deciding when to hold such meetings with suppliers may be negotiable, especially if the vendor must research several alternative technologies before sharing concrete results with the buyer. For this reason the team may grant a delay if the vendor's technology is of great value to the project and the proposal due date is too soon to permit adequate presentation. Any changes that the team makes to the original specifications may also justify a supplier's delay and request for an extension. However, once the bidding is under way, the buyer may issue a *performance bond* to guarantee the seller's ability to complete the work. Subsequent changes to the specifications after the contract has been issued must be readily reviewed, accepted, and documented by both the buyer and seller using a change control process (see Chapter 10).

As we have stated, a company will sometimes enter into noncompetitive procurement and issue to a single preferred vendor sole-source procurement documents. Although there are no regulations prohibiting organizations from proceeding in this manner, companies may not get the best deal for their projects' investment. Sometimes in the procurement of customized equipment and systems, only one vendor appears to qualify. The purchasing organization may therefore have to ask other vendors to disclose if they have quality improvements or revisions to their current offers that will be available commercially or ready when needed by the project. It is usually in the project's and company's best interest to insist upon competitive procurement.

14.3.3 Managing the Solicitation Response

In most large companies the centralized procurement organization receives all the bids and proposals submitted. A corporate quotation registrar time-stamps and opens the quotes and proposals in the presence of a witness. Procurement will either review the bids with the project team managers or distribute the proposals for approval to an advisory council or steering committee. In the case of large government agencies, a formal evaluation board composed of consultants who are experts in the technical subject matter may be assembled to review and score each section of the technical, cost, and management proposals. Each area is rated as *superior, acceptable, susceptible of being made acceptable,* or *unacceptable.* A vendor with unacceptable ratings in any area is usually disqualified. If several vendors submit technical proposals with high technical ratings, the project will usually give the award to the vendor with the lowest bid.

Here are a few well-founded techniques that we recommend using to effectively manage and evaluate vendor responses to formal solicitations on large and small projects in both the commercial and government sectors:

- Use formal channels to secure information about the vendor. Although many vendors would like access to project personnel at the engineering level, save this relationship for the project executives. Insist that the vendors deliver their corporate, management, and financial data to a central contact who will later negotiate the contract with the seller.

- Establish a preferred vendor policy. Have the procurement organization identify currently preferred vendors and their product lines and share the company rationale for using these preferred vendors with the project team.

- Document the selection criteria internally within the company prior to the solicitation to avoid bias selection. Make the selection process a standard way of doing business when evaluating vendor's proposals.

- Screen out unqualified competitors.

- Create a realistic timetable for the entire solicitation and evaluation process that allows vendors adequate time to prepare their proposals. Do not underestimate the amount of time needed to conduct a thorough evaluation of the vendors' responses and to secure the approval of senior management once a selection has been made.

- Use a formal vendor evaluation policy during a contract award that specifies how vendor evaluation will be conducted and states that the reviews of the vendor's technical, cost, and management proposals will be fair, confidential, and consistent. Include information in the RFP about the evaluation review and the weighting algorithms that will be used to make the selection.

- Complete a post-award vendor evaluation report. Share it with the procurement organization and other project management teams to let these teams understand the level of detail that needs to be included in future procurement documents for successful project execution.

- Conduct vendor debriefings with losing vendors if necessary to point out areas in their proposals that need improvement. Holding such reviews will help the project teams solicit future proposals from these vendors that are more responsive and competitive.

14.4 CONTRACT ADMINISTRATION

14.4.1 The Global Nature of Contracts

Today's world class organizations are aligning with suppliers around the globe and embracing electronic commerce technology, making possible a truly international marketplace. As project teams engage in the management of suppliers throughout the world, acquiring an appreciation of how the various legal systems and contractual relations differ becomes increasingly more important. However, to understand how the language, culture, and legal system in the countries where project organizations are doing business vary, team members first need to understand the fundamental framework of their own systems.

A contract is defined as an enforceable agreement or set of promises that when *breached* are remedied by action grounded in *civil* or *common law*. In general, contracts formalize the relationship between buyer and seller, and good procurement management integrates the principles of contract management. It is important for project teams involved in international development to understand that when a negotiation becomes enforceable may differ around the world. In most U.S. jurisdictions, contracts under a certain dollar amount can be oral. The U.S. Statutes of Fraud states that contract values over $500 must be in writing to be enforceable. Because statements or actions by either party may serve to defeat this statute and create an oral contract, it is best always to insist on written contracts. However, in other systems a written document is not needed to form a contract. Inadvertently, U.S. managers may enter into a contract if they do not stress from the outset with their international partners that no contract will result from any transactions unless a final written document is prepared.

Every project team needs to understand the legal framework behind international contracts in the global business environment of the twenty-first century. Foremost, there is no universal law of contracts. Most nations have adapted a civil law or statutory system of legal codes. Great Britain and her former colonies use a system of *common laws* or principles and rules that derive authority from usage, customs, judgments, and decrees that are recognized, affirmed, and enforced by the courts. Except for U.S. federal government contracting, commercial contracts in common law countries are governed by statues of state law and therefore differ from jurisdiction to jurisdiction. There is a hierarchy of parallel courts set out in the U.S. Constitution. Federal laws are prosecuted in federal courts; state laws, in state courts. However, in the United States a contract for the sales of goods is subject to the statues enacted by all states, called the *Uniform Commercial Code* (*UCC*). Elsewhere, some countries have accepted the U.N. Convention on Contracts for International Sales of Goods developed by the United Nations, which is a set of rules governing international sales similar to the UCC.

Unlike its common law counterpart, a commercial contract written under a civil law system does not give precedence to previous decisions of its courts. There is no reporting of previous cases in commercial contracts drafted under civil law because the courts interpret the terms and conditions. When preparing international commercial contracts, the procurement organization should seek in-country legal council to verify that the language used can be understood clearly without the benefit of how previous U.S. courts interpreted it since special U.S. case law interpretation has no meaning. Nor can a mistake of fact be appealed in civil law systems. Rather, all cases must be tried anew, making litigation unpredictable and slippage of tight time-to-market deadlines highly probable.

14.4.2   Basic Contract Principles

Since in the United States the basic principles of contracting law are common to all the states, an understanding of these principles and how they differ under civil law is needed for project teams simply to do business in today's global work environment. For a contract to be valid in the United States, it must contain five basic elements: (1) offer, (2) acceptance, (3) consideration/exchange, (4) capacity, and (5) legality. Not all these elements, however, are necessary under civil law.

1. *Offer.* To form an enforceable contract in the United States, there must be an offer, acceptance, and exchange of consideration. An offer is an invitation to make a deal or exchange a promise that is acceptable to another party.

2. *Acceptance.* The other party or its agent cannot withdraw an offer upon acceptance. For a contract to be enforceable, an acceptance must be communicated either orally, by silence, or by taking specific action. Under civil law, acceptance must be the mirror image of the offer, whereas under the U.S. Uniform Commercial Code, acceptance may be merely to the significant terms of the offer. Otherwise, it is considered a counteroffer, which automatically negates the original offer.

3. *Consideration.* Consideration binds the parties who enter into an agreement. Something of monetary value must be exchanged to make the contract enforceable, or a promise to do something or to refrain from doing

something not permitted by law must be made. Consideration, under the Uniform Commercial Code, cannot circumvent something already pledged or given up. The parties also have 72 hours to rescind an offer that has become a contract under mutual agreement. In a civil law jurisdiction, there is no concept of consideration. Formation of a contract requires only offer and acceptance.

4. *Capacity.* Under common law, capacity refers to the legal authority to enter into an agreement and bind a contract. The annual granting of authority through the formal corporate resolution of the board of directors to designated company officers is known as *de jure capacity*. All capacity flows through this document. *De facto capacity*, on the other hand, is not founded on law. It occurs when one party takes action to lead another to believe that the original has the authority to enter into an agreement. An agent acting on behalf of your organization would have de facto capacity to bind the organization, regardless of any agreements to the contrary. Consequently, this is an important concept for Americans to understand when setting up international agreements with oversees agents who will represent them. In parts of Europe and Asia, the names of specific officers of the corporation who are authorized to bind the corporation are kept in official court registries. A request to see the official papers should be a standard condition for continuing a negotiation. These employees, unless registered, cannot bind the corporation. Also note that agents whose employment is dictated by statues are not personally bound to any agreement made on behalf of a disclosed principle. Under common law, should the agent refuse in the negotiation to reveal the organization that he or she represents, he or she becomes personally bound to the agreement. This concept does not exist in jurisdictions with statutory agents, which means that in such negotiations your organization would be legally bound to execute the agreement the agent negotiated.

5. *Legality.* Contracts negotiated in both civil and common law countries share the requirement of legality of purpose. In the United States, a contract may not violate existing laws or statues of a specific state or the regulation of the U.S. Federal Acquisitions Regulations (FARs). Its terms and conditions must also be consistent with the Uniform Commercial Code.

14.4.3 Standard Clauses

As stated in Section 14.3, the formation and drafting of domestic and international agreements is rightfully the domain of the corporate legal department. However, it is in the project's best interest for the team leaders to acquaint themselves with the language used in drafting formal contracts and to become aware of how standard clauses may be interpreted under various legal systems. Contract clauses are extremely important because they serve to protect the issuer in an unintended dispute.

Most companies have a series of boilerplate contracts containing standard clauses that the legal department will review with the project team before drafting one with provisions specific to the project's requirements. Be aware, though, that standard domestic clauses drafted for the U.S. legal system may be misinterpreted outside American cultural settings. There is, however, an international source of guidance on the drafting of international contracts for the purchase and sale of goods and services that the company's legal advisors may consult to develop its own standard contract terms and conditions before tailoring them to match the needs of the project. The U.N. Convention on Contracts for the International Sale of Goods (CISG) has defined a set of rules governing contract formation that represents a compromise of laws between the civil and common law systems prevailing in many countries.

The following synopsis lists many of the clauses with which today's successful international development and production project managers need to be familiar.

Jurisdiction. In U.S. contract law, *jurisdiction* refers to the state and country in which a dispute will be resolved. Although a large company may be indifferent to this clause, it is important for small businesses to specify where lawsuits would be resolved to minimize the expense associated with out-of-state litigation. There may also be regional differences between how statues are interpreted, while certain states may be friendlier to business than others. The differences between state law and the versions of the Uniform Commercial Code that each state has adopted should also be considered when drafting the jurisdiction clause and determining the governing law for the contract. Usually, the choice of law under which the contract is written is also stated. Sometimes the court systems are stipulated as well. In the event that an international dispute arises over the failure to honor an agreement, litigation in a foreign country should be avoided.

Change Orders. Contracts should contain a *change orders clause* because the project will undoubtedly experience changes over the course of its development and production cycles. The change orders clause is used to specify: (1) the process for determining if a change order is warranted; (2) a method for quantifying the financial impact of the changes to the project, such as direct and indirect labor overruns; (3) the approval process specifying who has the authority to agree to, validate, and accept the changes; and (4) the process for formally documenting the changes in an addendum to the agreement.

Confidentially and Nondisclosure. A *confidentially clause* is written into a contract to protect businesses against the disclosure, copying, utilization, and outright theft of its products, trademarks, intellectual property, pricing, customer lists, and business operations. Often *confidentially* clauses are presented as separate *nondisclosure/noncompete agreements* to vendors for signature prior to the exchange of information that has market or strategic value. Generally, these agreements are broadly written.

Since patents, trademarks, and trade secrets outside the security of U.S. intellectual property protection are at greater risk, international contracts must be drafted with specific rules that define confidentiality and who will have access to the information in question. The contract should make it uneconomical for any vendor with whom project leaders are negotiating to benefit from the company's intellectual property. Similarly, if the seller is to retain control of the data rights to a software product, a separate licensee agreement must be negotiated for the software used in creating the product.

Use of Subcontractors. Vendors will often use subcontractors or nonemployees to complete some of the work on large contracts. A *subcontractor clause* should give the buyer the right to refuse their work unless (1) the vendor can certify that his or her subcontractor is in compliance with U.S. federal, state, or local laws, or (2) the subcontractors meet certain criteria negotiated on behalf of the buyer, such as adherence to certified business processes or audible software development methodologies that the Software Engineering Institute or the International Standards Organization (ISO) regularly reviews.

Effective Date of Contract. It is wise to include a clause that specifies exactly when the contract will be *enforceable*. Whereas under common law most contracts become effective when the last signature is obtained or a *condition precedent* is completed, the exact point of formation is not always clear in many international jurisdictions.

Termination. Just as when a contract begins is defined differently in various legal systems, the endpoint is also open to interpretation. In the United States, the *termination clause* provides for cancellation of a project before its term for *default* or *convenience*. *Termination for default* usually occurs when the supplier is unable to perform the schedule or is unwilling to comply with the quality parameters or technical standards specified in the contract. Under common law, if a supplier agrees to a termination for default, the vendor will typically try to correct the cause and seek the right to *remedy* or *cure*. The buyer has the responsibility to verify that the remedy matches the situation. In some instances, default may cause damage that is difficult to calculate. Therefore, the parties may agree in advance to a *liquidation of damages clause* stipulating the amount of damage to be assessed against the breaching party in the event of a termination for default.

Termination for convenience allows both a supplier and buyer to walk away from an agreement for any reason. This clause, often included in U.S. government contacts, allows the agency to terminate an agreement if it can no longer fund a project. However, the party terminating an agreement for convenience is responsible for reasonable costs, which are usually calculated according to a predetermined formula that is referenced in a *termination* or *settlement schedule*. Understandably, termination for convenience is not received well outside the United States. A project might seek the same effect by drafting a *limited liability clause* instead.

Disputes. It is important for the project team leaders to be unambiguously specific as to what events will trigger the termination of a contract, what *closeout* actions (see Section 14.4.5) will be expected, and what rights, if any, will survive the termination. The *termination clause* should also specify a method for adjudicating a *dispute*, the rules that will govern the proceeding, and the location of the adjudication. Few countries are as litigious as the United States. Companies that are involved in international projects will be expected to negotiate a resolution to disputes, even when a partner is flatly breaching a contract. For this reason, international contracts that contain provisions to settle a dispute before initiating litigation are looked upon more favorably.

A *choice of forum clause* can be used to dictate where a dispute will be resolved. Because of the amount of time consumed resolving disputes under the U.S. legal system and the difficulty enforcing U.S. judgments abroad, it may not be advantageous for companies involved internationally to want to hear the dispute in the United States. However, it may be equally disagreeable to have the litigation held in a foreign country. In this situation, *arbitration* would be an attractive alternative for resolving contractual disputes. Neutral parties called *arbiters* determine arbitration. It is generally speedier and less costly than U.S. litigation; it is confidential, and the awards are enforceable internationally. Usually, the courts cannot overturn a decision rendered by arbiters unless the award was rendered fraudulently. In the United States, third-party facilitators can also mediate *disputes* any time the buyer and supplier agree jointly to do so. However, the decision reached may not be binding. All that would be necessary for the ruling to be enforceable under the laws of the United States is for the agreement to include an unequivocal statement that disputes will be submitted to mandatory, binding arbitration.

Warranties. Contracts for the sale of goods or services generally include a statement by the seller that documents or represents the quality of goods and/or services for sale. This is called a *warranty clause*. It is extremely important for project leaders to have the procurement organization negotiate warranties and identify the implications—financial and legal—to the organization. An *expressed warranty clause* states specifically the performance parameters (i.e., 3 years/36,000 miles) under which the product or service will perform. An *implied warranty clause* is less specific and represents only what the expected performance may be. When a supplier makes an implied warranty about what a product can or cannot be used for, it is called a *warranty of fitness for a particular purpose*. If a statement is made concerning how it will perform under specific circumstances, it is called a *warranty of merchantability*. A supplier uses the warranty clause to limit the liability of the firm. Overseas, however, many countries have statutory warranties that cannot be waived, and a supplier may find that certain limitations may not be permissible under local law. Also, warranties may be assumed to be transferable or assignable to someone else if not stated in the United States. Under civil law, this must be stated. Procurement should also assure that any remedies for a breach are specified, are adequate, and are not exclusive.

Limitation of Liability and Indemnification. In U.S. contracting law, when the term *hold harmless* is written into an agreement's *indemnification clause*, it ensures against liability for possible loss, damage, or hurt from gross negligence or willful misconduct. However, when used in international contracts, the term does not have the meaning applied in U.S. law. If specific relief or compensation is sought for damages, loss, or bodily harm, this amount should be built into the price of the contract.

Notices. The *notices clause* specifies how communications between the buyer and the supplier will take place. This includes (1) the vehicle for communications (which may be by registered letter), (2) the required time length (e.g., 10 or 30 days), and (3) who specifically can receive the communications (names of the U.S. federal contracting officer, company technical representative, purchasing officer, etc.). We recommend keeping a log and maintaining the contract notices in the project archive. Remember to choose an appropriate form for communications when negotiating international contracts; mail may be the slowest, and outside the United States, certified mail loses its meaning.

Payment. The *payment clause* establishes (1) the method (cashier's check, wire transfer, letter of credit), (2) the timing (advanced and/or progress payments), and (3) the frequency (weekly, monthly, etc.) of the payments received for the exchange of goods or services. In international contracts payment clauses must specify the denominations of the currencies. A *price adjustment clause* or *maintenance of value clause* may also be needed to adjust the price based on fluctuations in exchange rates or other economic factors. Restrictions on the company's ability to repatriate payments or profits and newly imposed taxes and fees that also affect the value of the payments may require that the contract also have an *escape clause*.

To finance large international purchases, the seller will often open a documentary *letter of credit* with one or more banks. To receive payment the supplier presents to the bank the letter of credit along with certain documents. The bank acts as an intermediary and establishes a separate agreement with the buyer. To protect itself and both parties, the bank honors the letter of credit only when all the documentation conforms to its terms and conditions. Abroad, the International Chamber of Commerce's Uniform Customs and Practices for Documentary Credit governs international letters of credit. The creditworthiness of the contract will usually depend on the financial integrity of the issuing bank and the quality of the drafting of the letter.

Authority. The *authority clause* establishes who has the capacity to bind a company to the contract. Outside the United States it is standard procedure for the person purporting to bind a company to be asked to establish his or her authority to do so through independent means. Requesting authorization papers before the contract is finalized would do this.

14.4.4 Types of Contracts and Pricing Arrangements

Besides understanding the significance of the terms and conditions that appear in project contracts, it is equally important for project team leaders to understand how to choose a pricing arrangement that matches specific project situations. The type of contract chosen for the project and its pricing arrangement depends primarily on the resources that the team has available to perform the tasks outlined in the statement of work and the risk associated with the work's completion. A project with scarce internal resources that is also fairly important to the company's year-end business plans is more risky than one where resources can readily be assigned or has a fairly long lead time before the planned market introduction. Estimating the cost of procuring the needed resources and the parameters associated with making the completion estimate overly pessimistic or too optimistic to allow a vendor a reasonable profit is information that will determine the price of the vendor's contract.

The vendor will want to make as high a profit as the market will bear. The strategies and negotiation tactics need to align with the desired outcome and include the type of contract and pricing arrangements preferred by the buyer. For instance, to minimize the risk of agreeing to a higher-than-necessary price to cover the supplier's cost plus a reasonable profit, a risk-avoiding buyer will want to begin negotiating the cost of the work from the more optimistic range of possibilities. The risk-avoiding supplier, on the other hand, will want to avoid agreeing to a price that does not cover actual performance and push the price toward the more pessimistic range to cover all possible costs. Typically, actual costs exceed estimated costs, and neither party will want to assume the responsibility for the overruns. The vendor's and buyer's determination to minimize this expense is ultimately an important consideration in the type of contract written.

Table 14.1 illustrates the relationship of the three fundamental types of contracts in the marketplace. As we have just explained, contract type is generally set in the commercial marketplace by the economic law of supply and demand and is typically negotiated. In government acquisitions, the procurement method used by the agency's contracting officer prescribes the contract type. In the United States, federal agency contracts are held to a requirement for *full and open competition*. Therefore, as with private enterprise, most government procurement is competitive. It also uses the

TABLE 14.1 TYPES OF CONTRACTS AND RISK

Contract Type	High Risk to Buyer	Low Risk to Buyer	High Risk to Seller	Low Risk to Seller
Cost-reimbursement	×			×
Time and materials	×			×
Fixed-price		×	×	

procurement methods of competitive proposals and sealed bidding. However, noncompetitive contracts can be issued for urgent agency needs that compromise national security by securing a waiver of the requirement for full and open competition.

Fixed-Price (FP) Contracts. *Fixed-priced (FP) contracts* are the simplest and most common form of standard business contracts that a project can negotiate. Because they place the responsibility for performance and financial risk on the supplier, buyers usually prefer them. Fixed-price contracts also offer the contracted vendor greater opportunity to secure a substantial profit. Generally, there are four types of fixed-price contracts: (1) firm fixed-price (FFP), (2) fixed-price with economic price adjustment (FP/EPA), (3) fixed-price incentive (FPI), and (4) fixed-price redeterminable (FPR).

In a *firm fixed-price* (FFP) contract, the vendor agrees to render a service or sell a specified quantity of goods either for a lump-sum payment or for a fixed unit price. Because the price is fixed and not subject to change based on the vendor's actual expenses, the supplier sets the price based on previous experience. This type of contract is more appropriate in commercial transactions where the costs associated with the contract are generally within acceptable marketplace limits.

In situations where there are limiting factors beyond the supplier's immediate control, an *economic price adjustment* (EPA) clause may be added to the fixed-price contract to provide for the upward and downward adjustments of the price based on specified contingencies. The price adjustment in this arrangement, however, is not firm fixed-price. Since the vendor's actual performance determines the amount of the adjustment that the project will allow, the EPA clause protects a supplier from the risk of certain increases in variable costs.

If a project chooses to use a *fixed-price incentive* contract, both the buyer and vendor will need to negotiate (1) a target cost, (2) a target profit that is equal to the target price, and (3) a formula for sharing cost over- and underruns. When performance on the contract is complete, its cost is calculated. To determine if the vendor is entitled to a profit, the negotiated formula is applied to any underrun and overrun. In general, a project may want to use a fixed-price incentive contract for the initial production of complex technology. However, both the buyer's and supplier's lack of previous experience will often make it difficult for the parties to agree on the costs associated with the production.

A *fixed-price redeterminable* (FPR) contract is used infrequently in the commercial section and never in government procurement. A FPR contract provides for the contract price to be redetermined at various times.

Cost Reimbursement Contracts. Cost reimbursement (CR) contracts are also known as *cost-plus contracts* and require that the project leaders include in them (1) estimates of project cost, (2) provisions for reimbursing the supplier's expenses, and (3) provisions for paying a fee as profit. Cost reimbursement contracts are more favorable to the seller than to the buyer in that the buyer agrees to accept the risk of having to reimburse according to an acceptable allocation the seller's costs that may occur while the agreement is in force. This type of contract is used primarily in high-technology R&D projects where there is high cost uncertainty, no definitive specifications, or the scope of work is vague. They are also preferred in the international market, where the risks of an unstable political environment make cost uncertain.

Cost reimbursement contracts are categorized as follows (1) cost plus fixed fee (CPFF), (2) cost plus incentive fee (CPIF), (3) cost plus award fee (CPAF), (4) cost contract with no profit, and (5) cost sharing. The *cost plus fixed fee* (CPFF) *contract* is the most common type of CR contract. The supplier is reimbursed for all allowable costs and is paid a fixed fee that does not change in response to the cost overages or underruns that the project may actually experience. CPFF contracts virtually insulate the vendor from all financial risks while guaranteeing receipt of a fee. As a result of both the performance and the fee being fixed, the supplier may have little incentive to perform in a cost-effective way.

In contrast, a *cost plus incentive fee* (CPIF) *contract* provides a supplier with incentives to come in under cost. In this type of contract both parties negotiate a target cost, a target profit, a price ceiling, and an adjustment formula for establishing the profit that will be paid. Riskier that the fixed-price incentive contract, there is no guarantee that the supplier will be paid an actual profit; in addition, the supplier must pay the actual costs that exceed the ceiling. If costs are uncertain when project leaders negotiate the targets in these contracts, they should specify an option to set successive targets based on performance experience later in the project.

A *cost plus award fee* (CPAF) *contract* is a variation of the CPIF contract and is used by projects to provide incentives based on subjective measures when precise measures of cost and technical performance cannot be obtained. Even though there is more administrative overhead associated with this type of contract for the buyer, who must document how the supplier's performance is evaluated, the advantages of having the ability to determine whether the supplier has earned his or her fee are very satisfying. Projects use cost plus award fee contracts primarily to secure services after a long-term relationship between the buyer and the supplier has been established. They are frequently used to purchase software-engineering resources for systems maintenance and for the design and fabrication of microelectronics hardware as well as construction services. The U.S. government developed cost plus award fee contracts to pay the Army Corps of Engineers during the 1930s and has used this type of contract to procure services for the U.S. National Aeronautics and Space Administration.

A *cost contract with no profit* is utilized by nonprofit institutions.

A project would utilize a *cost sharing contract* to formalize a joint research alliance where both the seller and buyer are looking to share the benefits of a research or development effort. Cost would be allocated according to a prorated formula or ratio. Each party assumes some costs and forgoes profit.

Time and Materials and Unit Price Contracts. *Time and materials (T&M) contracts* require project leaders to negotiate hourly rates for specified labor and to obtain agreement on the cost of all parts and materials. This type of contract is frequently used to procure equipment maintenance and other support services, particularly when the time estimate to complete the setup, repair, or overhaul of the equipment is uncertain. The buyer receives a bill for the number of labor hours at the agreed-on hourly rate and for the cost of materials and parts.

Be careful when negotiating T&M contracts because each hourly rate includes a component for overhead costs, which include both *fixed* and *variable rate costs.* Although variable rate costs can be calculated rather easily by estimating how many hours will be performed during the contract period, a share of the fixed costs must be allocated to each hour. Suppliers often do not recover all their fixed costs, and the buyer benefits. Less frequently the hours are underestimated, and the supplier experiences a windfall profit. A technique that may be used to avoid this problem is to include in the contract a step-up discount plan that increases as the number of hours billed by the vendor increases.

A *unit price contract* is an arrangement by which the supplier is paid based on units of measurable output. A base floor and ceiling can be set and adjustments made to reflect price changes in the marketplace. These types of contracts are advantageous to both the seller and buyer because they are based on measurable costs; however, projects must establish fair rates and prices for the costs.

14.4.5 Closure

As a project draws to an end, the team will be involved in the closure with the centralized procurement organization of the many contracts negotiated on its behalf. Closure is primarily an administrative function. However, if either party has exercised its rights to discontinue or terminate the contractual performance either completely or partially, remedies due to the other party's omission or failure may be sought. Consult your organization's attorney immediately. Do not put your project and company at risk by attempting to understand the applicable law(s) or make a judgment under the terms of the contracts yourself!

Normally, closure will proceed smoothly. The supplier's payment is usually contingent upon receipt and acceptance of all deliverables by the delegate project representative(s). The buyer records the exchange of contract closeout documentation in a project log, and the data files and physical materials are transferred to a storage facility or archive for retention. The procurement organization will also issue a request for a signed statement from the supplier indicating that all contract terms and conditions have been met.

We also encourage project managers and team leaders to hold a project retrospective to record and capture both the positive and less positive lessons learned by the team members. The focus of this closure activity is to preserve the project's "learnings" and to share the best practices with other project teams so that potential problems and risks can be mitigated in the future.

14.5 CONCLUSIONS

In this era of intense globalization, more and more project teams will be required to outsource nonessential project activities and to procure the least-cost components from suppliers who are located around the globe. Understanding how to work with centralized procurement organizations and the company's legal advisors is rapidly becoming an essential project management skill that team leaders and members alike must cultivate. Managing contractual negotiations and arranging the pricing associated with materials and resources for the project is a team-based activity involving many different specialists from across the larger project organization.

BIBLIOGRAPHY

Arnavas, D. P., and W. J. Ruberry, *Government Contract Guidebook*. Washington, DC: Federal Publications, 1994.

Bonnell, M. J., *An International Restatement of Contract Law*. Irvington, NY: Transnational Juris Publications, 1994.

Garrett, G., *World-Class Contracting*. Arlington, VA: ESI International, 1997.

Mukulski, F. A., *Managing Your Vendors: The Business of Buying Technology*. Upper Saddle River, NJ: Prentice Hall, 1993.

EXERCISES

14.1. There are six processes involved in project procurement management. List each of them, and describe in detail the process activities that involve the project management team.

14.2. Explain the purchasing department's role in source qualification and how the project management team may become involved in selecting and qualifying vendors to appear on the purchasing department's preferred vendor list.

14.3. Name at least four advantages to the project organization of centralized contracting.

14.4. What is the process called when prior to awarding a contract, the buyer meets with each potential seller after receipt of his or her proposals?

14.5. For a contract to be valid in the United States it must contain five basic elements. Which of the basic elements is not necessary in a contract formulated under civil law?

14.6. Describe why cost-reimbursement contracts are more favorable to the seller than to the buyer.

14.7. In fixed-price and incentive-type contracts, is the responsibility for performance and the financial risks associated with nonperformance or delay placed on the buyer or the seller?

14.8. The purpose of contract administration is to ensure that both parties who negotiated the contract perform in accordance with the contract's terms and conditions. Explain why establishing and implementing a contract change control system is an important project team activity for effective contract administration.

14.9. What precautions should the procurement organization take on behalf of project team members who during the course of establishing a contract relationship with a vendor intend to share important proprietary data?

14.10. Explain how a project may submit a contract dispute to arbitration even though the contract has no mandatory binding arbitration clause.

14.11. Explain why the contracting party who wants the right to terminate an international agreement without the other party's being in default should include in the agreement both a voluntary termination and a limited liability clause.

14.12. A buyer has negotiated a fixed-price incentive contract with the seller. The contract has a target cost of $500,000 a target profit of $75,000, and a target price of $575,000. The seller's actual expenses are $450,000. How must profit would the buyer owe the seller if the negotiated ceiling price was $600,000 and the share ratio was 70:30?

15

PROJECT MANAGEMENT IN THE INFORMATION AGE

The Future has a way of arriving unannounced.

GEORGE WILL

15.1 INTRODUCTION

There are two great events of the twentieth century that will influence how projects are managed today and well into the twenty-first century. The first is the advent in the early 1980s of the personal computer (PC). This tool brought to the desktop the processing power of the first room-sized computer, which engineers at the University of Pennsylvania unveiled for the U.S. government in 1946—a behemoth that was made up of 30 separate units, weighed 30 tons, occupied 1800 square feet, and had 17,468 vacuum tubes. The second came a decade later when a British computer scientist envisioned the World Wide Web at the European Laboratory for Particle Physics in Geneva and created a program that has become today's familiar Web browser. Both these events, along with the technology that later followed, have had a profound influence on how project teams are able to communicate and share project information across multiple sites spanning the globe.

Quietly, the innovations of the past two decades are helping to fuel a revolution in how today's distributed workforce can collaborate and work together as a team. In the twenty-first century, information technology will become more strategic. It will permit the virtual project team to access

512

groupware applications, multimedia technology, and enterprise-wide project management software to manage more collaboratively and proficiently, making it easier to achieve project goals and hence the organization's objectives.

As we have seen throughout this book, the computational complexity required to analyze progress and produce scheduling and tracking reports makes manual computation very difficult. The advantages of project management software that permits, for example, individual project specialists to upload a personal time line of key deliverables for inclusion into the overall project baseline plan are all too numerous. The demand for increased control of project costs also requires that the project's management software be efficiently linked to the enterprise's cost management systems, making such pertinent cost information accessible to the management team when needed.

Similar tools that save time and space as well as maximize human interaction and permit simultaneous real-time access to information are becoming more and more prevalent. Such tools help teams to maintain control of the decision-making process and archive important documentation. In this chapter we explore some of the key factors influencing today's project management team in its choices concerning the latest information technology. The chapter illustrates how these innovative software tools and technology further promote greater collaboration and teamwork and serve as a conduit of programs that teams may use to plan, schedule, and control the costs, risk, and quality of their projects.

15.2 COMMUNICATING ACROSS THE ORGANIZATION

15.2.1 Conferencing and Collaboration Technology

Every project in our networked era can benefit from using *conferencing and collaboration tools* that enable distance-bound project teams to participate in real-time meetings electronically. As discussed in Chapter 11, these new tools are essential for helping teams to collaborate more effectively when distance precludes face-to-face decision making. The plethora of documentation that project teams use to manage both the business and technical aspects of a project must be accessible, cost-efficiently stored, and centrally available. Today's information-age projects may choose from among any number of *groupware applications* to satisfy these requirements. The convergence of electronic messaging technology with these software systems has also made the storage and retrieval of project information more efficient.

Team members are now able to store and extract their plans, schedules, and meeting notes from a central, customized, and project-exclusive database.

Unfortunately, implementing this new technology to augment the team's ability to communicate in a timely and productive way may not be easy. If members are to proceed comfortably with the applications, project leaders must set achievable expectations for them and explain how to use to their advantage both the technology and the tools described in subsequent sections of this chapter. All of the technology has its place in the project environment and the organization's computing infrastructure. To make best use of these applications, it may be beneficial to invite specialists from the corporate telecommunications or information systems (IS) department to assist in facilitating the setup and teardown of the team's first videoconference or electronic meeting. It may also be necessary to ask these specialists to teach the team how to access the groupware applications on their computers. The team's comfort and success with these new tools is very dependent on each member's acceptance and eventual familiarity with the new technology.

Electronic Meeting and Data Conferencing Systems. Project teams whose members are remotely located in virtual offices or in geographically distributed sites are finding simple collaborative technologies, such as an audio-teleconference and a shared computer screen, helpful for reducing overall product development cycle times and increasing productivity. Because what is discussed at project meetings often does not require that the team members be seen, this type of conferencing system is more easily affordable than other electronic meeting technology. Until recently, however, data conferencing software available to project teams over corporate local area networks (LANs) added voice traffic onto an often heavily utilized corporate data infrastructure. The frustration caused by the congestion all too frequently prevented the team from hosting successful electronic meetings to get the message across. The alternative often selected was the simple audioconference.

Today, however, software is available for electronic data conferencing that connects team members without adding traffic to the corporate data network. Remote team members are linked to a meeting host via the corporate intranet or extranet and the telephone network. Since no reservations are required, meetings can be spontaneous, a necessity in today's busy project environments. Electronic meeting and data conferencing software typically feature a shared screen or *white board* where a document, diagram, or picture can be viewed by the entire team. A *chat window* is left open for participants to provide real-time comments. Usually, the software applications require that the conference participants select a host to facilitate or moderate the meeting. In the best systems, the host is able to leave at any time, so his or her presence will not affect the overall conference.

Desktop Real-Time Video- and Audioconferencing. Digital computing and its tight integration of real-time video and audio allow geographically dispersed project teams to conduct two-way video- and audioconferences from their desktops, thereby saving participants the time and expense of having to travel to a common meeting site. Were a project organization to undertake a cost–benefits analysis for deploying this multimedia technology, the cost may prove to be very affordable, with the benefits far outweighing current transmission limitations.

Two-way, real-time video- and audioconferencing also promotes more collaborative work and group productivity than audio alone. To really experience these benefits, it is important for an organization to scale up its infrastructure and make the technology available to all its projects. This new ability to make decisions collectively helps to resolve and bring closure to several project issues. All too often, project issues and concerns remain unresolved for many business days, until a face-to-face meeting is held to understand what is impeding progress!

While Internet Protocol (IP)–based videoconferencing and collaborative technologies have already emerged as a standard configuration for the desktop computer, sometimes the images transmitted over this technology are blurry and jerky. Other times, synchronization of the audio and video is not perfect. Dissatisfied with the current quality of the transmission, many virtual teams will continue hosting their project meetings in video-conference rooms where rollout units can be connected to dedicated-telephony networks and programmed to handle multiple sites at various transmission speeds.

In an effort to bring the best of the new digital technology into their videoconference rooms, many organizations are purchasing *Web presentation units*. This is a form of technology that permits more interactive team-member participation in project meetings. With only a few minutes of preparation, team members can forward over the Internet the materials that they will be using in the meeting to the room's Web presentation unit. When hooked up to an *LCD projector* and the Internet, the unit is used to share documents with remote meeting sites. Mobile team members, who join the meeting either from home or from a virtual office, are able to view the documents by addressing the unit's home page. The unit automatically streams the pages live to the remote participants' desktop or laptop computers, where they are displayed just as they are in the conference room. As remote team members join the videoconference, they are provided with separate passwords giving them access to the unit's tool bar, so they too can take control of the meeting materials. Any of the documents shown in the meeting of interest to team members when they physically return to their offices can be accessed on their own PCs and printed out as hard copy.

15.2.2 Groupware Applications

The term *groupware* was first coined by social scientists in academia and industry, who worked to develop computer systems to increase work group productivity. The earliest groupware systems provided an interface to two or more users who were engaged in a common task or goal. Although there may be some confusion caused by how the computer press currently uses the term to describe the electronic technologies that support person-to-person collaboration via electronic multiuser databases and electronic mail, the recent and rapid advancement of this technology has simply extended the original meaning assigned to the term.

As discussed in Chapter 11, groupware applications allow virtual teams to manage document-centric project activity more productively. These tools enable projects to share information more efficiently in today's highly distributed workplace while creating a more collaborative team environment. The convergence of these applications with messaging and the World Wide Web technology, however, has made this an infrastructure that even collocated project teams have embraced.

The many groupware applications in the marketplace share five basic building blocks: messaging, calendars, discussion database, document management, and electronic forms. When project teams evaluate these applications, they quickly realize that the various vendors, depending on their technical prowess, emphasize different combinations of these five components.

1. *Messaging* is the underlying logical infrastructure used to manage the communications between the users of the groupware application. Group messaging allows team members to keep in touch by querying a stored directory of addresses and routing e-mail messages intelligently to these team members. A product that can't perform well at handling messaging will probably fail at effective collaboration. Many of the applications permit team members to filter mail messages and then select what to do when these messages arrive — copy them, for example, to a folder; delete them; or change the message's importance.

2. *Calendars* assist the project teams in dynamically scheduling status meetings, technical reviews, and other project functions. Time is an essential element of any organization's productivity management plans. The integration of this groupware component with publishing tools that can be accessed over the World Wide Web are very valuable to the time-restricted project team leader who cannot personally check every team members availability before posting an agenda for the ad hoc meeting that has to be scheduled. The calendar also lets team members control and manage their own personal "to do" lists.

3. *Discussion databases* or *bulletin boards* are the groupware forums that house the information needed by team members to manage the project. Information is not sent to any particular person, but rather, is deposited into a common repository that all team members can access with just a mouse click. In general, these databases are not relational and the information stored in them is unstructured. Theoretically, the project team can populate the database with documents that store just plain text, numerical data and graphics, audio clips, or even, full motion video, depending on the project's needs. The databases are expandable into multiple views and can be sorted typically by date order or message category. Initially, a pioneering technology known as *replication* created copies or replicas of the databases and kept them synchronized even though the computers on which they resided were not connected. This was of particular importance to mobile team members who logged onto different servers to access the databases while disconnected from their data network.

4. *Document management* provides a logical way to organize documents for retrieval. Libraries are created for team members to access, track, and provide version control of their documents. Document management has recently expanded to include *knowledge management*. Now that groupware providers have adapted their applications to the open standards of the Web browser technology, they offer an interface to the World Wide Web within their applications. This permits users to subscribe to accessible databases that they can bookmark for reference later. Extractable information can even be filtered and altered when a team member initiates a search that uncovers potentially valuable information to enhance how other team members manage project knowledge.

5. *Electronic forms* are reusable software modules that the team can use to input, display, output, or manage variable project information such as spreadsheets, charts, and drawings. A team can drop a spreadsheet into a mail message or create elaborate customized forms by lacing the features of

the different form modules together. There is even the capacity to generate electronic signatures for managing document sign-offs and approvals.

Before a project team selects any specific vendor's groupware application, consult with representatives from the corporate information systems department. The IS department must evaluate how the application will scale across the computing platforms, without sacrificing functionality, and ensure that the application will be secure and easy to administer. Some of the issues that may be addressed are desktop orientation, integration, application development effort, database strategy, current infrastructure, and the product's features and functions. It is also important to consider how seriously providers have invested in the convergence of the newest Internet technology.

Without a doubt, by adapting document-centric, knowledge-based groupware applications, project team members, both virtual or colocated, will improve their productivity, work more collaboratively, and enhance overall group morale. The return on the investment for this technology is quite high, when teams are taught how to use the applications. Organizations that make good use of the technology will also see the results where it counts, on the bottom line.

15.2.3 Using a Home Page

Information technology is truly at the center of nearly every aspect of project communication and information sharing. Sometimes project organizations that are highly technical will choose to create Web sites for their projects. These may be available over a corporate intranet to the larger, cross-functional team. These sites often are used in tandem with the broader groupware applications that provide point-and-click access to the team's *home page*.

Business rules are changing today in a way that requires business teams to become Net-savvy in order to survive. Building an on-line project Web site is an excellent way for these teams to become familiar with this technology. A project home page can be used to share key information that in the past was rarely available to the entire team. Sharing documents such as product announcements, launch schedules, market research, marketing collateral, press releases, event publications, or comments from product trials with customers would be of extreme value to the project team managing the product introduction as well as the development or engineering team building the next release of the product.

Most corporate IS organizations set the standards for intranet sites with the domain of home page creation rapidly becoming a haven for the graphic design departments whose artists and writers often are available to consult on the design of a project's site. However, teams in small and medium-sized businesses without access to a dedicated graphic arts department needn't be tech-savvy or have prior Web-building experience to create their own successful project home page. Freelance artists and programmers who understand this technology are always available for hire. Although, today, commercial programs can be used to guide a team through the process of setting up a home page, including prompts to enter text and menus to direct the uploading of images and other content. Even though the software packages are intuitive, the best way to make this experience positive and productive is for a project administrator to attend any number of classes that teach the ins and outs of the programs and tips concerning how to connect the page to the Web browser technology.

15.3   TOOLS FOR ENTERPRISE-WIDE PROJECT MANAGEMENT

15.3.1   Project Estimation Tools

Project teams who bring products and services to market in the Information Age will be required to improve on their ability to estimate the number of resources and time required for application development. Whether the project is small or larger, it will be insufficient to continue estimating software development schedules based on programmers' hopeful guess-timates that may have failed in the past. The inaccuracy and lack of reliability of using such techniques, to determine the workload and release schedule for both development and testing, takes its toll on the entire project team. No one enjoys working on a project that management perceives as always being late!

Given the tight demands of today's marketplace, the benefits of using a *knowledge-based estimation tool* to overcome the frustration of always working on a late software project are readily understood. The lack of any analytical methodology to rationalize the size of the effort necessary to address new requirements can even prohibit a software application from coming to market as planned. Projects estimates based on the programmers' "gut feel" are typically susceptible to the marketing organization's desire to satisfy pent-up demand for new features or functionality in whatever version of the application is about to come to market. Simply being unable to justify any trade-offs for taking on the additional effort makes an already late software project even later.

To help project teams overcome this seemingly traditional controversy, we recommend applying a knowledge-based estimation tool that can be used in tandem with the standard *project management information systems* that we discuss later in the chapter. We define knowledge-based estimation tools as any tool using a credible internal or external knowledge base to develop a detailed task-level plan that considers project size, complexity, classification, languages, and other attributes. The input into these tools is derived from a statistical method called *function point analysis*, which is used to measure the number of lines of code that a project will require. The estimating methodology involves analyzing a project's function points, including the sizing of nearly every aspect of the software that is related to the end-user interfaces, such as data sources and screen functions. For this reason, many noted software professionals consider estimation methodologies that count function points to be an excellent indicator of project scope. (For more detail about the underlying concepts used in the sizing of software development projects using function points, see Chapter 6.)

As project team leaders investigate using these knowledge-based tools to improve their estimating ability, they immediately should anticipate experiencing the following benefits:

· Timesaving over manual methods calculating overall estimates.
· Planning results that are more accurate, which may, in turn, lead to fewer canceled projects.
· Improved results in the marketplace due to (1) reduced time to market, (2) improved end-user or customer satisfaction, (3) improved product quality because adequate resources were assigned to test the product, and (4) lower costs required to provide maintenance or warranty support to customers resulting from improved software reliability and fault tolerances.

The more project teams that use these new tools, the greater the efficiency gains for the organization. Experts estimate that companies can experience efficiency gains as high as 80% from using standard procedures for estimating costs; tracking times, resources assignments, and document changes; and avoiding project cancellations.[1] Project teams who plan better will also experience higher morale, retain employees longer, and save on the inefficiencies and hidden expense of needing to train new employees frequently. Deployment of this class of tool also helps to weed out the

---

[1] C. Douglis, "Cost Benefit Discussion for Knowledge-Based Estimation Tools," *Project Management Journal*, Vol. 29, No. 2, June 1998, pp. 8–10.

riskier projects before they get started, because each team must identify major risk factors and manage them. Few organizations today can absorb the cost of coming to market late due to delays that can't be managed or worse yet, due to catastrophic results that were not anticipated. Like most other industries, the software development community is rapidly approaching the era when cowboy programming and undisciplined project management can no longer be tolerated.

15.3.2 Risk Analysis Tools

Project managers do not need to be experts in decision theory or statistics to bring the advantages of the science of risk analysis to the project environment. A number of excellent risk modeling tools are available today to assist teams in creating a *Monte Carlo simulation* of the activities that represent significant risk to the overall project plan. Because change is inevitable in the project environment and nearly all project activities have some degree of associated uncertainty, the objective of risk analysis tools is to help the project team quantify this uncertainty and interpret the distribution of the probable values of the output.

The concept of risk assessment, discussed in Chapters 5 to 7, first requires project teams to identify specific events as risky and then to quantify their probable impact, for example, on the final cost or end date of the project. Sold as add-ons to the most popular *project management information systems*, these powerful tools are helpful in letting a project team determine all the possible variables the riskiest activities could possibly take on and the likelihood of each value. The mathematical concepts behind the power of the Monte Carlo simulation recalculates a project schedule hundreds, even thousands of times, selecting random values from the distribution functions (i.e., beta PERT) entered into the program. The analysis results in a risk model of a finite number of possible values that tells the team not only what could happen on the project, but *how likely* it is that these events will occur.

The simulation technology used by these analytical tools also makes the team leader's interpretation of project risk events clearer. Each tool graphically presents the range of outcomes that result from the simulation. Most of the tools permit teams to display these results as customized graphs with overlaid titles. In addition, cumulative graphs can be integrated into other word- processing or spreadsheet applications and used to display risk over a range of simulation outputs.

Many of these tools also provide additional *sensitivity* and *scenario analysis* to help teams gain insight about their projects that is not readily apparent without the aid of a computer. Project leaders can use sensitivity analysis to rank individual risk factors in order of their contribution to the overall project risk and to determine, for instance, which factors have the most impact on a project's bottom line, completion date, and so on. *Scenario,* or *"what if" analysis* is used to determine which situations project leaders will want to achieve or which they should avoid. Conducting a scenario analysis lets the teams identify which combinations of variables are most important in causing a given output to occur and can serve as a powerful warning sign, for instance, that the project is about to go over budget.

Learning how to use the more popular risk analysis tools is fairly straightforward. Most programs include on-line tutorials and hot-line support. Many vendors also offer lectures and hands-on seminars designed to teach decision makers how to use the project tools and the underlying mathematical concepts supporting the programs. Non–English language versions of the software are available from some vendors as well.

15.3.3 Project Management Information Systems

How projects are managed today is rapidly becoming an activity that involves not only the corporate IS department but also every member of the project team. With the need to gather information about the status and completion of project activities from individual project specialists, an increasing number of *project management information systems* are introduced today as desktop applications linked to corporatewide business systems. This permits the project team to control a mariad of projects centrally in a multiuser environment.

The shift today from having just a few specialized technicians who use these tools to manage high-tech projects to a wider group of project team specialists who access them from across the entire business will continue well into the new millennium. This trend requires project teams to understand thoroughly the benefits these systems offer their organization to offset the cost that will be incurred to procure and maintain them.

The project management market has an expected annual grow rate of 20% for the foreseeable future,[2] making numerous software packages for every make of personal computer and a proliferation of Web-based, client-server, and mainframe applications available to project managers. The advantages experienced from using these tools far outweigh the tedious

---

[2] K. M. Carrilo, "Is It All a Project?" *Information Week*, No. 670, February 23, 1998, pp. 100–104.

task, for instance, of manually calculating and recalculating an early- or late-finish activity schedule. Here are a few of the reasons why a project team may benefit from using any one of today's computerized project management information systems to process project data.

1. *Time-phase the project baseline plans.* Computerized software can ease the work required to formulate multiple iterations of project baseline plans as well as document the initial planning process for further reference.

2. *Generate the project work and organizational breakdown structures WBS/OBS).* Quick and easy access to project files, whether for consultation, review, or updating, is a necessity. Graphical portrayal of both project structures also proves to be a helpful tool. Often, senior management will want to review the organizational structure proposed for a project and understand where common resources can be drawn from in the event of a shortage. The graphic portrayal of the two structures highlighting the common work packages lets the project management team show senior management where the linkages are.

3. *Compute the project's schedule and early and late finish or end dates.* R&D projects comprised of as few as 10 activities will benefit from a software solution. But on efforts where the number of activities is several hundred, the storage and retrieval by computer of project data leading to these calculations is a necessity. In addition, in the typical R&D multiple-project environment, there is generally a need to create:

- Subnetwork schedules of project activities that roll up into a subsuming hierarchy for processing separately or as an integral subset of the master network.
- Networks that require multiple startup dates.
- Schedules that are computed based on assigned times of different durations (i.e., hours, days, weeks, months, etc.) worked by individual project resources.

4. *Centralize project data collection.* For virtual team members who interface with other project specialists who reside in geographically dispersed locations, centralized maintenance of project schedules and cost reports is essential. This is also true for project personnel who work simultaneously on a number of different small or medium-sized projects. By centralizing project files, project personnel can electronically transfer, between corporate databases, information needed from other MIS systems (i.e., corporate accounting and workflow management systems).

5. *Track and update the project schedule.* Both the master schedule and cost variance reports require frequent and periodic updating to help determine when project work is to be completed and at what cost. Because of their frequency of occurrence, changes cannot be readily incorporated into these reports unless a computerized processing system is available.

6. *Search, sort, and display project data.* Searching for and sorting project data constitute primary tasks the project manager is concerned with in order to generate timely and meaningful project reports. Because project control is accomplished by analyzing a variety of plans and reports, the output capacity of computerized project management systems helps to direct this management activity. Production of graphic reports of the project plan is another valuable tool provided by computerized programs.

7. *Forecast and manage project resources.* Most medium- to large-scale R&D projects have finite but defined resources (personnel, equipment, materials, etc.) that must be scheduled and tracked against original plans and budgets. The capacity to code resources assigned to specific activities and tasks facilitates their allocation to a variety of projects. Another important requirement is the ability to smooth out or level resources used over time. Unpredicted constraints may therefore be accounted for by simple automatic adjustment of the project schedule.

8. *Integrate, at the project manager's fingertips, a number of sophisticated techniques for advancing the flow of project information across all the organizations that contribute to the realization of the project effort, such as business planning, marketing, sales, manufacturing, and customer support.* Combining on office workstations, for example, project planning and scheduling systems with computer-aided engineering (CAE) and materials requirement planning (MRP) software would satisfy engineering's simultaneous need to monitor design changes as well as to track the status of parts and materials on the factory floor while scheduling shipments of completed systems.

15.4 MANAGING EXPECTATIONS

Most project management software packages offer managers and project teams flexible control of project plans, schedules, costs, and resources. In fact, considering the many choices available today requires that the project's various management requirements be considered carefully before final selection and implementation of the program chosen. Most important, it is the user's needs that dictate what software is selected, not the reverse.

15.4.1 Understanding System Limitations

As a preliminary step in actually selecting a desktop, server-based, or mainframe project management information system, one should analyze senior management's exact expectations and detail them in writing. In this way, the features, capacity, cost, and limitations of the solution considered will be readily available for everyone's review. The analysis should consider the following list of questions. These questions will help team members who are investigating the various system choices to concentrate on several key factors that must be considered when choosing not only a project management information system, but also the hardware infrastructure on which it will run.

- What features are offered that will enable the project team to plan and to control project activity? How do the programs handle the creation of project plans, resource allocation, the optimal use of scheduled activities, and the comparison of actual versus planned progress?
- What constraints does the system impose on the size of the project network?
- What standard presentation graphics are included in the program? Do they illustrate the full integration of schedule, resource, and cost information as well as denote the dependency relationships among project activities.
- What level of quality is offered for presentation graphics? Can network (precedence) diagrams be generated directly without the need for additional programming, specialized sorting, or coding of project information? Are there provisions for add-on tools that present the network by activity names and WBS/OBS codes for specific intervals? Can the graphics be easily edited for presentation to senior management without converting the output to other graphic packages?
- Does the system generate easily accessible, on-line printed or plotted planning charts? Can professional planners and project team members obtain a plan baseline without having to produce presentation-quality charts?
- Are project team members allowed the flexibility of working interactively with the system to manipulate project data while ensuring the integrity of the project's plan?
- How difficult is the software to learn and to use for both novice and experienced project personnel? Does the vendor's staff need to be retained as a support service or to train other project personnel? Is an on-line tutorial included in the purchase price?

- Is the program's documentation straightforward? Is the graphical user interface intuitive? Is the menu structure easy to operate? Can user-specific data-entry screens be tailored to match the different requirements of a variety of projects? Are command-driven sequences and program macros available for flexible data entry as program users become more sophisticated? Are experienced programmers given access to the system through standard developer tool kits?

- What are the initial installation charges? How much capital outlay is required? Is specialized equipment required? Are there hidden charges for continuous vendor troubleshooting or hot-line support? Is user documentation and vendor-conducted training included in the price of the system?

- How frequently does the vendor update the product? Are product enhancements for new system releases or documentation updates supplied for a nominal fee?

- Are a wide variety of drivers available for the printers and plotters supported by the system proposed? Can these peripheral devices be located at sites remote from the computer?

- Does the system under investigation support an open architecture for horizontal integration of data from corporate CAE systems, on-line inventory control, or order-entry systems?

- How easily can data from the project management system be exchanged with other corporate MIS systems, publication software, spreadsheets, or ODBC data sources?

- Can documents, spreadsheets, graphics, or video be attached to reports or graphics generated in the project management system?

- Does the system architecture support two-way exchange of project information over standard mail systems? Is there a system interface to a simple Web browser? Do users have access to a Web wizard for Internet/intranet publishing of reports and graphics?

- Is the software flexible enough to accommodate demands for additional processing capacity and increased sophistication in the types of reports requested or the number of projects tracked? Can today's system grow to handle tomorrow's activities?

The next step in this analysis is to consider the availability and distribution of the computing infrastructure, including mobile accessibility for remote users. We recommend working with the resources available in your organization's IS department to determine the proper configuration. This will facilitate exchanges of files between different operating systems over interoperable communication networks that link the desktop computers with one another as well as the more powerful servers.

The tasks associated with system administration also need to be account-ed for and assigned to either project personnel or an IS support organi-zation. A project may also choose to engage the software vendor to set up the initial configuration of the more complex server-based project manage-ment software applications. This would ensure that (1) only members of the project management office have access to project files, and (2) the software system is installed specifically for data exchange between existing management systems in operation in the organization.

Another important consideration is to determine whether the potential vendor will service the project management system once it is installed, or make available skilled programmers and analysts to extend, tailor, and further develop application programs at its customer's request. It is also important to consider whether the vendor supplies remote diagnostic assistance or on-premises maintenance and repair of either the software system or any turnkey devices.

15.4.2 Installing a Solution That's Right for Your Project

Table 15.1 illustrates a number of the features that project teams will want to have in today's tools, however fundamental knowledge of the concepts of project planning, scheduling, and control is an important prerequisite for satisfactory selection. The advantages offered by one system over another are sometimes subtle. Therefore, the manager or project specialist embark-ing on a comparative shopping review can best judge the benefits provided by a prospective system when he or she understands the tasks the software must perform and has a determined preference for how they ought to be performed. Also, it is important to have realistic expectations because not all the software available commercially is sufficiently sophisticated to meet every manager's requirements or even support all the analytical techniques espoused in this book. Experienced R&D managers report very real limitations despite the fact that the features of many desktop systems have improved dramatically in recent years.

Some commonly experienced problems with commercial systems are as follows:

- Data entry
  - On screen production and manipulation of a large project network with smaller systems can be quite cumbersome.
  - Inability to assign user-specified activity codes in a system of any size makes tracking similar activities for different WBS levels diffi-cult.

TABLE 15.1 IMPORTANT ATTRIBUTES AND FEATURES TO LOOK FOR IN PROJECT MANAGEMENT SOFTWARE SYSTEMS

Data Entry	User Interface	Data Manipulation/Communication	Scheduling Techniques
1. Point and click interface	1. Intuitive graphical user interface	1. Automatic retrieval and display of historical project data when updating existing data	1. Critical path (CPM) scheduling with availability of the following:
2. Multiple dialogue boxes, pull-down menus, and screen buttons for easy data input	2. On-line help facilities, interactive tutorials, wizards, and user training that expedite learning	2. Ability to scroll over the full-screen network diagram and pick up and move individual nodes	• AOA and AON linear and variable time-scaled networks
3. Standard tool bars	3. Complete error checking and data validation		• PERT networks with milestone notations
4. Full drag and drop between the system and other applications using the same operating system	4. Copy commands that let users save project data for later reuse or move portions of the project elsewhere	3. Ability to zoom into sections of the network and graphical displays	• As-soon-as-possible and as-late-as-possible activity-based schedules
5. Full-screen forms that users can customize	5. Provision to display subsequent levels of detail in the project plan, WBS, or activity sequences	4. Ability to display networks across many pages or compress onto one page	• Early start/early finish and late start/late finish activity reports
	6. Allowance for multiple users to update, analyze, and report concurrently on portions of a project	5. Capacity to draw tree diagrams which can be used to display work and organizational breakdown structures	2. Multiple-workweek calendars and maximum calendar lengths for projects spanning several years
		6. Ability to preview reports and plots on the computer screen prior to printing or plotting	3. User control over scheduling algorithms with task and resource prioritization
		7. Ability to save reports to disk and in standard communications protocols readable by other systems	4. Flexible coding structures that allow users to code project tasks by WBS/OBS numbers or alphanumeric descriptors
		8. Project import/export to other project management systems, publication software, spreadsheets, etc.	5. Progress reporting and full summarization of scheduled activities by:
			• User-defined status dates
			• Actual start

Resource Management

1. Ability to schedule multiple projects from a common resource pool and to monitor changes in resource availability on project completion dates and costs
2. Ability to level resources by resource availability and and override constraints imposed by time-limited activities

Cost and Performance Measurement

1. Automatic conversion of resource usage to cost
2. Full integration of resources and cost with schedule
3. Cash-flow analysis
4. Detailed budgets by activity and resources
5. Reporting by:
   • Actual costs to date (ACWP)
   • Estimate at completion (EAC)

9. Ability to retrieve/export data to and from ODBC data sources for manipulation and reporting
10. Ability to exchange data from other corporate MIS/business systems
11. Web publishing wizard or HTML page and template publishing
12. E-mail integration wizard that embeds e-mail addresses of project specialists into the system for easy exchange of project information

Graphics and Displays

1. Availability of the following:
   • Work breakdown structure
   • Network diagrams
   • XY charts
   • Bar (Gantt) charts
   • Logic diagrams
   • Resource histograms
   • Pie charts
2. Ability to display multiple curves, histograms, and plot points on the same graph to

• Actual finish
• Duration complete
• Duration remaining
• Percent complete
6. Multiple-level project processing and provisions for linking project activities for multiple-project reporting
7. Resource pooling feature that allows resources to be shared across multiple projects

Other Features

1. Password protection for project and resource files
2. General and context-specific help facility; easily accessible and informative error messages
3. Vendor support and maintenance agreements that include:
   • On-line hot-line support
   • Telephone consultation
   • User training
   • Documentation updates

529

TABLE 15.1 (Continued)

Resource Management	Cost and Performance Measurement	Graphics and Displays	Other Features
3. Capability to level resources for a single resource class, particular activity, or the entire project 4. Ability to display resource usage histograms directly on the computer screen to assess the effect of variable resource availability over time 5. Multiproject resource leveling	• Budget to date (BCWS) • Budget at completion (BAC) • Earned value (BCWP) 6. Summarization of cost data across multiple projects	determine variances and predict trends 3. Provision to edit and generate multiple-size plots 4. Ability to specify sort sequences by activity name or WBS/OBS codes 5. Ability to retain in a library or file, user specifications for subsequent regeneration of bar charts, network diagrams, etc. 6. Device-independent interface for printing and plotting graphics and other displays 7. Ability to summarize and selectively print sections of the bar charts and network diagrams for reporting to senior management 8. Ability to integrate into the graphics text and user-defined symbols 9. Standard reports 10. Report wizard to create user-definable or custom reports 11. Ability to define a wide variety of reports, including time-distributed reports 12. Macro reporting	• Maintenance and repair 4. Support services for very large enterprise-wide systems that include: • Tool kits for applications program development • Consulting contracts 5. Comparison products: • Drivers for plotter graphics systems • Communications software that links the system with other MIS/business systems

- Absence of input procedures that provide users with an alternative to menu-driven input screens is time consuming for larger projects.
- Limitations on how data can be input over an on-line e-mail or Web connection makes the program inflexible for geographically dispersed users.
- Scheduling techniques
  - Limitations as to the types of precedence relationships or lead/lag times accepted by the software may result in an inaccurately scheduled project.
  - Confusion over the difference between PERT and CPM scheduling algorithms, as well as which methods the commercial systems use, poses difficulties for less advanced users first learning this terminology.
  - Inability to integrate a project's schedule, resource, and cost information seamlessly within one management information system means that extra time must be spent exchanging these data to and from other software programs.
- Graphics
  - Printed graphics that are merely a snapshot of the computer screen require extensive manipulation to generate legible, time-scaled networks that are suitable for presentation to senior-level management.
  - Limitation on the number of standard graphics reports or an inability to customize reports is a very real problem. Users should be given enough flexibility for generating as many different kinds of displays as wanted and for designing report formats that meet their own customer's needs.
- System design
  - Lack of a fully integrated, modular design creates havoc for project personnel when everything cannot be updated at once. This nightmare is compounded when separated programs are utilized for the storage and retrieval, for instance, of corporate financial records.
- Documentation
  - Poorly written user manuals, on-line help screens, wizards, or performance aids often confuse system users.
  - Overly complex or extremely simplified users' manuals often fail to serve the requirements of either novice or advanced users. Clear and concise multilevel documentation is sorely needed.
  - Documentation may not be available in more than one language.
- Vendor support

- Purchasers of the many smaller systems are frequently disappointed with the limited support provided by software vendors.
- On-line consultation or phone support may not be available 24 hours a day, 7 days a week.
- Security
  - Password protection and restricted data access, in earlier days, were the domain of the corporate IS department. This issue raises very serious concerns for users of desktop systems that permit project data to be accessible over the Internet to nonproject personnel.

The first step in installing a solution that's right for your project — no matter how small or large the project effort — is to evaluate and weight thoroughly a feature list such as the one given here to determine whether a match with the project team's requirements exists.

To install an appropriate solution successfully, the approach taken to select the most effective hardware–software combination must be systematic and methodical. Understanding your own project management requirements is key to making the best choice. Before embarking on your management's charge for selecting a system to automate project planning, tracking, and decision making, be sure that you have thoroughly analyzed *all* appropriate management requirements as well as determined what size computer configuration is needed to run the applications under consideration.

Begin by reviewing a written list of criteria to make this decision — in other words, the functionality needed to satisfy senior management's and the project management team's specific planning, scheduling, or tracking requirements. If both stakeholders are satisfied with your analyses, proceed with a discussion of how to physically configure the various systems, either on the desktop or in the corporate IS center. This lets you decide whether your choices impose any difficulties. Then determine, as accurately as possible, what each software program will cost to operate, and compare these charges to the projected performance savings.

To really understand the choices available, read current articles in trade journals about software systems. Periodical publications such as *PC World* that rate desktop and server-based systems annually are excellent references. The Project Management Institute conducts and publishes an annual review of the various systems. Additional references concerning the applications and limitations of these systems are also recommended in the Bibliography at the end of the chapter. Then to help ensure that the software under consideration has the potential to meet all your project needs, contact specialists involved in similar project management activity and conduct surveys among other corporate planning and project offices. Spell out your team's requirements in writing so that the features, capacity, costs, and limitations of several potentially selected systems can be studied in more detail.

Ultimately, you'll narrow down the list of accessible systems to a possible two or three, whose features and performance you can appropriately assign a weight and then rate accordingly. Information that colleagues share about these systems will enable your project team to understand any major drawbacks and to avoid common pitfalls. Ask to review sample written reports that represent the entire gamut or scope of their project work. Raising such interesting questions as whether the software was ever configured to handle input from other corporate MIS systems reveals a great deal. Also ascertain how acceptable a system's standard output is for senior- and executive-level briefings. If you intend on uploading data from a variety of mobile computers, determine how your colleagues will handle the problem of multiple access to data files. How secure is the computing environment? What protection schemes does the vendor offer to restrict user access to different levels of data?

You may also want to determine how your colleague's software handles the allocation of common resources that work on multiple projects. How adequately did their systems track resource use and expenses across these projects? Were they able to publish daily work assignments as well as track their completion? Also, how difficult was it to coordinate user training? Did a corporate computing group or the software vendor provide it? How instructionally sound was the training; is there an optimum number of students who should be scheduled at any one time?

Knowing the answers to these and many other questions will enlighten the project about the most commonly encountered pitfalls as well as help set more realistic expectations for implementing the software. Remember that final selection of a good project management system depends not only on whether but also how it meets the project's required criteria. Initially, many software systems may appear to satisfy many, if not all, of the team's general management requirements. For instance, until a more thorough evaluation is made by extensive trialing one or two demonstration systems obtained from the program vendors, it is not possible to judge the speed with which data are processed or the adequacy of on-line reporting. Benchmark project data taken from similar efforts can be used to evaluate a number of program parameters: capacity, power, versatility, user friend- liness, and graphic output.

Now your team may use its list of requirements to complete a decision matrix and assess comparable program features. This is accomplished by assigning priority weights to all essential features and then ranking their performance. We suggest a score from 1 to 5, where a value of 5 means that the software's performance is outstanding, and a 1 equals poor performance. Then multiply the preassigned numeric weights for each feature by the assigned ranking value, and calculate a grand total. As your team populates the matrix, it will become evident that for any of the software systems to be selected as candidates for further analysis or purchase, they must receive high scores in your most important categories. Rework the matrix more than once to validate the analysis. This is especially important whether you've decided to purchase and install the software in a single site without further trialing or have determined that leasing it under a multisite license is the best way to go.

Suggested steps included in this systematic selection procedure and some precautions that the project team should take follow:

Step 1. Determine your own project management requirements.
- Who will use the software program?
- What limitations or restrictions do the computer hardware, system requirements, and so on, impose?
- What features are absolutely essential and which would be nice to have in the future but not necessary now?
- How does the program's performance compare with its cost to operate?
- Who must review the program's output?
- Are presentation graphics required?
- What applications does the software need to perform?

Step 2. Search product/buyer's guides to project management software; contact other users to discuss general impressions and precautions.

Step 3. Review the many project entries to eliminate programs that fall short of your list of requirements.

Step 4. Narrow the list of accessible software systems to a minimum of two or three.

Step 5. Visit various system vendors' Web sites to obtain product literature, evaluation kits, or complementary passwords; conduct a system demonstration or trial using a case study of data taken from your own projects.

Step 6. Test each of the features and performance categories for each vendor's system; analyze problems encountered; validate the performance by testing each system more than once.

Step 7. Compare the results of these evaluations; a performance-weighted ranking technique is recommended.

As you and your team embark on evaluating and installing one or more of these potentially excellent project management software systems, good luck!

15.5 CONCLUSIONS

In an era where business changes at a phenomenal rate and the availability of new programs to enhance how project teams communicate and manage globe-spanning projects increases exponentially, we cannot review in this book the features of any one software application that the major vendors offer the marketplace. Nor is it our intention to replicate the results of recent testing conducted by the computer press. Our objective, instead, is to heighten your awareness of what technologies are available to teams today in the dawn of the information age. Undoubtedly, the product portfolios in each of these categories will evolve over time. We trust that the new features and functionality added to these products will be in direct response to your needs as managers of mobile project teams with an ever-increasing need to communicate on a daily basis and to keep the team informed of the project's changing status.

In Appendix 15.1 we list the Internet addresses of professional associations that are good sources of information as well as the URLs for the most popular vendors of groupware applications, project management information systems, add-on risk, and project estimation tools at the time this publication went to press. Before rushing out and procuring any of these systems for the project's tool shelf, we suggest that you first research these sources for further product information and then visit the vendors' Web sites. Your understanding of the basic concepts of the project management discipline from your work with this book should help your team make a more informed decision.

BIBLIOGRAPHY

Buchan, D. H., "Risk Analysis: Some Practical Suggestions," *Cost Engineering,* Vol. 36, No. 1, January 1994, pp. 29–34

Coleman, D., *Groupware: Collaborative Strategies for Corporate LANs and Intranets.* Upper Saddle River, NJ: Prentice Hall, 1997.

Gordon, P., "To Err Is Human, to Estimate Divine," *Information Week,* No. 717, January 18, pp. 65–72.

Guide to @ RISK for Project, Newfield, NY: Pallisade Corporation, 1997.

Hapgood, F., "Tools for Teamwork," *CIO Web Business,* Vol. 12, Sec. 3, November 1, 1998, pp. 68–74.

Hecht, A., "A Web-Based Project Management Framework," *PM Network*, Vol. 12, No. 12, December 1998, pp. 40–44.

Jossi, F., "Creating Presentations as a Team," *Presentations*, Vol. 10, No. 7, July 1996, pp.18–26.

"100 Years of Innovation," *Business Week,* Summer 1999.

Pendergast, M., and S. Hayne, "Groupware and Social Networks: Will Life Ever Be the Same Again?" *Information and Software Technology*, Vol. 41, No. 6, April 25, 1999, pp. 311–318.

Raskin, A., "Task Masters," *Inc. Technology*, Vol. 21, No. 4, 1999, pp. 62–72.

Santosus, M., "Working Smart," *CIO*, Vol. 12, No. 11, March 15, 1999, p. 80.

Williams, L., "Evaluating Project Management Software," *PM Network*, Vol. 12, No. 12, December 1998, pp. 27–30.

APPENDIX 15.1: PROFESSIONAL PROJECT MANAGEMENT ASSOCIATIONS AND REPRESENTATIVE SOFTWARE AND HARDWARE VENDORS*

Vendor	System
Conferencing and Collaboration	
Avaya Inc. Basking Ridge, NJ *www.avaya.com*	DEFINITY®AnyWhere MultiMedia Communications eXchange MultiPoint Conference Unit
Microsoft Corporation Redmond, WA *www.microsoft.com*	NetMeeting
PictureTel Corporation Andover MA *www.picturetel.com*	Venue 2000 PictureTel 240/330 PictureTel 550 Swift Site Suitcase
Polycom, Inc. San Jose, CA *www.polycom.com*	Polycom Global Management System The MeetingSite 5000™ Conferencing System SoundPoint® and SoundPoint PC® SoundStation Premier® and SoundStation Premier with Extended Mics SoundStation® and SoundStation®EX

*Comments from individual readers concerning additional software and hardware systems are welcome. Please bring to the authors' attention the names of any vendors whose products should be included in this list.

Vendor	System
	SoundStation Premier Satellite®
	ShowStation®IP
	SteamStation™
	ViewStation™FX
	The VS4000
	ViewStation™MP
	ViewStation™V.35
	WebStation™
Sony NY, NY *www.sony.com/videoconference*	Trinicom 5100PLUS PCS DS150 VID-P110 VID-P50 Contact™
Tandberg Reston, VA *www.tandbergvision.com*	TANDBERG 7000 TANDBERG 6000 TANDERG 5000 TANDBERG 2500 TANDBERG 800 TANDERG 600
	Groupware
Instinctive Technology Cambridge, MA *www.instinctive.com*	eRoom
Lotus Development Corporation Cambridge, MA *www.lotus.com*	Notes/Domino Family Server/SMART SUITE Sametime
Microsoft Corporation Redmond, WA *www.microsoft.com*	Microsoft® Exchange Server Outlook 2000
Novell Orem, UT *www.novell.com*	GroupWise NetWare
Open Text Corporation Minneapolis, MN *www.opentext.com*	Livelink®
The Ventana East Corp. Johnson City, NY *www.ventana-east.com*	GroupSystems

Vendor	System

Web Authoring

Microsoft Corporation Redmond, WA *www.microsoft.com*	FrontPage® 2000

Project Estimation

Marotz Inc. Jamut, CA *www.marotz.com*	Cost Xpert
Price Systems, L.L.C. Mt. Laurel, NJ *www.pricesystems.com*	PRICE Estimating Suite
Software Productivity Research Burlington, MA *www.spr.com*	KnowledgePlan®

Risk Analysis

Palisade Corporation Newfield, NY *www.palisade.com*	@RISK Professional for Project
Primavera Bal Cynwyd, PA *www.primavera.com*	Monte Carlo™ for Primavera®

Project Management Information Systems

ABT Corporation New York, NY *www.abtcorp.com*	ABT Publisher™ 5 ABT Planner™ 5 ABT Workbench™ 5 ABT Connect™ 5/ABT Team™ 5 ABT Integrator™ ABT Resource™ ABT Repository™
Advanced Management Solutions Los Angeles, CA *www.amsusa.com*	AMS REALTIME Product Suite AMS REALTIME Projects AMS REALTIME Costs AMS REALTIME Solo AMS REALTIME Resources AMS REALTIME Server

Vendor	System
Angel Group LLP Coppell, TX *www.angelgroup.com*	Tracking DataBase (TRDB)™
Artemis Management Systems Boulder Corporation *www.artemispm.com*	Artemis Views
Computer Associates International Islandia, NY *www.cai.com*	CA-Super Project CA-Super Project Net
Critical Tools Austin, TX *www.criticaltools.com*	PERT Chart EXPERT WBS Chart for Project
IMSI San Rafael, CA *www.imsisoft.com*	TurboProject Professional
Microsoft Corporation Redmond, WA *www.microsoft.com*	Project 2000
PlanView, Inc. Austin, TX *www.planview.com*	PlanView Software
Primavera Bal Cynwyd, PA *www.primavera.com*	DATASTORE™ FOR PRIMAVERA Expedition® Expedition Analyzer® Expedition Express™ Primavera Project Planner (P3®) Enterprise Resource Planning System Integration SureTrak Project Manager TEAMPLAY™ Webster *for PRIMAVERA*™
SAS Institute Inc. Cary, NC *www.sas.com*	SAS/OR Software
Scitor Corporation Menlo Park, CA *www.scitor.com*	Project Scheduler 8™

Vendor	System
Welcome Solutions Technology Houston, TX *www.welcom.com*	Open Plan Suite

Professional Project Management Associations

AACE International
Morgantown, WV
www.aacei.org

PMWebRing
www.pmforum.org

Project Management Institute (PMI)
Upper Darby, PA
www.pmi.org

Software Engineering Institute
Carnegie Mellon University
Pittsburgh, PA
www.sei.cmu.edu

Software Program Managers Network
www.spmn.com

GLOSSARY

Acceptable quality level (AQL): maximum number of defects per hundred units or maximum percent defective that, for the purpose of sampling inspection, can be considered satisfactory as a process average.

Acceptance: agreement to the terms offered in a contract. An acceptance must be communicated, and in common law, it must be the mirror image of the offer.

Accountability: being answerable to a higher power. Accountability is generally tied to integrity and flows up through the various levels of an organization.

Active empathetic listening: listening in which the receiver paraphrases both the content and the feeling in the message received.

Activity on arrow (AOA): the graphical notation used to represent the start and finish of work tasks in a schedule or network where the arrow used to denote the specific work activity has a time duration and the geometric circles or rectangles it connects to do not.

Actual cost of work performed (ACWP): in the earned value cost accounting method, the total costs incurred (direct and indirect) in accomplishing work during a given time period. The calculation answers the question, How much did the completed product cost?

ACWP: *see* Actual cost of work performed.

Adaptive organization: the more traditional organization that tends to cope with or adjust to present situations and encourages the maintenance of the status quo.

AOA: *see* Activity on arrow.

AQL: *see* Acceptable quality level.

Arbiter: a person appointed to judge or decide a disputed issue.

Arbitration: the process by which parties in a dispute submit to settlement or judgment by an impartial party appointed by mutual consent or statutory provision.

Assessment: a planned activity to measure the effectiveness of a process improvement of an organization's quality system.

Attribute control chart: a chart displaying the percentage of defective characteristics, the number of defective items, or actual defects in a manufactured product sample that makes the quality of a lot unacceptable.

Audio conferencing: holding a conference among people at remote locations where the communication is sent as audio signals over telephone lines. *See also* Teleconferencing.

Audit: the examination of products or processes conducted to determine compliance with specifications, usually performed by personnel from the quality assurance organization, who randomly sample and inspect a product, document, or process to ensure that it meets the needs and requirements of both its internal and external customers.

Authority: a legal term for the power and capacity, either expressed specifically in writing or implied through neglect, granted by a principal to a third party that allows an agent to perform acts that are customary and necessary on his or her behalf.

Autocratic: political term referring to a more directive, unilateral approach to management and leadership.

BAC: *see* Budget at completion.

Back charges: incurred expenses that are overdue or in arrears.

BCWP: *see* Budgeted cost of work performed.

BCWS: *see* Budgeted cost of work scheduled.

Beta test: the first test performed in a customer environment to demonstrate that a software system meets its specified requirements.

Bid invitation: a formal statement issued by the buyer to a contractor asking what he or she will charge for supplies or labor.

Bill of lading: in commercial law, the receipt that a common carrier supplies to a shipper for goods given to the carrier for transportation. The bill evidences the contract between the shipper and the carrier and can also serve as a document of title, creating in the person possessing the bill ownership of the goods shipped.

Boundary spanner: the project manager's role of liaison across various organizational entities.

Boundary spanning: using political and interpersonal skills to cut across organization lines and relationships.

BPR: *see* Business process reengineering.

Break-even time: interval that a project requires to reach a point where revenues equal costs. The point is located by a break-even analysis, which determines the volume of sales needed to cover fixed and variable project costs. Sales over the break-even point produce profits for the firm; sales below produce losses.

Browser: a computer program providing access to sites on the World Wide Web. *See also* World Wide Web.

Budget at completion (BAC): in the earned value cost accounting method, may refer to the total baseline cost of the project at the end or at the completion of a task, if partial budgeting is done. The calculation answers the question, What is the total job supposed to cost?

Budgeted cost of work performed (BCWP): in the earned value cost accounting method, the sum of the approved cost estimates of any completed work (including overhead) or portions of work budgeted to be done during a given period of time; known as the project's earned value. The calculation answers the question, How much work is done?

Budgeted cost of work scheduled (BCWS): in the earned value cost accounting method, the sum of the approved cost estimates of any scheduled work (including overhead) or portions of work planned to be spent during a given period of time; known as the project performance measurement baseline or budget to date. The calculation answers the question, How much work should be done?

Bureaucracy: an organization characterized by formal rules, strict procedures, and detailed reporting relationships.

Business process reengineering (BPR): fundamental rethinking and radical redesign of business processes to bring about dramatic improvements in critical, contemporary measures of performance, such as cost, quality, service, and speed.

Capacity: the ability or power created by a law or an operation to give consent.

Cause-and-effect diagram: a graphical tool used to visually isolate and to diagram possible related causes. Also known as a fishbone or Ishikawa diagram, it is often used to identify root causes or problems. *See also* Root cause.

Cash flow (CF): the money coming into and out of an organization or project as income and expenditures. Net cash flow or net income from operations is established after charges such as depreciation, interest, and other noncash charges are considered.

Centralized organization: a firm or business in which decision-making authority and responsibility is held by a few within a centralized location, such as home headquarters.

CF: *see* Cash flow.

Change control: methodology adopted by software and hardware development projects to understand the consequences of making changes to project baselines.

Change control board: a formally constituted group of stakeholders responsible for approving or rejecting changes to the project baselines.

Change management plan: a high-level plan adopted by a software or hardware development project that provides the structure for (1) assessing the impact of proposed changes on the project's schedule, resources, and budget; (2) notifying all affected parties of each proposed change, its acceptance, or rejection; and (3) providing an audit trail of the decisions related to the product content.

Change orders clause: a written statement directing the seller to make changes according to the provisions of the contract document.

Change request (CR): that part of the audit trail structure outlined in a project's change management plan that serves to identify and record a proposed change to the software or hardware product.

Checkpoint: a point in the new product development process where all activities are stopped and inspected to determine progress complying to previously established criteria to continue as planned.

Choice of forums clause: a clause in an agreement in which the parties stipulate what judicial body or assembly will be engaged to resolve disputes arising between them.

Civil law: Roman law embodied in the Justinian Code and presently prevailing in most Western European states.

Code walkthrough: a review of software code presented by the programmer to peer personnel to (1) detect errors, omissions, and other flaws; and (2) appraise the quality of a product.

Coding phase: the new product development phase during which a programmer codes, tests, and debugs modular units of a software application program.

Commercial-off-the-shelf (COTS) components: an enumeration of software elements obtained commercially and included in a product's design (i.e., commercial database, customer action request system, etc.).

Commodity prices (CP): the price of bulk goods such as metal, grain, and foods traded on a spot market for cash or at a commodity exchange for an agreed-upon sum.

Common law: the system of jurisprudence originating in England that was later applied in the United States. It is based on judicial precedent (court decisions) rather than on legislative enactment (statutes) and is therefore derived from principles rather than rules.

Compliance review: the inspection of the final output from any project activity by internal experts, who are typically members of the project team, to determine compliance with established project management procedures, user requirements, design specifications, development processes, or quality standards.

Component development architecture: an enumeration of software elements that may be reused with no changes in a range of other product designs.

Concept phase: the early new product development planning phase during which customer requirements for a new product or service are translated into written technical specifications.

Concept to customer time (CTC): the time from conception of a product to its introduction to the market.

Confidentiality clause: a stipulation or provision in an agreement, often issued as a separate formal document, indicating that it is unauthorized for the party signing the agreement to convey or communicate information disclosed in discussions with the issuer.

Configuration management system: the formal management system established during the development life cycle of a software program to (1) control the definition of the functional and physical characteristics of each computer program configuration item, which has a specific end-use function, (2) maintain a continuous accounting of the computer program's design and configuration, and (3) control the adoption of changes to project baselines.

Consideration/exchange: an act, forbearance, or promise give by one party for the act or promise of another. Under common law, except in the state of Louisiana, consideration is a necessary element to the creation of a contract.

Constructive conflict: differences that are approached with a win–win attitude and occur in a relatively open and supportive environment.

Contingency fund: a reserve separately planned for in a project's cost and schedule baselines that is set aside to manage uncertain events that have a high probability of occurring. *See also* Management reserves.

Contingency theory of leadership: a theory of leadership first proposed in the 1950s, that encourages the leader to match his or her leadership style with such variables presented in the work environment as power position, task structure, and leader–member relations.

Contract administration: managing the relationship with the seller.

Contract work breakdown structure (CWBS): a presentation included in a contract transacted between two parties showing the corresponding major elements that the vendor has promised to deliver or supply. *See also* Work breakdown structure.

Control chart: a graphical representation of a product attribute or process characteristic showing a data plot of statistics gathered from the process which are used for real-time control.

COQ: *see* Cost of quality.

Core team: an integrated, cross-functional team of project leaders responsible for managing the development and launch of a commercial product or service.

Cost and price proposal: a document presented to the buyer by the selling organization in response to a formal procurement solicitation request that indicates how the cost of the contract is divided among direct and indirect expenses, between personnel or labor categories, between tasks by month, and so on.

Cost of quality (COQ): A key financial tool that can be used to measure the effects of process control on the total quality management effort. The sum of the costs to ensure quality including the cost of prevention (quality planning and control), appraisal (quality assurance), and failure (rework).

Cost reimbursement (CR) Contract: a type of contract that usually includes an estimate of the project costs, a provision for reimbursing the seller's expenses, and a provision for paying a fee as profit.

Cost schedule: a schedule illustrating how much money has been allocated and spent for each project activity as a function of time, permitting the project manager to determine whether or not the project will finish within budget.

Cost schedule status report: the standard output of the U.S. Department of Defense planning and control reporting system, intended to foster greater uniformity and provide early warning of any impending schedule or budget overruns.

Cost variance (CV): the difference between the estimated cost of an activity and the actual cost of an activity. In the earned value cost accounting method, CV is calculated as the budgeted cost of work performed (BCWP) minus actual cost of work performed (ACWP).

COTS: *see* Commercial-off-the-shelf components.

CP: *see* Commodity price.

CPM: *see* Critical path method.

CR: *see* Change request *and* Cost reimbursement contract.

Crashing: an action to decrease the duration of an activity or project by increasing the expenditure of resources.

Critical design reviews: a detailed review of the logic, components, or some portion of the proposed software system's design (such as a database), conducted by technical experts who are members of the project team, to ensure its integrity.

Critical path: the longest sequential series of project activities, which determine the length of the project. The critical path is the network path with no slack and usually zero total float.

Critical path method (CPM): a scheduling technique using either activity-on-arrow (AOA) or precedence (PDM) diagrams to graphically display project activities.

Cross-functional: referring to the many disciplines across an organization.

CTC: *see* Concept to customer time.

Customer–market phase: the preliminary planning phase of a new product development project that analyzes the feasibility of pursuing product and service concepts identified by market research.

Customer value analysis (CVA): management science that provides quantifiable data for monitoring and measuring market-perceived quality, customer satisfaction, and competitive information.

CV: *see* Cost variance.

CVA: *see* Customer value analysis.

CWBS: *see* Contract work breakdown structure.

Daily build and smoke tests: activities that compile, link, and integrate software modules into executable programs and routinely test them.

Decentralized organization: a firm or business in which decision making and empowerment are dispersed and encouraged at all levels throughout the organization.

Decision-path flow diagram: a representation in flowchart form of the multiple conditions and corresponding actions to be taken to solve a problem.

Defect tracking: the process of measuring the imperfections or flaws that make a unit of nonconforming or defective product or output.

Definition phase: the new product development phase during which the product or service design is decomposed into definable units of activities and all associated cost and production time estimated.

Design phase: the new product development phase during which all aspects of a product or service are simulated.

Design reviews: the inspection by internal experts who are typically members of the project team to determine whether the elaboration by the software or hardware developers of a product design matches the logic used to satisfy the requirement.

Destructive conflict: disagreements that are characterized by a win–lose stance and a lack of team spirit.

Detailed process-flow diagram: a representation in flowchart form of the conditions and corresponding actions to be taken to solve a problem in which symbols are used to represent operations, data, flow, equipment, and so on.

Detailed requirements/specifications: documented statements of valued technical performance objectives required of the product or service to comply satisfactorily with customer needs.

Development phase: the new product introduction phase, during which the product concept is operationalized and early models are constructed and tested according to specifications supplied by the customer or interpreted from market requirements.

Development time (DT): the time elapsed that occurs from the initial funding of a new research and development concept, through its transfer to a manufacturing site, to eventual market introduction.

Discount rate (DR): the interest rate used when calculating the present value of an investment's future cash flows.

Divestment phase: the new product development phase during which postproject evaluations are conducted to improve future project management activities, and where customer acceptance of the newly launched product or service is carefully monitored to enhance market position.

Downsize: to eliminate layers of management and/or personnel within an organization.

DR: *see* Discount rate.

DT: *see* Development time.

Dummy: an activity in a network diagram that requires no work, signifying a precedence condition only.

Early finish (EF): in a network diagram, the earliest time that an activity can be completed, depending on its early start time.

Early start (ES): in a network diagram, the earliest time that an activity can be started, depending on the duration of its predecessors.

Earned value analysis (EVA): a method for measuring project performance that compares work that was planned with what was actually accomplished to determine if cost and schedule performance are proceeding on plan.

ECAC: *see* Estimated cost at completion.

EF: *see* Early finish.

Electronic mail (e-mail): messages sent and received electronically (as between terminals linked by telephone lines or microwave relays).

E-mail: *see* Electronic mail.

Empathy: understanding the emotions of others vicariously.

Empowerment: the concept of delegating sufficient authority to an organization, position, or team, so as to create a feeling of ownership and pride in one's accomplishments.

Environmental test: a special series of tests whereby the equipment is subject to different conditions of operation, such as vibration, shock, and cold.

ES: *see* Early start.

Essential work: the required work that sustains the organization but is not valued by customers.

Esteem or status value: the nonfunctional features that prompt a customer to purchase a product or service.

Estimated cost at completion (ECAC): the expected total cost of an activity, group of activities, or project, when the defined scope of work has been completed. In earned value analysis the formula for calculating the ECAC equals the actual cost of work performed (ACWP)/budgeted cost of work performed (BCWP) × budget at completion (BAC). The calculation answers the question, What do we expect the total job to cost now?

Estimated time at completion (ETAC): the expected additional cost needed to complete an activity, a group of activities, or the project. In the earned value cost accounting method, the formula for calculating the ETAC equals the budgeted cost of work scheduled (BCWS)/budgeted cost of work performed (BCWP) × time at completion (TAC).

ETAC: *see* Estimated time at completion.

EVA: *see* Earned value analysis.

Events: an identified point in time highlighted in reports and plots.

Executive committee: a group of key executives within the organization that have decision-making authority and responsibility for instituting corporate change and direction. *See also* Steering committee.

Expected value: a statistical term used in probability theory to represent the average value in the long run.

Expressed warranty: a written statement arising out of a sale to the consumer of a consumer good, pursuant to which the manufacturer, distributor, or retailer undertakes to preserve or maintain the utility or the good's performance or provide compensation if there is a failure.

Extranet: the computer network outside an organization's firewall where communications are not screened. *See also* Intranet.

Facimile: a system of transmitting and reproducing graphic matter (as printing or still pictures) by means of signals sent over telephone lines.

FAR: *see* Federal Acquisitions Regulation.

Federal Acquisitions Regulation (FAR): a U.S. federal government-wide procurement regulation mandated by Congress and issued by the Department of Defense, the General Services Administration, and the National Aeronautics and Spare Administration.

Finish node: the node in a network diagram at which two or more subsequent activities finish.

Fishbone diagrams: *see* Cause-and-effect diagrams.

Fitness for use: the concept whereby products and services perform as specified in technical requirements when introduced into the market. *See also* Usable.

Fixed-priced (FP) contract: a contract form in which the price and fee are predetermined and not dependent on cost.

Float: the amount of working time by which a project can be delayed and still meet the specified completion date. Float occurs when the required completion date is specified later than the critical path completion date.

Foreign exchange (FX): an instrument employed in making payments between countries (i.e., notes, paper currencies, checks, bill of exchange, electronic notifications of international debits and credits).

FP: *see* Fixed-priced contract.

Front end: the concept and planning phases for a product or service development during which customers needs are translated into a product definition.

Function points: a qualification of the functionality provided to the user of a software system helpful in estimating the size of the development project.

FX: *see* Foreign exchange.

Gantt chart: a graphic representation of project activities shown by a time-scaled bar line named after a World War II industrialist, Henry Gantt.

Gemba kaizen: Masaaki Imai's concept of continuous improvement to a process or organization, by involving everyone in implementing incremental changes in all aspects of creating a quality product.

Generative organization: the newer, more learning organization that encourages creativity, innovation, and change, so as to improve competitiveness.

Groupware: an umbrella term describing the electronic technologies that support person-to-person collaboration, which include e-mail, electronic meeting systems, desktop videoconferencing, and messaging and workflow applications.

Hierarchical organizations: an organization characterized by several reporting levels and bureaucracy.

Histogram: a graphical tool that displays a frequency distribution. Histograms use vertical bars to indicate the number of occurrence (frequencies) of each value of a response variable.

Home page: the entry or starting point for a presentation available on the worldwide Internet or corporate intranet.

How: a QFD term for a mechanism that fulfills a *What*; the first-level technical attributes of a product or service that explicitly address customer needs or wants.

Implied warranty: a promise arising from operations of the law that something which is sold (1) can be offered for sales free from security interest (implied warranty of title); (2) must be fit for the special purpose of the buyer (implied warranty of fitness); and (3) conforms to the promises or affirmations of fact made on the container or label (warranty of merchantability).

Indemnification clause: a contract clause exempting a party from incurred penalties or liabilities against hurt, loss, or damage resulting from an act or forbearance by another.

Information revolution: the extensive changes in today's culture, communication, and business brought about by the widespread use of computer technology.

Installation phase: the new product development phase during which an early market version of the product or service is delivered to a customer for testing in a controlled environment.

Integrated manufacturing and production phase: the new product development phase during which the technology underlining the fabrication of a commercial product or service is transferred from the R&D laboratory to a firm's production site.

Integration and test phase: the new product development phase during which the software modules are integrated together (compiled, linked, and tested) and then integrated with any system hardware.

Integration test: an internal laboratory test whereby software components, hardware components, or both are combined and tested to evaluate the interaction between them.

Integrative planning process (IPP): a technique for building successful cross-functional teams that involves everyone in the preparation of the project plan, thereby ensuring that the entire team understands project objectives, cost, and quality parameters.

Interest rate (IR): the cost of credit expressed as a percentage rate, computed from the ratio of interest to principal.

Internal rate of return (IRR): the true annual rate of earnings on an investment.

Internet: the electronic communications network that connects computer networks and organizational computer facilities around the world using the Internet Protocol (IP) to provide file transfer, remote login, electronic mail, news, and other services.

Interoperability test: a test whereby software systems are combined and the interaction between them evaluated.

Intersystem test: a test conducted outside the laboratory whereby software systems, hardware systems, or both are combined with existing field systems and tested to evaluate the interactions among them.

Intranet: generally, any collection of distinct computer networks working together as one in which a firewall, or server, is used to regulate and screen communications. *See also* Extranet.

IPP: *see* Integrative planning process.

IR: *see* Interest rate.

IRR: *see* Internal rate of return.

Ishikawa diagram: *see* Cause-and-effect diagram.

Job classifications: the documented functions, processes, and responsibilities assigned to each position within an organization.

Jurisdiction: the power, right, or authority to interpret, apply, or declare the law. Questions of jurisdiction arise regarding quasi-judicial bodies such as administrative agencies in their decision-making capacities.

Knowledge worker: a contemporary employee who characterizes today's more information processing-oriented versus production-oriented organizations.

Late finish (LF): the latest time an activity may be completed without delaying the project completion date.

Late start (LS): the latest time an activity may begin without delaying the project finish date.

Launch strategy: a detailed plan for achieving product financial and customer satisfaction objectives when introduced to the market.

Learning organization: a contemporary corporation, agency, or firm that encourages innovation, transformation, and change.

Letter of credit: an instrument or document issued by a bank guaranteeing payment of a customer's drafts up to a stated amount for a specified period. Used extensively in international trade, it substitutes the bank's credit for the buyer's and eliminates the seller's risk.

Level of effort: a vendor's accomplishment of support activities not easily measured but typically characterized by a uniform rate of activity over a specified period.

LF: *see* Late Finish.

Linear responsibility chart (LRC): a graphical or chartlike portrayal of the members of a project team, which indicates their primary and secondary project responsibilities.

Liquidation for damages clause: a stipulation in an agreement of the amount of monies to be awarded to compensate an injured party reasonably as reparation for loss in the event of a breach.

Lower control limit: the lower line on a control chart that is used to judge whether the process is statistically out of control. *See also* Upper control limit.

LRC: *see* Linear responsibility chart.

LS: *see* Late Start.

Maintainable: the concept whereby customer acceptance of a software application program's initial life is extended in the market through the introduction of upgrade releases.

Maintenance development: software development performed to (1) make a computer usable in a changed environment, (2) correct faults in the software, and (3) prevent problems from occurring.

Maintenance of value clause: a stipulation in an agreement that permits both parties to reappraise the estimated fair market or monetary return of the contract to adjust for changes in the price of commodities, foreign exchange rates, and so on.

Maintenance/operations phase: the new product development phase following the deployment of a software system to customers during which programmers actively attend to change requests resulting from errors, faults, and omissions of desired functionality or features.

Management proposal: a document presented to a buyer by the selling organization in response to a formal procurement solicitation request to demonstrate that the buyer has won other related contracts and performed satisfactorily.

Management reserve: a reserve separately planned for in a project's cost and schedule baselines that is set aside to manage unknown events. *See also* Contingency fund.

Master budget: an estimate of projected expenditures for resources and anticipated capital outlays over the duration of a project.

Master test plan: a document describing the scope, approach, resources, and schedule of intended test activities as well as the technical and management approach to be followed for testing a system or component. Typical contents include items to be tested, tasks to be performed, required resources, responsibilities, and schedules for the testing activities.

Mean time to first failure (MTFF): the mean value of the lengths of time between initial failures under stated conditions.

Messaging technology: the strategies and solutions for managing incoming electronic mail, providing remote access, and integrating e-mail with other desktop and mobile applications over proprietary LAN-based e-mail systems and the latest Internet e-mail standards.

Milestones: key events or critical accomplishments planned at time intervals throughout a project, used to gauge overall progress toward completing the schedule.

Milestone schedule: a summary-level schedule which identifies the significant events that are usually associated with the completion of the major deliverables in a project.

Monte Carlo simulation: a schedule risk assessment technique that performs a project simulation many times in order to calculate a distribution of likely results.

MTFF: *see* Mean time to first failure.

Multimedia: using, involving, or encompassing several media, usually sound, still pictures, and full-motion video to communicate.

Net present value (NPV): a method of determining whether the expected financial performance of a proposed investment promises to be adequate.

New product development (NPD) process: the design, development, manufacturing, and product introduction actions required to take a product or service from its initial concept to its delivery to customers.

NIH: *see* Not-invented-here attitude/syndrome.

Nodes: graphical notation used in the precedence diagramming method network to indicate the start, completion, and time frame of an activity.

Nonessential work: activities performed by an organization that do not sustain the business. *See also* Non-value-adding work.

Non-value-adding work: activities performed by an organization that a customer does not value. *See also* Non-essential work.

Notices clause: the stipulation in an agreement of how one party will give another communications either of a fact, claim, demand, or a proceeding.

Not-invented-here (NIH) attitude/syndrome: a slang expression that refers to the belief that only the ideas or concepts generated by one's own department or unit are worthy of consideration.

NPD: *see* New product development.

NPV: *see* Net present value.

Object-oriented development (OOD): a software development methodology focused on the object-orientation approach to specifying functional requirements and data structures as instances of types of objects they must contain. *See also* Object-oriented program.

Object-oriented program: a software program that possesses characteristics associated with object orientation. Programs are organized as cooperative collections of objects, each of which represents an instance of a class whose members are united in a hierarchy of inherited relationships. *See also* Object-oriented development.

ODBC: *see* Open database connectivity.

OOD: *see* Object-oriented development.

Open database connectivity (ODBC): an interface that makes it possible to access a variety of database management system platforms with a common language. ODBC is based on the CLI (Call Level Interface) of the Windows Open Service Architecture standard developed by Microsoft Corp. There are ODBC drivers and development tools for Windows, Macintosh, UNIX, and OS/2.

Operational pilot: a functional test of a fully integrated hardware and software system conducted at the operational site where it is expected to perform under normal operating procedures.

Operational plans (planning): a plan that specifies details on how the overall objectives are to be achieved.

Organizational culture: the norms, rules, attitudes, and acceptable and unacceptable behaviors that typify a particular company or firm.

PACE: *see* Product and cycle-time excellence.

Parallel organization/structure: a method of implementing change whereby a steering committee is established in conjunction with the current organization to (1) handle activities that deviate from the norm and (2) discover technologies for improving the organization and changing the manner of operation. The primary purpose of the parallel organization is to explore new methods of operating and to institute change.

Pareto chart: a graphical tool used to rank problems by number of occurrences, to give an idea of their severity relative to each other. It is especially useful in identifying the "critical few" and the "trivial many."

Payment clause: a statement in an agreement that establishes the method, timing, currency, and frequency for the receipt of payment for the exchange of goods and services between a buyer and seller.

PDM: *see* Precedence diagramming method.

P/E Ratio: *see* Price/earnings ratio.

Perceived quality: a subjective, user-based approach to quality that lies in the eyes of the beholder.

Performance schedule: a schedule illustrating when each project activity is expected to begin and end, allowing the project manager to determine whether the overall project will finish as estimated.

Personnel schedule: a schedule illustrating which project resources are allocated to work on project tasks, when, and for how long, permitting the project manager to distribute the workload so that each resource is able to work at optimal capacity.

Phased development: the new product development methodology that separates projects into distinctly financed phases where approval to continue working is sought at the end of each project phase.

Planning audit: an examination of the records kept by a project that is performed after project planning is completed to assess the adequacy of the project plan and project management system to meet customer expectations.

Postmortem review: an informal evaluation conducted at the end of a project to discover findings and present recommendations for improvement, where project members are interviewed to capture their personal experience and output documented in technical or quality assessments that were conducted during the project are reviewed.

Precedence diagramming method (PDM): a network representation in which project activities or work tasks are represented by a node and dependency is shown by a line or arrow.

Precedence relationship: the basic relationship between the start and/or finish of one project activity and the start and/or finish of another, which includes:

- *Finish to finish*: restriciting the finish of an activity for some specified duration until the finish of another following activity.
- *Finish to start*: letting an activity start as soon as another is finished.
- *Start to finish*: restricting the finish of an activity for some duration before another can start.
- *Start to start*: restricting the start of an activity for some specified duration until the proceeding activity has started.

Price/earnings (P/E) ratio: the relationship between a firm's current stock price to its earnings over a specified earnings period. P/E is calculated by dividing the current stock price by the earnings per share.

Probability distribution: a mathematical devise used for presenting the uncertainty of the qualified project risks. There are a wide variety of distributions, each of which describes a range of possible values and the likelihood of their occurrence on the project.

Process capability: the performance level of a process after it has been brought under control, which satisfies a customer's tolerance limits for an acceptable product.

Process control: the activities involved in ensuring that a process performs consistently at some target level of performance with only normal variation; the state of stability, normal variation, and predictability.

Process-flow diagram: a representation in flowchart form of an associated set of activities defined by a specific sequence of events with distinct start and end points.

Process quality: the system of metrics and management practices that are implemented to monitor, control, and improve continually not only the end products delivered to customers but also the operations used to develop and produce them.

Product advantage: a product or service feature built into a product that is perceived by customers to differentiate it from other competitive offers in the market.

Product and cycle-time excellence (PACE): the approach to new product development where project activities occur in phases and the decision to proceed between each is established by passing predetermined criteria documented in an operational plan.

Production or manufacturing phase: the new product development phase where a previously hand-assembled product from tool-made samples is fabricated and tested in a production facility.

Product scope: a statement describing the degree of functionality developed and included in the product. Product scope is specified and communicated to project team members in a product requirements document.

Project executive sponsor: a person serving on the executive steering committee who is an advocate for the project.

Project familiarity: an advantage created by introducing new products to the marketplace with some similarity to those already known by customers.

Project handbook: a repository of project documentation that references the standards and procedures that are used to manage the project.

Project management audit: an assessment conducted by an experienced group of professionals to help a particular project evaluate its current plan. It is a formal, confidential evaluation of the adequacy of the project's adherence to guidelines, policies, and procedures and the adequacy and implementation of executed project controls.

Project management information system: a software application program that permits project specialists to create project plans to track progress and to manage and communicate information about a project, as well as integrate and analyze data at the enterprise reporting level.

Project management requirements analysis: a technique for outlining project methodology and procedures that match project needs to project management strategies.

Project organization: the structure imposed on a team of people engaged in the temporary endeavor of realizing a product or service.

Project review meeting: a periodical gathering of project team members that is conducted by project officials for the purpose of comparing progress with baseline data.

Project scope: a statement of the products and services to be delivered or provided by a project.

Project scope management: the subset of project management activities associated with defining, planning, and controlling the realization of major project deliverables and objectives that assure the project contains only the work required to complete it successfully.

Project status report: a written report prepared periodically by task team leaders for management in a format requested by the sponsor that presents control-oriented project information in summary form.

Project summary plan: a formal, approved document used to execute and control the realization of project deliverables and objectives. The document captures critical assumptions used to guide decision making and facilitate communications among project team members. *See also* Summary Plan.

QFD: *see* Quality function deployment.

QMS: *see* Quality management system.

Quadruple constraints: the four parameters that define projects: (1) time schedules, (2) budget or costs, (3) performance specifications, and (4) quality. Often, projects are constrained by these seemingly conflicting objectives since they must be accomplished simultaneously.

Quality: the ability of a product or service to meet and exceed customer expectations consistently.

Quality assurance: traditionally, the role of auditing conformance of manufactured product, performed by staff specialists independent of line operations. More recently, the role of quality assurance has evolved to focus more on reliability and maintainability, supplier certification and rating, and product and quality system audits.

Quality circles: a group of employees linked together with the common purpose of identifying problems within the organization and participatively defining solutions.

Quality control: the process of monitoring specific project results to (1) determine if they comply with relevant quality standards and (2) identify ways to eliminate causes of unsatisfactory performance.

Quality function deployment (QFD): the methodology that recognizes customer needs as the most important input into the successful definition of a product or service.

Quality goal: the targeted purpose toward which an organization directs all its quality improvement endeavors to ensure their attainment.

Quality improvement: a focused effort to introduce change in a specifically defined area.

Quality management: an applied process that helps to assure the attainment of quality goals.

Quality management plan: a project-specific plan that guides management of the activities used to predict or determine the ability of a product, service, or process to meet customer expectations.

Quality management system (QMS): the process-oriented structure within an organization used to ensure that the development processes are operating in an environment of continuous control.

Quality management system plan: an organization-wide plan that documents the processes and methodologies all projects in a specific area use to carry out quality functions and track their implementation. Many organizations include or cross-reference the content of the QMS plan in the project summary plan.

Quality policy: a program that guides the management of the processes used by an entire business to maximize customer satisfaction at the lowest overall cost to the organization.

Quality standards: the recognized criteria, often published by an acknowledged body such as a government agency or international professional organization, that an organization or project team must satisfy to obtain an achievable level of excellence.

R: the range value that equals the difference between the minimum and maximum values in a sample of measurements.

$\bar{R}$ (pronounced "R bar"): the value that equals the average range of a series of R values.

R Chart: a type of control chart for variable data that plots the scatter in the process at a particular point in time; also called a *range chart. See also* $\bar{X}$ chart.

Reengineering: a method used for systematic overhaul or revamping of an entire process.

Regression test: a test conducted in the laboratory to verify that a software system works after changes are made.

Reliability: the probability that an item will perform a required function under stated conditions for a specified time.

Request for information (RFI): a document similar to an RFP, except that the desired items are stock or catalog items, so the recipients need to provide only information about the items in the form of standard documentation.

Request for proposal (RFP): a document issued by an organization to others describing work that the issuer would like the recipients to undertake by inviting them to respond with a proposal.

Request for quotation (RFQ): a document similar to an RFP except that the desired items are stock or catalog items, so the recipients need to propose only prices and delivery times.

Restructure: to reorganize or regroup the reporting relationships, units, and departments within an organization.

Reuse: the systematic design and implementation of a software component for a range of products.

RFI: *see* Request for information.

RFP: *see* Request for proposal.

RFQ: *see* Request for quotation.

Risk analysis: identifying the components of a project, its risks, and the controls that need to be put in place to manage the risks.

Risk response plan: the program of activities identified by the project team and agreed to by senior management that will be implemented to avoid and/or control the occurrence of the associated project risks.

Roof of the house of quality: the vertex of the QFD matrix referred to as the house of quality, where the relationships between two conflicting or correlated customer requirements are displayed graphically.

Root cause: the underlying reason for nonconformance of a product, process, or service. *See also* Cause-and-effect diagram.

Run chart: a quality control chart, where criteria have been applied to three or more consecutive points to detect unnatural patterns.

Scatter diagram: a graphic tool used to study the possible relationship between the response variable and a regression explanatory variable.

Scenario/what if analysis: a technique used to determine the effect generated on a targeted output variable, such as the end date of the project, by combinations of a specific input variable: for example, length of specific programming tasks.

Schedule variance (SV): an estimate of the amount of money needed to put a late project back on track. In the earned value cost accounting method, SV is calculated as the budgeted cost of work performed (BCWP) minus the budgeted cost of work scheduled (BCWS).

SEI CMM: *see* Software Engineering Institute Capability Maturity Model.

Simulation: a technique whereby a model, such as an Excel worksheet, is calculated many times over with different input values with the intent of arriving at a representation of all the possible scenarios that may occur in an uncertain situation.

Situational leadership model: a leadership theory that stresses the importance and interaction of leadership behavior with the needs and maturity of the followers.

Slack: the amount of time a noncritical activity may be delayed without altering the early start of its immediate successor.

Software Engineering Institute Capability Maturity Model (SEI CMM): a federally funded research and development initiative sponsored by the U.S. Department of Defense which is dedicated to helping organizations improve the maturity of their software engineering practices and long-term business performance.

Software process assessment (SPA): the collection, analysis, and reporting of information necessary to baseline (i.e., thoroughly define) and manage the process associated with software development.

Sole-source procurement: limiting, for justifiable reasons, the awarding of a contract in a competitive marketplace to a single source among others that may be worthy.

SOW: *see* Statement of work.

SPA: *see* Software process assessment.

SPC: *see* Statistical process control.

Stage/gate model: the new product development process managed by a fully integrated cross-functional core team that must seek senior management's approval to proceed with project work beyond certain quality gate checkpoints.

- *Stage 0, concept generation:* the stage of the product development cycle during which the ideas for a product or service first surface.
- *Stage 1, project definition:* the stage of the product development cycle during which the strategy and target markets for a product or service

offer along with the technical specifications and customer requirements are identified.

- *Stage 2, development*: the stage of the product development cycle during which research and development of the actual physical product or service deliverables are completed.

- *Stage 3, integrated manufacturing and production*: the stage of the product development cycle during which laboratory prototypes are replicated and released for sale from a production facility.

- *Stage 4, commercialization*: the stage of the product development cycle dedicated to full-scale production and/or assembly of the product or the components of a service offer.

Stakeholders: people outside the project team that have an interest in the success (or failure) of the project.

Start node: the node in a network diagram at which two or more subsequent activities start.

Statement of work (SOW): that portion of a proposal or the resulting contract that states exactly what will be delivered and when.

State-of-the-art: an adjective used to describe modern, contemporary approaches.

Statistical process control (SPC): applying statistical methods to analyze, study, and monitor the process capability and performance of data samples.

Steering committee: a group or task force composed of senior-level executives charged primarily with corporate strategic decision-making responsibilities. *See also* Executive committee.

Strategic business unit: a division or entity within the organization with responsibility for profit and loss.

Strategic management: the process by which a business's senior-level management determines the long-range direction and performance of an organization by ensuring the continuous evaluation, formulation, and effective implementation of its goals.

Strategic plan (planning): a plan that is organizational-wide, establishes overall objectives, and positions the organization in terms of its environment.

Summary plan: the entire plan for a project, which includes, at a minimum, the project WBS, network diagram, and task budgets. *See also* Project summary plan.

SV: *see* Schedule variance.

Sympathy: experiencing the emotion of pity.

Systems approach: applying the concepts of integration and synergy to understand the workings of an organization, business, or technology.

Systems design review: a process whereby representatives from an organization who determine what components and logic to use in a proposed software system meet with others who have decision-making authority concerning the needs of customers in the marketplace to ensure that the system is acceptable.

Systems requirements review: a process whereby representatives from an organization who develop the components of a proposed software system and others who have decision-making authority about what components and logic to use meet to ensure that the system satisfies all specified user requirements.

System test: an internally conducted laboratory test to determine whether a software program is capable of performing as required by the design specifications or system verification plan.

System verification plan: a plan describing the tests that must be conducted to determine whether the software program is capable of performing functionally as required by the design specifications.

TAC: *see* Time at completion.

Tactical plan (planning): a program or project specific plan that guides the daily execution of objectives laid out in a higher-level plan.

T&Cs: *see* Terms and conditions.

T&M contract: *see* Time and materials contract.

Target costing: an engineering design technique whereby the costs associated with the realization of new features are based on benchmarked targets established in the marketplace. *See also* Value analysis *and* Value engineering.

Technical proposal: a document proposing the means of accomplishing specific technical objectives or other requirements presented by the selling organization to the buyer in response to a formal procurement solicitation request.

Technical prospectus: a statement of customer expectations for a product or service concept specified in terms understood by the profession engaged in its development.

Technical reviews: a routine, in-depth evaluation by an independent group of auditors of all or part of the project's proposed solutions (including the architecture, requirements, and high-level design) to ensure that the problems that the project was asked to solve are addressed. The opinion of the review team is based on its experience and knowledge since no standards exist against which to judge the correctness of the solution.

Teleconferencing: holding a conference among people at remote locations where the audio or video portion of the communications is sent over telephone lines. *See also* Audioconferencing *and* Videoconferencing.

Termination for convenience: to bring a contract to a definite end before its natural conclusion if agreed to suitably by both parties.

Termination for default: to bring a contract to a definite end before its natural conclusion due to either party's failure to perform or comply with its terms for which damages may be recoverable under specific contract provisions.

Terms and conditions (T&Cs): the clauses in a contract that describe particular events and the effect their occurrence or nonoccurrence has on the parties that signed the agreement.

Testable: the concept whereby the performance of a software application program is verifiable against initial design specifications.

Time and materials (T&M) contract: a contractual form in which the customer agrees to pay the contractor for all time and material used on the project, including a fee as a percentage of costs.

Time at completion (TAC): the total time expected to complete an activity, a group of activities, or the project.

Total project quality management (TPQM): the participation of all members of the project organization in the continuous improvement of its business culture, process methodologies, products, and services to manage the long-term success of the business through customer satisfaction.

Total time (TT): the total elapsed time from when the ideas for a product or service first emerge to its full-scale introduction to the market.

TPQM: *see* Total project quality management.

Transaction costs: the costs associated with investing in the business, buying equipment or supplies, paying bills, withdrawing money from the business, paying rent, and so on.

Transaction exposure: the potential for an organization to suffer a gain or loss in net worth due to fluctuations in the foreign exchange markets.

Trend chart: a chart that displays information in sequence of time to reveal long-term changes.

TT: *see* Total time.

UCC: *see* Uniform Commercial Code.

Uniform Commercial Code (UCC): a law developed to standardize commercial contracting law among the U.S. states. It has been adapted by all 49 states except in significant portions of Louisiana.

Unit price contract: a category of contracts under which the seller is paid based on variable units of measurable output, such as dollars per hour, although a base floor and ceiling, which are set in ranges that are adjustable, are typically used.

Unit test: a test conducted in the laboratory to determine whether the logically separable parts of a computer program are capable of performing as required by the design specifications.

Upper control limits: the upper line on a control chart that is used to judge whether the process is statistically out of control. *See also* Lower control limit.

Usable: the concept whereby customers are readily able to access with minimal effort the functionality they require in a software application program. *See also* Fitness for use.

Value: the extent to which a product or service meets a customer's needs or wants measured in willingness to pay.

Value-adding work: activities or work performed in an enterprise that is done to satisfy customer requirements.

Value analysis: a cost-reduction technique directed at analyzing and redesigning certain functions and components of a service enhancement or a product design without degrading performance or quality. *See also* Target costing *and* Value engineering.

Value chain mapping: a flowchart from the customer's perspective of all steps performed by an organization to provide a service or produce a product.

Value engineering: a problem-solving technique directed at analyzing the total costs associated with the realization of a product. The effort may include evaluation of the systems, equipment, facilities, procedures, and supplies used to assemble a product design for the purpose of achieving the lowest cost of fabrication without degrading performance or quality. *See also* Value analysis *and* Target costing.

Variable control chart: a chart displaying measurements that are capable of showing actual degrees of variation in a manufactured product sample that makes the quality of the lot unacceptable.

Videoconferencing: holding a conference among people at remote locations by means of transmitted audio and video signals. *See also* Teleconferencing.

Virtual teams: a work team that routinely crosses boundaries through an array of interactive electronic communication technologies.

Voice of the customer: the specific needs and desires of a customer segment in the marketplace.

Voice mail: an electronic communication system in which spoken messages are recorded or digitized for later playback to the intended recipient.

WBS: *see* Work Breakdown Structure.

WBS branch manager: the supervisor or task team leader of a subdivision of a project's WBS.

What: a QFD term for a statement of need or want explicitly expressed by a customer, fulfilled by a *How*.

Widget: an imaginary or hypothetical product.

Work breakdown structure (WBS): a deliverable-oriented representation of project elements that depicts the total project scope graphically.

Work group: a composite of employees or personnel, consisting of specialists with similar job functions and/or tasks.

Work package: a description given to the lowest elements in the project WBS prepared by task team leaders to inform management what work must be performed, by whom, and in what time frame.

World Wide Web: a part of the Internet designed to allow easier navigation of the network through the use of graphical user interfaces and hypertext links between different addresses. *See* Browser.

$\bar{X}$ **(pronounced "X bar")**: the average value of a sample of measurements.

$\bar{\bar{X}}$ **(pronounced "X double bar")**: the average of a series of $\bar{X}$ values.

$\bar{X}$ **chart**: a type of control chart for variable data that plots sample means; also called an average chart. *See also* R chart.

Zero defects: (1) long-range quality goal, first defined by Joseph Juran, which defines quality as continuous improvement; (2) literally, a defect-free product.

INDEX

Printed in the United Kingdom
by Lightning Source UK Ltd.
120062UK00001B/25